GSLIB

Geostatistical Software Library and User's Guide

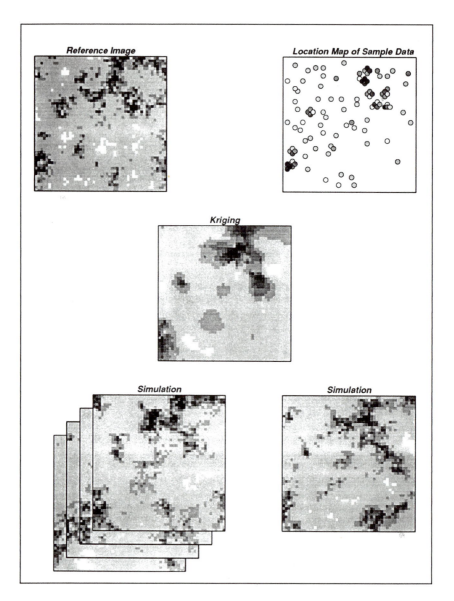

Reference Image

Location Map of Sample Data

Kriging

Simulation

Simulation

GSLIB

Geostatistical Software Library and User's Guide

CLAYTON V. DEUTSCH

Department of Applied Earth Sciences
Stanford University

ANDRE G. JOURNEL

Department of Applied Earth Sciences
Stanford University

New York   Oxford
OXFORD UNIVERSITY PRESS
1992

Oxford University Press

Oxford   New York   Toronto
Delhi   Bombay   Calcutta   Madras   Karachi
Kuala Lumpur   Singapore   Hong Kong   Tokyo
Nairobi   Dar es Salaam   Cape Town
Melbourne   Auckland

and associated companies in
Berlin   Ibadan

Library of Congress Cataloging-in-Publication Data
Deutsch, Clayton V.
GSLIB [computer file] : geostatistical software library and user's guide / Clayton V.
Deutsch, André G. Journel.
2 computer disks ; 3 1/2 in. + 1 manual.
System requirements; IBM AT or compatible; 640K RAM; MS-DOS 2.0 or higher;
FORTRAN compiler; 1-1.44M floppy disk; monochrome or color monitor.
Title from disk label.
Copy-protected.
Audience: Geoscientists.
Summary: A collection of geostatistical routines for providing a source code that serves
as a starting point for custom programs, advanced applications, and research. Powerful,
flexible, and documented programs not confined to friendly menus. The most advanced
methods in the field.
ISBN 0-19-507392-4
1. Geology—Statistical methods—Software. I. Journel, A. G.
II. Title.
QE33.2.S82 <MRCRR>
550—dc12   92-17261

9 8 7 6 5 4 3

Printed in the United States of America
on acid-free paper

Preface

The primary goal of this work is to present a geostatistical software library known as GSLIB. An important prerequisite to geostatistical case studies and research is the availability of flexible and understandable software. Flexibility is achieved by providing the original Fortran source code. A detailed description of the theoretical background along with specific application notes allows the algorithms to be understood and used as the basis for more advanced customized programs.

The three main chapters of this guidebook are based on the three major problem areas of geostatistics: quantifying spatial variability (variograms), generalized linear regression techniques (kriging), and stochastic simulation. Additional utility programs and problem sets with partial solutions are given to allow a full exploration of GSLIB and to check new software installations.

This guidebook is aimed at graduate students and advanced practitioners of geostatistics; it is not intended to be a theoretical reference textbook. Proofs, lengthy theoretical discussions, and heavy matrix notation are omitted as much as possible. Instead, this guidebook contains many footnotes, application notes, brief warnings, and multiple cross references.

The GSLIB source code has been assembled from programs developed and used at Stanford University over the course of 12 years. These programs are constantly questioned and modified to handle new algorithms. GSLIB is not a commercial product and carries no warranties, software support, or maintenance. Undoubtedly, there will be "bugs" left in the published version of the programs, most of them introduced during the rewriting of the code and thus the sole responsibility of the authors of this guide. Even though the main avenues of these programs have been tested and used extensively there are simply too many possible combinations of input data and parameters to ensure bug-free programs.

We would like to acknowledge the many graduate students who have contributed to GSLIB during their stay at Stanford University. This text is dedicated to them and to future generations who will continue to make GSLIB a living, evolving collection of programs.

Stanford, California   C.V.D
June 1992   A.G.J

Contents

GSLIB
*Geostatistical Software Library
and User's Guide*

Chapter I

Introduction

GSLIB is the name of a directory containing the geostatistical software developed at Stanford. Generations of graduate students have contributed to this constantly changing collection of programs, ideas, and utilities. Some of the most widely used public-domain geostatistical software [50,54,62] and many more in-house programs were initiated from GSLIB. It was decided to open the GSLIB directory to a wide audience, providing the source code to seed new developments, custom-made application tools, and, hopefully, new theoretical advances.

What started as a collection of selected GSLIB programs to be distributed on diskette with a minimal user's guide has turned into a major endeavor; a lot of code has been rewritten for the sake of uniformity and the user's guide looks more like a textbook. Yet, our intention has been to maintain the original versatility of GSLIB, where clarity of coding prevails over baroque theory or the search for ultimate speed. GSLIB programs are development tools and sketches for custom programs, advanced applications, and research; we expect them to be torn apart and parts grafted into custom software packages.

The strengths of GSLIB are also its weaknesses. The flexibility of GSLIB, its modular conception, and its machine independence may assist researchers and those beginners more eager to learn than use; on the other hand, the requirement to make decisions at each step and the lack of a slick user interface will frustrate others.

Reasonable effort has been made to provide software that is well documented and free of bugs; but, there is no support or telephone number for users to call. Each user of these programs assumes the responsibility for their *correct* application and careful testing as appropriate.[1]

---

[1] Reports of bugs or significant improvements in the coding should be sent to the authors at Stanford University.

3

Trends in Geostatistics

Most geostatistical theory [55,102,163,167] was established long before the availability of digital computers allowed its wide application in earth sciences [94,113,118]. The development of geostatistics since the late 1960s has been accomplished mostly by application-oriented engineers: they have been ready to bend theory to match the data and their approximations have been strengthened by extended theory.

Most early applications of geostatistics were related to mapping the spatial distribution of one or more attributes, with emphasis given to characterizing the variogram model and using the kriging (error) variance as a measure of estimation accuracy. Kriging, used for mapping, is not significantly better than other deterministic interpolation techniques that are customized to account for anisotropy and other important spatial features of the variable being mapped [67,80,121]. The kriging algorithm also provides a measure of variance. Unfortunately, the kriging variance is independent of the data values and cannot be used, in general, as a measure of estimation accuracy.

Recent applications of geostatistics have de-emphasized the mapping application of kriging. Kriging is now used to build models of uncertainty that depend on the data values in addition to the data configuration. The attention has shifted from mapping to conditional simulation, also known as *stochastic imaging*. Conditional simulation allows drawing alternative, equally probable realizations of the spatial distribution of the attribute(s) under study. These alternative stochastic images provide a measure of uncertainty about the unsampled values taken *altogether* in space rather than one by one. In addition, these simulated images do not suffer from the characteristic smoothing effect of kriging (see frontispiece).

Following this trend, the GSLIB software and this guidebook emphasize stochastic simulation tools.

Book Content

Including this introduction, the book consists of 6 chapters, 5 appendices including a list of subroutines given, a bibliography and an index. Diskettes containing all the Fortran source code are provided.

Chapter II provides the novice reader with a review of basic geostatistical concepts. The concepts of random functions and stationarity are introduced as models. The basic notations, file conventions, and variogram specifications used throughout the text and in the software are also presented.

Chapter III deals with covariances, variograms, and other more robust measures of spatial variability/continuity. The "variogram" programs allow the simultaneous calculation of many such measures at little added computational cost. An essential addition to GSLIB would be a graphical, highly interactive, variogram modeling program (such as available in other public-domain software [50,54]).

Chapter IV presents the multiple flavors of kriging, from simple and ordinary kriging to factorial and indicator kriging. As already mentioned, kriging can be used for many purposes in addition to mapping. Kriging can be applied as a filter (Wiener's filter) to separate either the low frequency (trend) component or, conversely, the high frequency (nugget effect) component of the spatial variability. When applied to categorical variables or binary transforms of continuous variables, kriging is used to update prior local information with neighboring information to obtain a posterior probability distribution for the unsampled value(s). Notable omissions in GSLIB include kriging with intrinsic random functions of order k (IRF-k) and disjunctive kriging (DK). From a first glance at the list of GSLIB subroutines, the reader may think that other more common algorithms are missing, e.g., kriging with an external drift, probability kriging, and multiGaussian kriging. These algorithms, however, are covered by programs named for other algorithms. Kriging with an external drift can be seen as a form of kriging with a trend and is an option of program **ktb3d**. Probability kriging is a form of cokriging (program **cokb3d**). As for multiGaussian kriging, it is a mere simple kriging (program **ktb3d**) applied to normal score transforms of the data (program **nscore**) followed by a back transform of the data (program **backtr**).

Chapter V presents the principles and various algorithms for stochastic simulation. Major heterogeneities, possibly modeled by categorical variables, should be simulated first, then, continuous properties are simulated within each reasonably homogeneous category. It is recalled that fluctuations of the realization statistics (histogram, variograms, etc.) are expected, particularly if the field being simulated is not very large with respect to the range(s) of correlation. Stochastic simulation is currently the most vibrant field of research in geostatistics, and the list of algorithms presented in this text and coded in GSLIB is not exhaustive. GSLIB includes only the most commonly used stochastic algorithms as of 1991. Deterministic process simulations [155], the vast variety of Boolean algorithms [147], and spectral domain algorithms [18,64,68] are not covered here, notwithstanding their theoretical or practical importance.

Chapter VI proposes a number of useful utility programs including some to generate graphics in the PostScript page description language. Seasoned users will already have their own programs. Moreover, similar utility programs are available in classical textbooks or in inexpensive statistical software. These programs have been included, however, because they are essential to the practice of geostatistics. Without a histogram and scatterplot, summary statistics, and other displays there would be no practical geostatistics; without a declustering program most geostatistical studies would be flawed from the beginning; without normal score transform and back transform programs there would be no multiGaussian geostatistics; an efficient and robust linear system solver is essential to any kriging exercise.

Each chapter starts with a brief summary and a first section on methodology (II.1, III.1, IV.1, and V.1). The expert geostatistician may skip this first section, although it is sometimes useful to know the philosophy underlying the software being used. The remaining sections in each chapter give commented lists of input parameters for each subroutine. A section entitled "Application Notes" is included near the end of each chapter with useful advice on using the programs.

Chapters II through V end with sections suggesting one or more problem sets. These problems are designed to provide a first acquaintance with the most important GSLIB programs and a familiarity with the parameter files and notations used. All problem sets are based on a common synthetic data set provided with the GSLIB distribution diskettes. The questions asked are all open-ended; readers are advised to pursue their investigations beyond the questions asked. Limited solutions to each problem set are given in Appendix A; they are not meant to be definitive general solutions. The first time a program is encountered the input parameter file is listed along with the first and last 10 lines of the corresponding output file. These limited solutions are intended to allow readers to check their understanding of the program input and output. Working through the problem sets before referring to the solutions will provide insight into the algorithms. Attempting to apply GSLIB programs (or any geostatistical software) directly to a real-life problem without some prior experience may be extremely frustrating.

About the Source Code

The source code provided in GSLIB adheres as closely as possibly to the ANSI standard Fortran 77 programming language. Fortran was retained chiefly because of its familiarity and common usage. Some consider Fortran the dinosaur of programming languages when compared to newer object-oriented programming languages; nevertheless, Fortran remains a viable and even desirable scientific programming language due to its simplicity and the introduction of better features in new ANSI standards or as enhancements by various software companies.

GSLIB was coded to the ANSI standard to achieve machine independence. We did not want the use of GSLIB limited to any specific PC, workstation, or mainframe computer. Many geostatistical exercises may be performed on a vintage IBM XT clone. Large stochastic simulations and complex modeling, however, would be better performed on a more powerful machine. GSLIB programs have been compiled on machines ranging from an XT clone to the latest multiple processor Cray Y-MP. No executable programs are provided; only ASCII files containing the source code and other necessary input files are included in the enclosed diskettes. The source code can immediately be dismantled and incorporated into custom programs, or compiled and used without modification. Appendix B contains detailed information about software installation and troubleshooting.

The 1978 ANSI Fortran 77 standard does not allow dynamic memory allocation. This poses an inconvenience since array dimensioning and memory allocation must be *hardcoded* in all programs. To mitigate this inconvenience, the dimensioning and memory allocation have been specified in an "include" file. The include file is common to all necessary subroutines. Therefore, the array dimensioning limits, such as the maximum number of data or the maximum grid size, has to be changed in only one file.

Although the maximum dimensioning parameters are hardcoded in an include file it is clearly undesirable to use this method to specify most input parameters. Given the goal of machine independence, it is also undesirable to have input parameters enter the program via a menu driven graphical user interface. For these reasons the input parameters are specified in a prepared "parameter" file. These parameter files must follow a specified format. An advantage of this form of input is that the parameter files provide a permanent record of run parameters. With a little experience users will find the preparation and modification of parameter files faster than most other forms of parameter input. The non-standard *namelist* facility available with some compilers would be one logical alternative.

Further, the source code for any specific program has been separated into a main program and then all of the necessary subroutines. This was done to facilitate the isolation of key subroutines for their incorporation into custom software packages. In general, there are four computer files associated with any algorithm or program: the main program, the subroutines, an include file, and a parameter file.

Most program operations are separated into short, easily understood, program modules. Many of these shorter subroutines and funtions are then used in more than one program. Care has been taken to use a consistent format with a statement of purpose, a list of the input parameters, and a list of programmers who worked on a given piece of code. Extensive in-line comments explain the function of the code and identify possible enhancements. A detailed description of the programming conventions and a dictionary of the variable names are given in Appendix C.

Chapter II

Getting Started

This chapter provides a brief review of the geostatistical concepts underlying GSLIB and an initial presentation of notations, conventions, and computer requirements.

Section II.1 establishes the estimation of posterior probability distribution models for unsampled values as a principal goal. From such distributions, best estimates can be derived together with probability intervals for the unknowns. Stochastic simulation is the process of Monte Carlo drawing of realizations from such posterior distributions.

Section II.2 gives some of the basic notations, computer requirements, and file conventions used throughout the GSLIB text and software.

Section II.3 considers the important topic of variogram modeling. The notation and conventions established in this section are consistent throughout all GSLIB programs.

Section II.4 considers the different search strategies that are implemented in GSLIB. The super block search, spiral search, two-part search, and the use of covariance look-up tables are discussed.

Section II.5 introduces the data set that will be used throughout for demonstration purposes. A first problem set is provided in section II.6 to acquaint the reader with the data set and the utility programs for exploratory data analysis.

II.1 Geostatistical Concepts: A Review

Geostatistics is concerned with "the study of phenomena that fluctuate in space " and/or time ([125], p. 31). Geostatistics offers a collection of deterministic and statistical tools aimed at understanding and modeling spatial variability.

The basic paradigm of predictive statistics is to characterize any unsampled (unknown) value z as a random variable (RV) Z, the probability distribution of which characterizes the uncertainty about z. A random variable is a variable that can take a variety of outcome values according to some probability (frequency) distribution. The random variable is traditionally denoted by a capital letter, say Z, while its outcome values are denoted with the corresponding lower case letter, say z or $z^{(l)}$. The RV model Z, and more specifically its probability distribution, is usually location-dependent; hence the notation $Z(\mathbf{u})$, with $\mathbf{u}$ being the location coordinates vector. The RV $Z(\mathbf{u})$ is also information-dependent in the sense that its probability distribution changes as more data about the unsampled value $z(\mathbf{u})$ become available. Examples of continuously varying quantities that can be effectively modeled by RV's include petrophysical properties (porosity, permeability, metal or pollutant concentrations), and geographical properties (topographic elevations, population densities). Examples of categorical variables include geological properties such as rock types or counts of insects or fossil species.

The cumulative distribution function (cdf) of a continuous RV $Z(\mathbf{u})$ is denoted:

$$F(\mathbf{u}; z) = Prob\left\{Z(\mathbf{u}) \leq z\right\} \qquad (\text{II.1})$$

When the cdf is made specific to a particular information set, e.g., (n) consisting of n neighboring data values $Z(\mathbf{u}_\alpha) = z(\mathbf{u}_\alpha), \alpha = 1, \ldots, n$, the notation "conditional to (n)" is used, defining the conditional cumulative distribution function (ccdf):

$$F(\mathbf{u}; z|(n)) = Prob\left\{Z(\mathbf{u}) \leq z|(n)\right\} \qquad (\text{II.2})$$

In the case of a categorical RV $Z(\mathbf{u})$ that can take any one of K outcome values $k = 1, \ldots, K$, a similar notation is used:

$$F(\mathbf{u}; k|(n)) = Prob\left\{Z(\mathbf{u}) = k|(n)\right\} \qquad (\text{II.3})$$

Expression (II.1) characterizes the uncertainty about the unsampled value $z(\mathbf{u})$ prior to using the information set (n); expression (II.2) characterizes the posterior uncertainty once the information set (n) has been accounted for. The goal, implicit or explicit, of any predictive algorithm is to update prior models of uncertainty such as (II.1) into posterior models such as (II.2). Note that the ccdf $F(\mathbf{u}; z|(n))$ is a function of the location $\mathbf{u}$, the sample size and geometric configuration (the data locations $\mathbf{u}_\alpha$, $\alpha = 1, \ldots, n$), and the sample values (the n values $z(\mathbf{u}_\alpha)$'s).

From the ccdf (II.2) one can derive various optimal estimates for the unsampled value $z(\mathbf{u})$ in addition to the ccdf mean, which is the least-squared error estimate. One can also derive various probability intervals such as the 95% interval $[q(0.025); q(0.975)]$ such that

$$Prob\left\{Z(\mathbf{u}) \in [q(0.025); q(0.975)]|(n)\right\} = 0.95$$

with $q(0.025)$ and $q(0.975)$ being the 0.025 and 0.975 quantiles of the ccdf, e.g.,

$$q(0.025) \text{ is such that } F(\mathbf{u}; q(0.025)|(n)) = 0.025$$

Moreover, one can draw any number of simulated outcome values $z^{(l)}(\mathbf{u}), l = 1, \ldots, L$, from the ccdf. Calculation of posterior ccdf's and Monte Carlo drawings of outcome values are at the heart of stochastic simulation algorithms (see Chapter V).

In geostatistics most of the information related to an unsampled value $z(\mathbf{u})$ comes from sample values at neighboring locations $\mathbf{u}'$, whether defined on the same attribute z or on some related attribute y. Thus, it is important to model the degree of correlation or dependence between any number of RV's $Z(\mathbf{u}), Z(\mathbf{u}_\alpha), \alpha = 1, \ldots, n$ and more generally $Z(\mathbf{u}), Z(\mathbf{u}_\alpha), \alpha = 1, \ldots, n$, $Y(\mathbf{u}'_\beta), \beta = 1, \ldots, n'$. The concept of a random function (RF) allows such modeling and updating of prior cdf's into posterior ccdf's.

II.1.1 The Random Function Concept

A random function (RF) is a set of RV's defined over some field of interest, e.g., $\{Z(\mathbf{u}), \mathbf{u} \in$ study area$\}$ also denoted simply as $Z(\mathbf{u})$. Usually the RF definition is restricted to RV's related to the same attribute , say z, hence another RF would be defined to model the spatial variability of a second attribute, say $\{Y(\mathbf{u}), \mathbf{u} \in$ study area$\}$.

Just as a RV $Z(\mathbf{u})$ is characterized by its cdf (II.1), a RF $Z(\mathbf{u})$ is characterized by the set of all its K-variate cdf's for any number K and any choice of the K locations $\mathbf{u}_k, k = 1, \ldots, K$:

$$F(\mathbf{u}_1, \ldots, \mathbf{u}_k; z_1, \ldots, z_K) = Prob\left\{Z(\mathbf{u}_1) \leq z_1, \ldots, Z(\mathbf{u}_K) \leq z_K\right\} \quad \text{(II.4)}$$

Just as the univariate cdf of RV $Z(\mathbf{u})$ is used to characterize uncertainty about the value $z(\mathbf{u})$, the multivariate cdf (II.4) is used to characterize joint uncertainty about the K values $z(\mathbf{u}_1), \ldots, z(\mathbf{u}_K)$.

Of particular interest is the bivariate ($K = 2$) cdf of any two RV's $Z(\mathbf{u})$, $Z(\mathbf{u}')$, or more generally $Z(\mathbf{u}), Y(\mathbf{u}')$:

$$F(\mathbf{u}, \mathbf{u}'; z, z') = Prob\left\{Z(\mathbf{u}) \leq z, Z(\mathbf{u}') \leq z'\right\} \quad \text{(II.5)}$$

Consider the particular binary transform of $Z(\mathbf{u})$ defined as

$$I(\mathbf{u}; z) = \begin{cases} 1, & \text{if } Z(\mathbf{u}) \leq z \\ 0, & \text{otherwise} \end{cases} \quad \text{(II.6)}$$

Then, the previous bivariate cdf (II.5) appears as the non-centered indicator (cross) covariance:

$$F(\mathbf{u}, \mathbf{u}'; z, z') = E\left\{I(\mathbf{u}; z)I(\mathbf{u}'; z')\right\} \quad \text{(II.7)}$$

Relation (II.7) is the key to the indicator formalism as developed later in section IV.1.9: it shows that inference of bivariate cdf's can be done through sample indicator covariances.

Rather than a series of indicator covariances, inference can be restricted to a single moment of the bivariate cdf (II.5), for example the attribute covariance:

$$C(\mathbf{u}, \mathbf{u}') = E\left\{Z(\mathbf{u})Z(\mathbf{u}')\right\} - E\left\{Z(\mathbf{u})\right\} E\left\{Z(\mathbf{u}')\right\} \qquad \text{(II.8)}$$

The Gaussian RF model is extraordinarily convenient in the sense that any K-variate cdf of type (II.4) is fully defined from knowledge of the covariance function (II.8). This remarkable property explains the popularity of Gaussian-related models. That property is also the main drawback: these models allow reproduction only of the data covariance when other spatial properties may also be important.

II.1.2   Inference and Stationarity

Inference of any statistic, whether a univariate cdf such as (II.1) or any of its moments (mean,variance) or a multivariate cdf such as (II.4) or (II.5) or any of its moments (covariances), requires some repetitive sampling. For example, repetitive sampling of the variable $z(\mathbf{u})$ is needed to evaluate the cdf $F(\mathbf{u}; z) = Prob\left\{Z(\mathbf{u}) \leq z\right\}$ from experimental proportions. In many applications, however, only one sample is available at any single location $\mathbf{u}$ in which case $z(\mathbf{u})$ is known (ignoring sampling errors), and the need to consider the RV model $Z(\mathbf{u})$ vanishes. The paradigm underlying most inference processes (not only statistical inference) is to trade the unavailable replication at location $\mathbf{u}$ for another replication available somewhere else in space and/or time. For example, the cdf $F(\mathbf{u}; z)$ may be inferred from the cumulative histogram of z-samples collected at other locations, $\mathbf{u}_\alpha \neq \mathbf{u}$, within the same field, or at the same location $\mathbf{u}$ but at different times if a time series is available at $\mathbf{u}$. In the latter case the RF is actually defined in space-time and should be denoted $Z(\mathbf{u}, t)$.

This trade of replication or sample spaces corresponds to the hypothesis (rather a decision [1]) of stationarity.

The RF $\left\{Z(\mathbf{u}), \mathbf{u} \in A\right\}$ is said to be stationary within the field A if its multivariate cdf (II.4) is invariant under any translation of the K coordinate vectors $\mathbf{u}_k$, that is:

$$F(\mathbf{u}_1, \ldots, \mathbf{u}_k; z_1, \ldots, z_K) = F(\mathbf{u}_1 + \mathbf{l}, \ldots, \mathbf{u}_k + \mathbf{l}; z_1, \ldots, z_K), \qquad \text{(II.9)}$$

$$\forall \text{ translation vector } \mathbf{l}.$$

Invariance of the multivariate cdf entails invariance of any lower order cdf, including the univariate and bivariate cdfs, and invariance of all their moments,

---

[1] Stationarity is a property of the RF model, not of the underlying spatial distribution. Thus it cannot be checked from data. In a census, the decision to provide statistics per county or state rather than per class of age is something that cannot be checked or refuted. The decision to pool data into statistics across rock types is not refutable a priori from data; however, it can be shown inappropriate a posteriori if differentiation per class of age or rock type is critical to the undergoing study. For a more extensive discussion see [76,85].

including all covariances of type (II.7) or (II.8). The decision of stationarity allows inference. For example, the unique stationary cdf

$$F(z) = F(\mathbf{u}; z), \forall \, \mathbf{u} \in A$$

can be inferred from the cumulative sample histogram of the z-data values available at various locations within A. Another example concerns the inference of the stationary covariance

$$C(\mathbf{h}) = E\left\{Z(\mathbf{u} + \mathbf{h})Z(\mathbf{u})\right\} - \left[E\left\{Z(\mathbf{u})\right\}\right]^2 \qquad \text{(II.10)}$$

$$\forall \mathbf{u}, \mathbf{u} + \mathbf{h} \in A$$

from the sample covariance of all pairs of z-data values approximately separated by vector $\mathbf{h}$; see programs **gamv2** and **gamv3** in Chapter III.

The decision of stationarity is critical for the representativeness and reliability of the geostatistical tools used. Pooling data across geological facies may mask important geological differences; on the other hand, splitting data into too many sub-categories may lead to unreliable statistics based on too few data per category and an overall confusion. The rule in statistical inference is to pool the largest amount of *relevant* information to formulate predictive statements.

Stationarity is a property of the model; thus, the decision of stationarity may change if the scale of the study changes or if more data become available. If the goal of the study is global then local details can be averaged out; conversely, the more data available the more statistically significant differentiations become possible.

II.1.3 Variogram

An alternative to the covariance defined in (II.8) and (II.10) is the variogram defined as the variance of the increment $[Z(\mathbf{u}) - Z(\mathbf{u} + \mathbf{h})]$. For a stationary RF:

$$
\begin{aligned}
2\gamma(\mathbf{h}) &= Var\left\{Z(\mathbf{u} + \mathbf{h}) - Z(\mathbf{u})\right\} \qquad \text{(II.11)} \\
\gamma(\mathbf{h}) &= C(0) - C(\mathbf{h}), \ \forall \, \mathbf{u}
\end{aligned}
$$

with: $C(\mathbf{h})$ being the stationary covariance, and: $C(0) = Var\{Z(\mathbf{u})\}$ being the stationary variance.

Traditionally, the variogram has been used for modeling spatial variability rather than the covariance although kriging systems are more easily solved with covariance matrices ($\gamma(0)$ values are problematic when pivoting). In accordance with tradition,[2] GSLIB programs consistently require

---

[2] In the stochastic process literature the covariance is preferred over the variogram. The historical reason for geostatisticians preferring the variogram is that its definition (II.11) requires only second-order stationarity of the RF increments, a condition also called the "intrinsic hypothesis." This lesser demand on the RF model has been recently shown to be of no consequence for most practical situations; for more detailed discussions see [77,97,134].

*semi*variogram models which are then promptly converted into equivalent covariance models. The "variogram" programs of Chapter III allow the computation of covariance functions in addition to variogram functions.

A Note on Generalized Covariances:

Generalized covariances of order k are defined as variances of differences of order $(k + 1)$ of the initial RF $Z(\mathbf{u})$, see e.g. [35,162]. The traditional variogram $2\gamma(\mathbf{h})$, defined in relation (II.11) as the variance of the first order difference of the RF $Z(\mathbf{u})$, is associated to a generalized covariance of order zero. The order zero stems from the variogram expression's filtering any zero-order polynomial of the coordinates $\mathbf{u}$, such as $m(\mathbf{u})=$constant, added to the RF model $Z(\mathbf{u})$. Similarly, a generalized covariance of order k would filter a polynomial trend of order k added to the RF model $Z(\mathbf{u})$.

Unfortunately, inference of generalized covariances of order $k > 0$ poses severe problems since experimental differences of order $k + 1$ are not readily available if the data are not gridded. In addition, more straightforward algorithms for handling polynomial trends exist, including ordinary kriging with moving data neighborhoods [97].

Therefore, notwithstanding the theoretical importance of the IRF-k formalism, it has been decided not to include generalized covariances and the related intrinsic RF models of order k in this version of GSLIB.

II.1.4   Kriging

Kriging is "a collection of generalized linear regression techniques for minimizing an estimation variance defined from a prior model for a covariance" ([125], p. 41).

Consider the estimate of an unsampled value $z(\mathbf{u})$ from neighboring data values $z(\mathbf{u}_\alpha), \alpha = 1, \ldots, n$. The RF model $Z(\mathbf{u})$ is stationary with mean m and covariance $C(\mathbf{h})$. In its simplest form, also known as simple kriging (SK), the algorithm considers the following linear estimator:

$$Z^*_{SK}(\mathbf{u}) = \sum_{\alpha=1}^{n} \lambda_\alpha(\mathbf{u})Z(\mathbf{u}_\alpha) + \left[1 - \sum_{\alpha=1}^{n} \lambda_\alpha(\mathbf{u})\right] m \qquad (\text{II}.12)$$

The weights $\lambda_\alpha(\mathbf{u})$ are determined to minimize the error variance, also called the "estimation variance." That minimization results in a set of normal equations [82,105]:

$$\sum_{\beta=1}^{n} \lambda_\beta(\mathbf{u})C(\mathbf{u}_\beta - \mathbf{u}_\alpha) = C(\mathbf{u} - \mathbf{u}_\alpha), \qquad (\text{II}.13)$$

$$\forall \alpha = 1, \ldots, n$$

The corresponding minimized estimation variance, or kriging variance, is:

$$\sigma_{SK}^2(\mathbf{u}) = C(0) - \sum_{\alpha=1}^{n} \lambda_\alpha(\mathbf{u})C(\mathbf{u} - \mathbf{u}_\alpha) \geq 0 \qquad (\text{II.14})$$

Ordinary kriging (OK) is the most commonly used variant of the previous simple kriging algorithm, whereby the sum of the weights $\sum_{\alpha=1}^{n} \lambda_\alpha(\mathbf{u})$ is constrained to equal 1. This allows building an estimator $Z_{OK}^*(\mathbf{u})$ that does not require prior knowledge of the stationary mean m, yet remains unbiased in the sense that $E\{Z_{OK}^*(\mathbf{u})\} = E\{Z(\mathbf{u})\}$.

Non-linear kriging is but linear kriging performed on some non-linear transform of the z-data, e.g., the log-transform $\ln z$ provided that $z > 0$, or the indicator transform as defined in relation (II.6).

Traditionally, kriging (SK or OK) has been performed to provide a "best" linear unbiased estimate (BLUE) for unsampled values $z(\mathbf{u})$, with the kriging variance being used to define Gaussian-type confidence intervals, e.g.,

$$Prob\{Z(\mathbf{u}) \in [z_{SK}^*(\mathbf{u}) \pm 2\sigma_{SK}(\mathbf{u})]\} \cong 0.95$$

Unfortunately, kriging variances of the type (II.14), being independent of the data values, only provides a comparison of alternative geometric data configurations. Kriging variances are usually not measures of local estimation accuracy [85].

In addition, users have come to realize that kriging estimators of type (II.12) are "best" only in the least-squared error sense for a given covariance/variogram model. Minimizing an expected squared error need not be the most relevant estimation criterion for the study at hand; rather, one might prefer an algorithm that would minimize the impact (loss) of the resulting error; see [143]. This decision-analysis approach to estimation requires a probability distribution of type (II.2), $Prob\{Z(\mathbf{u}) \leq z|(n)\}$, for the RV $Z(\mathbf{u})$ [14,82].

These remarks seem to imply the limited usefulness of kriging and geostatistics as a whole. Fortunately, the kriging algorithm has two characteristic properties that allow its use in determining posterior ccdf's of type (II.2). These two characteristic properties are the basis for, respectively, the multiGaussian (MG) approach and the indicator kriging (IK) approach to determination of ccdf's:

(i) **The multiGaussian Approach:** If the RF model $Z(\mathbf{u})$ is multivariate Gaussian,[3] then the simple kriging estimate (II.12) and variance (II.14) identify the mean and variance of the posterior ccdf. In addition, since

---

[3] If the sample histogram is not normal, a normal score-transform can be performed on the original z-data. A multiGaussian model $Y(\mathbf{u})$ is then adopted for the normal score data. For example, the transform is the logarithm, $y(\mathbf{u}) = \ln z(\mathbf{u})$, if the z-sample histogram is approximately lognormal. Kriging and/or simulation are then performed on the y-data with the results appropriately back-transformed into z-values (see programs **nscore** and **backtr** in Chapter VI).

that ccdf is Gaussian, it is fully determined by these two parameters; see [9] and [157]. This remarkable result is at the basis of multiGaussian (MG) kriging and simulation (see program **sgsim** in Chapter V). The MG approach is said to be parametric in the sense that it determines the ccdf's through their parameters (mean and variance). The MG algorithm is remarkably fast and trouble-free; its limitation is the reliance on the very specific and sometimes inappropriate properties of the Gaussian RF model [92].

(ii) The Indicator Kriging Approach: If the value to be estimated is the expected value (mean) of a distribution, then least-squares (LS) regression, i.e., kriging, is a priori the preferred algorithm. The reason is that the LS estimator of the variable $Z(\mathbf{u})$ is also the LS estimator of its conditional expectation $E\{Z(\mathbf{u})|(n)\}$, i.e., of the expected value of the ccdf (II.2); see [94], p. 566. Now, instead of the variable $Z(\mathbf{u})$, consider its binary indicator transform $I(\mathbf{u}; z)$ as defined in relation (II.6). Kriging of the indicator RV $I(\mathbf{u}; z)$ provides an estimate that is also the best LS estimate of the conditional expectation of $I(\mathbf{u}; z)$. Moreover, the conditional expectation of $I(\mathbf{u}; z)$ is itself equal to the ccdf of $Z(\mathbf{u})$; indeed:

$$
\begin{aligned}
E\left\{I(\mathbf{u}; z)|(n)\right\} &= 1 \cdot Prob\left\{I(\mathbf{u}; z) = 1|(n)\right\} \\
&+ 0 \cdot Prob\left\{I(\mathbf{u}; z) = 0|(n)\right\} \\
= 1 \cdot Prob\left\{Z(\mathbf{u}) \le z)|(n)\right\} &\equiv F(\mathbf{u}; z|(n)), \text{ as defined in (II.2)}
\end{aligned}
$$

Thus, the kriging algorithm applied to indicator data provides LS estimates of the ccdf (II.2). Note that indicator kriging (IK) is not aimed at estimating the unsampled value $z(\mathbf{u})$ or its indicator transform $i(\mathbf{u}; z)$ but at providing a ccdf model of uncertainty about $z(\mathbf{u})$. The IK algorithm is said to be non-parametric in the sense that it does not approach the ccdf through its parameters, rather, the ccdf values for various threshold values z are estimated directly.

A Note about Disjunctive Kriging

Disjunctive kriging (DK) generalizes the kriging estimator to the following form:

$$
Z^*_{DK}(\mathbf{u}) = \sum_{\alpha=1}^{n} f_\alpha(Z(\mathbf{u}_\alpha))
$$

with $f_\alpha()$ being functions, possibly non-linear, of the data, see [94], p. 573, [117], [125], p. 18.

DK can be used, just like IK, to derive ccdf models characterizing the uncertainty about an unknown $z(\mathbf{u})$.

As opposed to the IK approach where the direct and cross indicator covariances are inferred from actual data, the DK formalism relies on parametric models for the bivariate distribution with rather restrictive assumptions.

The typical bivariate distribution model used in DK is fully characterized by a transform of the original data (equivalent to the normal score transform used in the MG approach) and the covariance of those transforms. The yet unanswered question is which transform to use. Moreover, if a normal scores transform is to be adopted, why not go all the way and adopt the well understood multivariate Gaussian model with its unequaled analytical properties and convenience?

In the opinion of these authors, DK is conveniently replaced by the more robust MG or median IK approaches whenever only one covariance model is deemed enough to characterize the spatial continuity of the attribute under study. In cases when the model calls for reproduction of multiple indicator (cross) covariances, the indicator kriging approach allows more flexibility. Therefore, it has been decided not to include DK in this version of GSLIB.

II.1.5   Stochastic Simulation

Stochastic simulation is the process of drawing alternative, equally probable, joint realizations of the component RV's from a RF model. The (usually gridded) realizations $\{z^{(l)}(\mathbf{u}), \mathbf{u} \in A\}\, l = 1, \ldots, L$ represent L possible images of the spatial distribution of the attribute values $z(\mathbf{u})$ over the field A. Each realization, also called a "stochastic image," reflects the properties that have been imposed on the RF model $Z(\mathbf{u})$. Therefore, the more properties that are inferred from the sample data and incorporated into the RF model $Z(\mathbf{u})$, the better that RF model.

Typically, the RF model $Z(\mathbf{u})$ is constrained by the sole z-covariance model inferred from the corresponding sample covariance, and the drawing is such that all realizations honor the z-data values at their locations, in which case these realizations are said to be conditional (to the data values). More advanced stochastic simulation algorithms allow reproducing more of the sample bivariate distribution by constraining the RF model to a series of indicator (cross) covariances of type (II.7).

There is another class of stochastic simulation algorithms that aim at reproducing specific geometric features of the sample data. The reproduction of bivariate statistics such as covariance(s) is then traded for the reproduction of the geometry, see literature on Poisson or Boolean algorithms, e.g., [13,46,65,66,135,147]. In a nutshell, Boolean algorithms consist of drawing shapes, e.g., elliptical shales in a clastic reservoir, from prior distributions of shape parameters, then locating these shapes at points along lines or surfaces randomly distributed in space. GSLIB offers a basic Boolean-type simulation program (**ellipsim**) that can be easily modified to handle any geometric shape. Note that programs such as indicator principal components simulation (**ipcsim**) and simulated annealing (**anneal**) in Chapter V represent alternatives for reproducing aspects of the geometry of categorical variables.

Conditional simulation was initially developed to correct for the smoothing effect shown on maps produced by the kriging algorithm. Indeed, kriging estimates are weighted moving averages of the original data values; thus they have less spatial variability than the data. Moreover, depending on the data configuration, the degree of smoothing varies in space entailing possibly artifact structures. Typical conditional simulation algorithms trade the estimation variance minimization for the reproduction of a variogram/covariance seen as a model of spatial variability. A smoothed map as provided by kriging is appropriate for showing global trends. On the other hand, conditionally simulated maps are more appropriate for studies that are sensitive to patterns of local variability such as flow simulations; see, e.g., [68,91]. A suite of conditionally simulated maps also provides a measure of uncertainty about the attribute(s) spatial distribution.

Increasingly, stochastic simulations are used to provide a numerical and visual appreciation of spatial uncertainty beyond the univariate ccdf $Prob\{Z(\mathbf{u}) \leq z|(n)\}$ introduced in relation (II.2). That univariate ccdf measures only the uncertainty prevailing at a given location $\mathbf{u}$; it does not provide any indication about the joint uncertainty prevailing at several locations. For example, consider the locations within a subset B of the field A, i.e., $\mathbf{u}_j \in B \subset A$. The probability distribution of the block average $\frac{1}{|B|} \int_B Z(\mathbf{u})d\mathbf{u}$, or the probability that all values $Z(\mathbf{u}_j)$ simultaneously exceed a threshold z, cannot be derived from the set of univariate ccdf's $Prob\{Z(\mathbf{u}_j) \leq z|(n)\}, \mathbf{u}_j \in B$. They can be derived, however, from a set of stochastic images.

II.2 GSLIB Conventions

The notation adopted in this user's manual and in the GSLIB source code adheres, as much as possible, to established geostatistical conventions used in textbooks such as [76,82,125].

Appendix C of this manual provides the GSLIB programming conventions and a dictionary of variable names. Appendix E provides a list of acronyms and notations.

Certain typographical conventions are used to distinguish files, program names, and variable names.

- The names of actual computer files, e.g., source code files, are enclosed in a box, e.g., $\boxed{\text{gam2.f}}$.

- Each program name is written in typographic script, e.g., **gam2**.

- When program variables begin a descriptive paragraph they are in bold script, e.g. **datafl**, otherwise, variables are written in italics, e.g., *datafl*.

II.2.1 Computer Requirements

The subroutines in GSLIB are written to adhere as closely as possible to ANSI standard Fortran 77. Some very obvious, and easily corrected, departures from the standard have been taken. For example, although the ANSI standard specifies upper case characters, the source code is in lower case characters. This could be corrected if necessary by a global change. The idea is to keep the software independent of a particular compiler or computer. The software has been compiled and executed, without modification, on a number of machines including IBM compatible PC's, DEC 5000, IBM 6000, SUN sparc, SUN Motorola, and Apollo 10000 workstations, and a Cray Y-MP. Therefore, the sole requirement should be the availability of a Fortran compiler.

For reasonably modest tasks the programs may be executed on a PC in an acceptable amount of time. Large kriging or simulation runs may require special consideration in terms of available memory (particularly RAM) and speed.

One short comment on the choice of Fortran: there was no need to retain Fortran as the programming language for this version of GSLIB; enough changes were planned to justify conversion to another language (C being a logical choice). Fortran was retained chiefly because of its common usage and familiarity. We also feel that structured programming is possible with Fortran and that extended capabilities (e.g., ENDDO's, dynamic memory allocation, etc.) available with the next ANSI standard or as non-standard extensions of currently available compilers will keep Fortran a viable programming language for the near future.

II.2.2 Data Files

Although users are strongly encouraged to customize the programs, workable main programs are useful starting points. The main programs read and write data with a format similar to the menu-driven packages Geo-EAS [50] and Geostatistical Toolbox [54]. The format, described below, is a simple ASCII format with no special allowance for regular grids or data compression.

The data file format is a simplified Geo-EAS format, hence with no allowance for the user to specify explicitly the input format. The data values are always read with free format. The accessibility of the source code allows this to be easily changed.

The following conventions are used by the "simplified Geo-EAS format" used by GSLIB data files:

1. The first line in the file is taken as a title and is possibly transferred to output files.

2. The second line should be a numerical value specifying the number of numerical variables *nvar* in the data file.

3. The next *nvar* lines contain character identification labels and additional text (optional) that describe each variable.

4. The following lines, from *nvar*+3 until the end of the file, are considered as data points and must have *nvar* numerical values per line. Missing values are typically considered as large negative or positive numbers

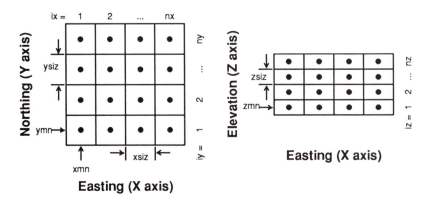

Figure II.1:   Plan and vertical cross section views that illustrate the grid definition used in GSLIB.

(e.g., less than -1.0e21 or greater than 1.0e21). The number of data will be the number of lines in the file minus $nvar + 2$ minus the number of missing values. The programs read numerical values and *not* alphanumeric characters; alphanumeric variables may be transformed to integers or the source code modified.

II.2.3   Grid Definition

Regular grids of data points or block values are often considered as input or output. The conventions used throughout GSLIB are:

- The X axis is associated to the east direction. Grid node indices ix increase from 1 to nx in the positive x direction, i.e., to the east.

- The Y axis is associated to the north direction. Grid node indices iy increase from 1 to ny in the positive y direction, i.e., to the north.

- The Z axis is associated to the elevation. Grid node indices iz increase from 1 to nz in the positive z direction, i.e., upward.

The user can associate these three axes to any coordinates system that is appropriate for the problem at hand. For example, if the phenomenon being studied is a stratigraphic unit, then some type of stratigraphic coordinates relative to a marker horizon could make the most sense [80,93]. The user must perform the coordinate transformation; there is no allowance for rotation or stratigraphic grids within the existing set of subroutines.

The coordinate system is established by specifying the coordinates at the center of the first block (xmn, ymn, zmn), the number of blocks/grid nodes (nx, ny, nz), and the size/spacing of the blocks/nodes $(xsiz, ysiz, zsiz)$. Figure II.1 illustrates these parameters on two 2-D sectional views.

Sometimes a special ordering is used to store a regular grid. This avoids the requirement of storing node coordinates or grid indices. The ordering is point by point to the east, then row by row to the north, and finally level by level upward, i.e., x cycles fastest, then y, and finally z. The index location of any particular node ix, iy, iz can be located by:

$$loc = (iz - 1) * nx * ny + (iy - 1) * nx + ix$$

Given the above one-dimensional index location of a node the three coordinate indices can be calculated as:

$$iz = 1 + int\left(\frac{loc}{nx * ny}\right)$$

$$iy = 1 + int\left(\frac{loc - (iz - 1) * nx * ny}{nx}\right)$$

$$ix = loc - (iz - 1) * nx * ny - (iy - 1) * nx$$

The uncompressed ASCII format is convenient because of its machine independence and easy access by visual editors; however, a binary compressed format would be most efficient and even necessary if large 3-D grids are being considered.

II.2.4 Program Execution and Parameter Files

The default driver programs read the name of a parameter file from standard input. If the parameter file is named for the program and has a ".par" extension, then simply keying a carriage return will be sufficient (e.g., the program **gam2m** would automatically look for $\boxed{\text{gam2.par}}$). All of the program variables and names of input/output files are contained in the parameter file. A typical parameter file is illustrated on Figure II.2. The user can have as many lines of comments at the top of the file as desired, but formatted input is started immediately after the characters "START" are found at the beginning of a line. Some in-line documentation is available to supplement the detailed documentation provided in this user's manual. Example parameter files are included with the presentation of each program in Chapters III through VI.

An interactive and "error checking" user interface would help make GSLIB programs more user friendly. A graphical user interface (GUI) has been avoided because there is no single GUI that works on all computers. A useful addition to each program would be an *intelligent* interpretation program that would read each parameter file and create a verbose English language description of the "job" being described by the parameters.

When a serious error is encountered an error message is written to standard output and the program is stopped. Less serious problems or apparent inconsistencies cause a warning to be written to standard output or a debugging file, and the program will continue.

```
                    Parameters for HISTPLT
                    **********************

START OF PARAMETERS:
cluster.dat                          \data file
3    0                               \column for variable and weight
histplt.out                          \output PostScript file
-0.99    999999.                     \trimming limits
0.0      20.0                        \histogram minimum and maximum
40                                   \number of classes
0                                    \1=log scale, 0=arithmetic
Clustered Data                       \title
```

Figure II.2: An example parameter file for histplt.

II.2.5 Machine Precision

When calculations are performed on a computer, each arithmetic operation is generally affected by roundoff error. This error arises because the machine hardware can only represent a subset of the real numbers [60]. Roundoff error is not a major concern because the size of geostatistical matrices is typically small. Moreover, large problems are often separated into a large number of smaller problems by adopting local search neighborhoods.

Roundoff error is noticed most in the matrix solution of kriging systems; storing the matrix entries in single precision causes the kriging weights obtained from two different solution methods to change as soon as the third decimal place. For this reason, all matrix inversion subroutines in GSLIB have been coded in double precision. To save on storage, however, the LU simulation program lusim (V.6.2) is coded in single precision. Roundoff error becomes significant only when simulating very large grid systems.

Another source of imprecision is due to numerical approximations to certain mathematical functions. For example, the cumulative normal distribution $p = G(z)$ and inverse cumulative normal distribution $y = G^{-1}(p)$ have no closed form expression and polynomial approximations are used [100]. Imprecision in these approximations is small relative to the error introduced by the floating point representation of real numbers.

II.3 Variogram Model Specification

This section can be scanned quickly the first time through. The conventions for describing a variogram model are explained here rather than within all 12 kriging and simulation programs. Most of the kriging and simulation subroutines call for covariance or pseudo-covariance values; however, a semivariogram model rather than a covariance model must be specified. This apparent inconsistency allows for the traditional practice of modeling variograms and also permits the straightforward incorporation of the power model which has no covariance counterpart.

An acceptable semivariogram model for GSLIB consists of an isotropic nugget effect[4] and any positive linear combination of the standard semivariogram models (some dimensioning parameters may have to be increased if more than four nested structures are being considered). The standard models are:

1. **Spherical** model defined by an actual range a and positive variance contribution or *sill* value c.

$$\gamma(h) = c \cdot Sph\left(\frac{h}{a}\right) \begin{array}{ll} = c \cdot \left[1.5\frac{h}{a} - 0.5\left(\frac{h}{a}\right)^3\right], & \text{if } h \leq a \\ = c & \text{if } h \geq a \end{array} \qquad (\text{II.15})$$

2. **Exponential** model defined by a parameter a (effective range $3a$) and positive variance contribution value c.

$$\gamma(h) = c \cdot Exp\left(\frac{h}{a}\right) = c \cdot \left[1 - exp\left(-\frac{h}{a}\right)\right] \qquad (\text{II.16})$$

3. **Gaussian** model defined by a parameter a (effective range $a\sqrt{3}$) and positive variance contribution value c. [5]

$$\gamma(h) = c \cdot \left[1 - exp\left(-\frac{h^2}{a^2}\right)\right] \qquad (\text{II.17})$$

4. **Power** model defined by a power $0 < a < 2$ and positive slope c.

$$\gamma(h) = c \cdot h^a \qquad (\text{II.18})$$

The type of variogram structure is specified by an integer code, which is the order in the above list, i.e., $it = 1$: spherical model, $it = 2$: exponential model, $it = 3$: Gaussian model, and $it = 4$: power model. The a and c parameter values, which correspond to the description in the above list, are also needed.

Each nested structure requires an additional two or five parameters that define its own geometric anisotropy in 2-D or 3-D. Figure II.3 illustrates the angle and anisotropy factor required in 2-D:

- The rotation angle *ang* corresponds to an azimuth angle measured in degrees clockwise from the positive Y or north direction. The range parameter a is applied directly to this principal direction. The distances along the minor direction, that is, at 90 degrees from the principal direction, are obtained by multiplying a by the second parameter *anis*.

---

[4] Anisotropic nugget effects can be modeled by setting some of the directional ranges of the first nested structure to a very small value.

[5] Note that matrix instability problems are often encountered with a Gaussian model with no nugget effect [131,146]. Also, a Gaussian variogram model with no nugget effect should not be used for categorical variables [114].

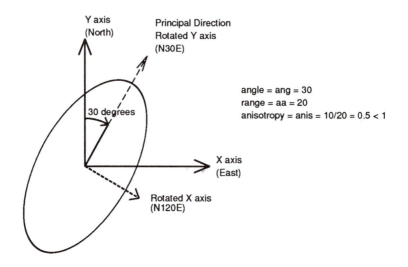

Figure II.3:   An example of the two parameters needed to define the geometric anisotropy of a variogram structure in 2-D.

- The anisotropy factor *anis* is the range in the minor direction divided by the range in the principal direction. Hence it is normally less than one. There is no requirement, however, that the *anis* parameter be less than one; for example, it may be set very large to model a zonal anisotropy. Note that a very large anisotropy factor will add the variogram structure in the principal direction and add nothing in any other direction, a feature known as "zonal anisotropy."

Within the software the actual distance is corrected so that it accounts for the specified anisotropy. That is, the distance component in the rotated X axis (see Figure II.3) is divided by *anis*. In other words, the anisotropy parameters do not apply to the *a* parameter of the variogram. Consequently, the anisotropy of the power model is handled in an intuitively correct manner; an anisotropic distance is calculated and the power *a* is left unchanged.

Figure II.4 illustrates the angles and anisotropy factors required in 3-D. Many software packages take a shortcut and only use two angles and two anisotropy factors. The added complexity of three angles is not in programming but in documentation. It is quite straightforward to visualize and document a phenomenon that is dipping with respect to the horizontal at a dip azimuth that is not aligned with a coordinate axis. The third angle is required to account for the geological concept of a plunge or rake. One example that requires a third angle is modeling the geometric anisotropy within the limbs of a plunging syncline.

The easiest way to describe the three angles and two anisotropy factors is to imagine the rotations and squeezing that would be required to transform a sphere into an ellipsoid, [6] see [57]. The outer shell of the ellipsoid consists of points at the same structural distance, e.g., if the ellipsoid is one half as large in one direction then the attribute is one half as continuous. We will refer to the original Y axis as the principal direction and consider the rotations such that it ends up being the actual principal structural direction (direction of maximum continuity):

- The first rotation angle *ang*1 rotates the original Y axis (principal direction) in the horizontal plane: this angle is measured in degrees clockwise.

- The second rotation angle *ang*2 rotates the principal direction from the horizontal: this angle is measured in negative degrees down from horizontal.

- The third rotation angle *ang*3 leaves the principal direction, defined by *ang*1 and *ang*2, unchanged. The two directions orthogonal to the principal direction are rotated clockwise relative to the principal direction when looking toward the origin. The rotation of the Third Step in Figure II.4 appears as counterclockwise since the view is away from the origin.

Zonal anisotropy can be considered as a particular case of a geometric anisotropy (see [76], p. 385-386). This can be handled by entering the anisotropy parameter *anis* as a very large number, which causes the implicit range in the minor direction to be infinity; the particular variogram structure is then added only to the major direction.

The program **vmodel** (see section VI.2.5) will write out the model semivariogram in any number of arbitrary directions and lags. This will help validate the correct entry of a semivariogram model.

Note that, whether in 2-D or 3-D, the anisotropy directions need not be the same for each nested structure, allowing for a great flexibility in modeling experimental anisotropy.

A variogram model should not be needlessly complex. Ideally, each variogram structure should have a physical interpretation. The more complicated the variogram model, the longer it takes to construct each kriging matrix; hence, the longer the kriging or simulation program will take.

II.3.1 A straightforward 2-D Example

Consider the semivariogram shown on Figure II.5. The dots are the experimental semivariogram points in two orthogonal directions. The semivariogram that reaches the sill first (at about 10-15 distance units) is in the

---

[6] The default is isotropy, i.e., no preferential direction of continuity; the five anisotropy parameters describe the transformations that an initially isotropic medium would have to go through to achieve the desired anisotropy. Conversely, we could have considered the transformations that an anisotropic medium would have to undergo to achieve isotropy.

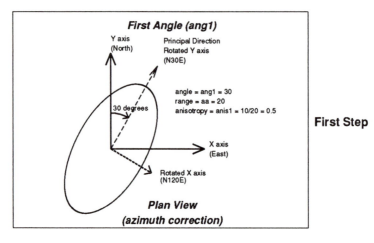

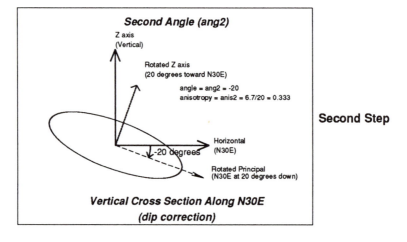

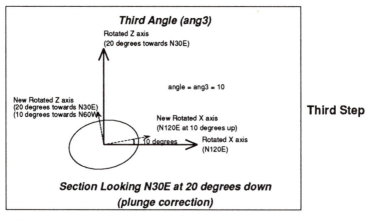

Figure II.4: An example of the five parameters needed to define the geometric anisotropy of a variogram structure in 3-D.

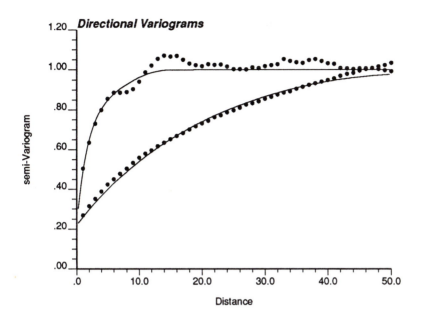

Figure II.5: Example semivariograms from a 2-D data set. The variogram with the longer range is in the east-west direction. The solid line is the fitted model (as described in the text).

north-south direction (an azimuth of 0.0) and the variogram with the longer range is in the east-west direction (an azimuth of 90 degrees). The solid line in both directions is the fitted semivariogram model. For an excellent discussion on variogram modeling refer to Chapter 16 of Isaaks and Srivastava [76].

The north-south model was fitted with a nugget effect of 0.22, an exponential structure with contribution 0.53 and range parameter a of 1.6, and a spherical structure with contribution 0.25 and range 15.0. The east-west model was fitted with nugget effect of 0.22, an exponential structure with contribution 0.53 and range parameter a of 16.0, and a spherical structure with contribution 0.25 and range 50.0. The semivariogram parameters required by the kriging or simulation programs would be specified as follows:

c0 = nugget = 0.22.

nst = number of nested structures = 2.

it(1) = type of structure 1 = 2 (exponential).

azimuth(1) = 90 degrees (the east-west direction).

cc(1) = contribution of structure 1 = 0.53.

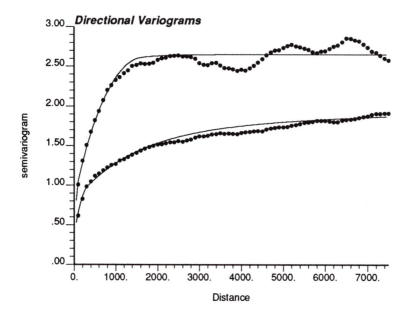

Figure II.6: Example semivariograms from a 2-D data set. The variogram with the higher sill is at azimuth 60 degrees (N60E). The solid line is the fitted model (as described in the text).

aa(1) = range of structure 1 in the direction **azimuth** = 16.0.

anis(1) = anisotropy of structure 1 = 1.6/16.0 = 0.10.

it(2) = type of structure 2 = 1 (spherical).

azimuth(2) = 90 degrees (the east-west direction).

cc(2) = contribution of structure 2 = 0.25.

aa(2) = range of structure 2 in the direction **azimuth** = 50.0.

anis(2) = anisotropy of structure 2 = 15.0/50.0 = 0.30.

II.3.2   A 2-D Example with Zonal Anisotropy

Consider the semivariogram shown on Figure II.6. The dots are the experimental semivariogram points in two orthogonal directions. The semivariogram that reaches the higher sill is at a direction N60E (an azimuth of 60.0) and the semivariogram with the lower sill is in the perpendicular direction (an azimuth of 150 degrees). The solid line in both directions is the fitted semivariogram model.

The semivariogram with the higher sill was fitted with a nugget effect of 0.40, a spherical structure with a contribution of 0.40 and range of 100.0, an exponential structure with contribution 0.95 and range parameter of 500.0, and a spherical structure with contribution 0.90 and range 1600.0. The semivariogram in the perpendicular direction (azimuth 150) was modeled with a nugget effect of 0.40, a spherical structure with a contribution of 0.40 and range of 300.0, an exponential structure with contribution 0.95 and range parameter 1500.0, and a spherical structure with contribution 0.90 and range 80000.0. Thus, for all practical purposes this last spherical structure does not contribute to direction N150E. This semivariogram model would be specified as follows:

c0 = nugget = 0.40.

nst = number of nested structures = 3.

it(1) = type of structure 1 = 1 (spherical).

azimuth(1) = 60 degrees (the direction with higher sill).

cc(1) = contribution of structure 1 = 0.40.

aa(1) = range of structure 1 in the direction **azimuth(1)** = 100.0.

anis(1) = anisotropy of structure 1 = 300.0/100.0 = 3.0.

it(2) = type of structure 2 = 2 (exponential).

azimuth(2) = 60 degrees (the direction with higher sill).

cc(2) = contribution of structure 2 = 0.95.

aa(2) = range of structure 2 in the direction **azimuth(2)** = 500.0.

anis(2) = anisotropy of structure 2 = 1500.0/500.0 = 3.0.

it(3) = type of structure 3 = 1 (spherical).

azimuth(3) = 60 degrees (the direction with higher sill).

cc(3) = contribution of structure 3 = 0.90.

aa(3) = range of structure 3 in the direction **azimuth(3)** = 1600.0.

anis(3) = anisotropy of structure 3 = 80000.0/1600.0 = 50.0.

II.4   Search Strategies

This section can be scanned quickly the first time through. The details of searching for nearby data are discussed here rather than with each program that calls for such a search.

Most kriging and simulation algorithms consider a limited number of nearby conditioning data. The first reason for this is to limit the CPU and memory requirements. The CPU time required to solve a kriging system increases as the number of data cubed, e.g., doubling the number of data leads to an eightfold increase in CPU time. The storage requirements for the main kriging matrix increases as the number of data squared, e.g., doubling the number of data leads to a fourfold increase in the memory requirements.

Furthermore, adopting a global search neighborhood would require knowledge of the covariance for the largest separation distance between the data. The covariance is typically poorly known for distances beyond one half or one third of the field size. A local search neighborhood does not call for covariance values beyond the diameter of the search ellipsoid.

A third reason for a limited search neighborhood is to allow local rescaling of the mean when using ordinary kriging, see section IV.1.2 and [97]. All of the data may have been pooled together to establish a reliable histogram and variogram; however, at the time of estimation it is often better to relax the decision of stationarity locally and use only nearby data.

A number of constraints are used to establish which nearby data should be considered:

1. Only those data falling within a search ellipsoid centered at the location being estimated are considered. This anisotropic[7] search ellipsoid is specified in the same way as an anisotropic variogram (see Figures II.3 and II.4): by a search radius, one or three angles specifying the orientation, and one or two anisotropy factors.

2. The allowable data may be further restricted by a specified maximum *ndmax*. The closest data, up to *ndmax*, are retained. In all the kriging programs, and in simulation programs where original data are searched independently from simulated nodes, closeness is measured by the Euclidean distance (possibly anisotropic). In the sequential simulation algorithm (discussed in V.1.3) the previously simulated grid nodes are searched by variogram distance.[8]

3. An octant search is available as an option to ensure that data are taken on all sides of the point being estimated. This is particularly important when working in 3-D with the data are often aligned along drillholes; an octant search ensures that data are taken from more than one drillhole.

---

[7]Beware that an artificially anisotropic search neighborhood may create artifact anisotropic structures in the resulting kriging and/or simulated maps.

[8]A small component of the Euclidean distance is added to the variogram distance so that the nodes are ordered by Euclidean distance beyond the range of the variogram.

An octant search is specified by choosing the number of data *noct* to retain from each octant.

If too few data (less than *ndmin*) are found then the node location being considered is left uninformed. This restricts estimation or simulation to areas where there are sufficient data.

Three different search algorithms have been implemented in different programs:

Exhaustive Search: the simplest approach is to check systematically all *nd* data and retain the *ndmax* closest that meet the three constraints noted above. This strategy is inefficient when there are many data and has been adopted only in the straightforward 2-D kriging programs (**okb2d** and **xvok2d**) and in the cokriging program (**cokb3d**).

Super Block Search: this search strategy (discussed in detail below) partitions the data into a grid network *super*imposed on the field being considered. When estimating any one location it is then possible to limit the search to those data falling in nearby *super blocks*. This search has been adopted for non-gridded data in most kriging and simulation programs.

Spiral Search: this search strategy (discussed in detail below) is for searching gridded data. The idea is to visit the closest nearby grid nodes first and spiral away until either enough data have been found or the remaining grid nodes are beyond the search limits. This search has been adopted in all sequential simulation progams.

Super Block Search

The *super block* search strategy is an efficient algorithm to be used in cases where many points are to be estimated, using local data neighborhoods, with the same set of original data. The algorithm calls for an initial classification and ordering of the data according to a regular network of parallelipedic blocks, see [94] page 361. This grid network is independent of any particular grid network of points/blocks being estimated or simulated. Typically, the size of the search network is much larger than the final estimation or simulation grid node spacing.

When estimating any one point, only those data within nearby super blocks have to be checked. A large number of data are thus quickly eliminated because they have been classified in super blocks beyond the search limits. This is illustrated in 2-D on Figure II.7 where an 11 by 11 super block grid network has been established over an area containing 140 data. When estimating a point anywhere within the dark gray super block, only those data within the dark black line need be considered. Note that all search resolution less than the size of a super block has been lost. Also note that the light gray region is defined by the search ellipse (circle in this case) with its center

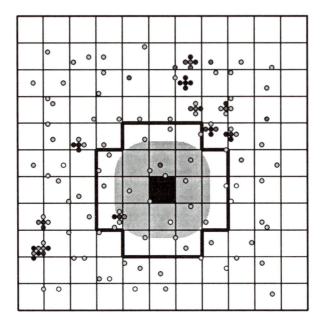

Figure II.7: An example of a super block search network.

translated to every node to be estimated within the dark gray super block. All super blocks intersected by the light gray region must be considered to ensure that all nearby data are considered for estimation of any node within the central dark gray superblock.

The first task is to build a template of super blocks, centered at the super block that contains the node being estimated. For example, the template is the relative locations of all 21 blocks enclosed by the solid line on Figure II.7. With this template, the nearby super blocks are easily established when considering any new location.

A second important computational *trick* is to sort all of the data by super block index number (the index is defined as in section II.2.3). An array (of size equal to the total number of blocks in the super block network) is constructed that stores the cumulative number of data for each super block and all super blocks with lesser block indices, i.e.,

$$cum(i) = \sum_{j=1}^{i} nisb(j)$$

where $nisb(j)$ is the number of data in super block j, and $cum(0) = 0$. Then, the number falling within any super block i is $cum(i) - cum(i-1)$ and their index location starts at $cum(i-1) + 1$. Therefore, this one array contains information on the number of data in each super block and their location in memory.

Spiral Search

The *spiral* search strategy is an efficient algorithm for cases when the data are (or have been relocated) on a regular grid. The idea is to consider values at grid nodes successively further away from the point being estimated. The ordering or template of nearby nodes to check is established on the basis of variogram distance. A spiral search template is constructed as follows:

1. Consider a *central* grid node at $ix = 0, iy = 0, iz = 0$

2. Programs that call for a spiral search require the maximum size of the search to be input as maximum dimensioning parameters *MAX-CTX*, *MAXCTY*, and *MAXCTZ* parameters which are odd integers.[9] The maximum search distance, in terms of grid nodes, is then $nx = MAXCTX/2$, $ny = MAXCTY/2$, $nz = MAXCTZ/2$ i.e.,

$$jx = -nx, \ldots, -1, 0, 1, \ldots, nx,$$

$$jy = -ny, \ldots, -1, 0, 1, \ldots, ny,$$

$$jz = -nz, \ldots, -1, 0, 1, \ldots, nz.$$

3. Compute the semivariogram between the fixed node ix, iy, iz and all nodes jx, jy, jz plus a small contribution of the corresponding Euclidean distance:

$$dis(jx, jy, jz) = \gamma(jx, jy, jz) + \epsilon \cdot h_{jx,jy,jz}$$

4. Sort the $dis(jx, jy, jz)$ array in ascending order.

5. Now, given a node kx, ky, kz the closest nodes are found by spiraling away according to the ordered jx, jy, jz offsets.

When building the spiral search template the covariance values for all offsets jx, jy, jz within the search ellipsoid are computed and stored to be later retrieved to construct the kriging matrices. This table look-up approach is much quicker than recomputing the covariance each time it is needed.

Two-part Search

Sequential simulation algorithms (see V.1.3) call for sequentially simulating the attribute at the grid node locations. A random path through all the nodes is followed, where the local distribution is conditioned to the original data and all previously simulated grid nodes. A common approximation is to relocate the original data to grid node locations before starting the simulation procedure. The advantage is that all conditioning data are then on a regular

---

[9] These parameters must be odd to allow n nodes to be searched on either side of a node being estimated, i.e., the search will be $2 \cdot n + 1$ nodes

grid which allows a spiral search and the use of covariance look-up tables for all covariance values needed in the kriging equations. The CPU time required is considerably reduced relative to the alternative which is to keep the original data separate from the previously simulated values, and use a two-part search strategy, that is, search for nearby original data with a superblock strategy and search for previously simulated nodes with a spiral search strategy.

Depending on the situation the more CPU-intensive two-part search may be appropriate. Considerations include the following:

- If the grid being simulated is coarse, then original data may have to be moved a significant distance from their actual locations; the final maps will honor the data at their relocated grid node locations and not their original locations. Also, when there are more than one data near a single grid node location only the closest is kept; thus, some data may not be used.

- There may be data outside of the grid network that contribute to the simulation. These data must be relocated to blocks at the edge of the grid network or a two-part search must be implemented to use these data.

- Commonly, the number of nodes to be simulated greatly exceeds the number of data; consequently, the original data will be dominated quickly by simulated values. Special patterns in the data, that are not captured by the variogram, will be reproduced better if there is a special effort to use the original data.

  Near the end of the simulation there will be so many close spaced previously simulated grid nodes that the long range variogram structure will not be used. By imposing the consideration of original data that are farther away, one can impart the long range structure by data "conditioning". Another approach to incorporate the long range structure is to consider a multiple-step simulation procedure; see section V.10.

All GSLIB programs that use the sequential simulation concept allow both options: relocation of original data to grid node locations or a two-part search strategy.

II.5   Data Sets

Problem sets are proposed at the end of each chapter to allow users to check proper installation of the programs and to check their understanding of the program scope and input parameter files. Some thoughts and *partial* solutions are given in Appendix A.

All problem sets refer to the same reference data set. Results based on this particular data set should not be used to draw general conclusions about the algorithms considered.

A complete 2-D gridded variable was created by simulated annealing where the first lag of a "low nugget" isotropic variogram was matched. The gridded data, defined below, provide reference data for all subsequent problem sets. It is important to note that most GSLIB programs work with 3-D data. Only 2-D data have been considered to keep the problems straightforward and easy to visualize.

The reference 2-D data file of 2500 values is characterized by the following geometric parameters:

xmn $=0.5$ (origin of x axis).

nx $=50$ (number of nodes in x direction).

xsiz $=1.0$ (spacing of nodes in x direction).

and similarly, **ymn**$=0.5$, **ny**$=50$, and **ysiz**$=1.0$.

Four sample data sets were derived from this full valued 2-D grid of 2500 values. The following data files may be found in the **data** subdirectory on the distribution diskettes:

1. The complete reference data set contained in $\boxed{\text{true.dat}}$.

2. A clustered sample of 140 values was drawn from the reference data set. The first 97 samples were taken on a pseudo-regular grid and the last 43 samples were clustered in the high-valued regions (as identified from the first 97 samples). This data set, found in $\boxed{\text{cluster.dat}}$ is used for the first two problem sets on exploratory data analysis and variogram analysis.

3. A subset of 29 samples was retained in $\boxed{\text{data.dat}}$ for selected kriging and simulation exercises. This smaller number of samples is more typical of many applications and also better illustrates the use of secondary variables.

4. An exhaustively sampled secondary variable (2500 values) was created to illustrate techniques that allow the incorporation of different attributes. These 2500 secondary data *and* the 29 primary sample data are contained in $\boxed{\text{ydata.dat}}$.

The histogram of the 2500 values in $\boxed{\text{true.dat}}$ is shown on Figure II.8. This histogram was created with the program **histplt** described in Chapter VI. The parameter file was presented earlier (Figure II.2).

A gray scale map of these reference values is shown on Figure II.9. The 140 clustered data shown on Figure II.10 were taken from these reference values.

A more exhaustive analysis of the sample data $\boxed{\text{cluster.dat}}$ is asked in the following problem set. A limited solution to this exploratory data analysis is given in Appendix A.

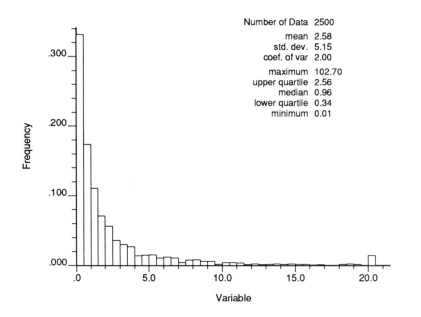

Number of Data	2500
mean	2.58
std. dev.	5.15
coef. of var	2.00
maximum	102.70
upper quartile	2.56
median	0.96
lower quartile	0.34
minimum	0.01

Figure II.8: Histogram of the complete 2-D gridded reference data in ⟨ true.dat ⟩.

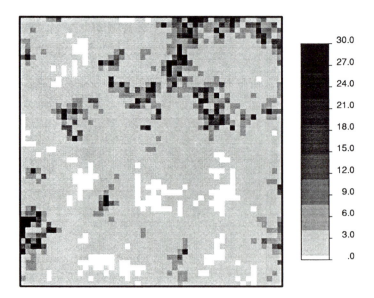

Figure II.9: The 2500 reference data values in ⟨ true.dat ⟩.

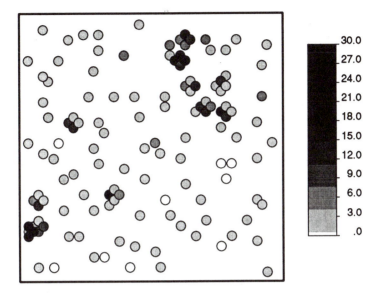

Figure II.10: Location map of the sample data in cluster.dat .

II.6   Problem Set One: Data Analysis

The goal of this first problem set is to learn the basics of GSLIB through experimentation with data analysis techniques. Also, the user will get familiar with the data sets used extensively in later chapters. For the sole purpose of exploratory data analysis, there is no need nor advantage in using GSLIB programs; any graphical-based software would be appropriate, e.g., Geo-EAS [50], Geostatistical Toolbox [54], SAS, S, your own custom programs, etc.

The GSLIB software must first be installed and compiled (see Appendix B). The specific programs called for in this problem set are documented in Chapter VI.

The data

The data can be interpreted as representing the sampling of a hazardous waste site. There are 97 initial samples taken on a pseudo-regular grid over a 2-D area 50 miles by 50 miles. An additional 43 samples were taken around those original sample locations with high concentration values. The total 140 samples are used to infer global statistics representative of the entire 2500 square-mile area. The location map of these samples is shown in Figure II.10; the corresponding data are stored in cluster.dat .

Questions

1. Comment on the sample data and the effect of clustering on the sample histogram and statistics (mean, variance, and high quantiles).

2. Create an equal-weighted sample histogram, a normal and lognormal probability plot. Comment on the results.

3. Can you think of a quick-and-easy way to correct the sample histogram for spatial clustering (in geostatistical jargon to "decluster" the sample histogram)? Try it.

4. Perform cell declustering using the original 140 data. Choose a reasonable range of cell sizes. What would be a "natural" cell size? Plot a scatterplot of the declustered cell mean versus the cell size. Check out the cell size providing the smallest mean and calculate the corresponding declustered histogram. Comment on the results. For more details about declustering algorithms see [76].

5. Generate the histogram and probability plot of the 2500 reference data stored in $\boxed{\text{true.dat}}$ (see Figure II.9). Comment on the effectiveness of the declustering algorithms you have tried.

6. Plot Q-Q (quantile-quantile) and P-P (cumulative probabilities) plots of the sample data (both equal-weighted and declustered) versus the reference data. Comment on the results.

Chapter III

Variograms

This chapter presents the variogram programs of GSLIB. These facilities extend well beyond the calculation of traditional variograms and cross variograms. A single subroutine call allows alternative measures of spatial continuity such as the correlogram or pairwise relative semivariogram to be computed in addition to the semivariogram. These alternative measures are known to be much more resistant to data sparsity, outliers values, clustering, and sampling error.

Section III.1 presents the ten measures of spatial variability/continuity considered by GSLIB with a brief discussion of their respective advantages and disadvantages. The principles underlying the four variogram subroutines are presented in section III.2. The detailed parameter specification of the programs for regularly spaced data is given in sections III.3 (2-D data) and III.4 (3-D data). Sections III.5 and III.6 handle the corresponding 2-D and 3-D cases for irregularly spaced data.

Some useful application notes are presented in section III.7. Finally, section III.8 proposes a problem set that explores the capability of the various variogram programs.

III.1   Measures of Spatial Variability

A variogram as defined in relation (II.11) or, more generally, a measure of spatial variability, is the key to any geostatistical study. In essence, the variogram replaces the Euclidean distance $\mathbf{h}$ by a structural distance $2\gamma(\mathbf{h})$ that is specific to the attribute and the field under study. The variogram distance measures the average degree of dissimilarity between an unsampled value $z(\mathbf{u})$ and a nearby data value. For example, given only two data values $z(\mathbf{u}+\mathbf{h})$ and $z(\mathbf{u}+\mathbf{h}')$ at two different locations; the more "dissimilar" sample value should receive lesser weight in the estimation of $z(\mathbf{u})$.

The distance measure need not be inferred from the sample semivariogram (III.1); the sole constraint is that the measure of distance be modeled by a

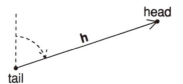

Figure III.1: An example of the *tail* and *head* naming convention.

positive definite model[1] (to ensure existence and uniqueness of solutions to
the kriging equations; see [24]).

The following nine experimental measures of spatial variability/continuity
can be computed with GSLIB subroutines:

1. *Semivariogram:* this traditional measure is defined as half of the av-
    erage squared difference between two attribute values approximately
    separated by vector $\mathbf{h}$:

$$\gamma(\mathbf{h}) = \frac{1}{2N(\mathbf{h})} \sum_{i=1}^{N(\mathbf{h})} (x_i - y_i)^2 \qquad (III.1)$$

    where $N(\mathbf{h})$ is the number of pairs, x_i is the value at the start or *tail*
    of the pair i and y_i is the corresponding end or *head* value. Figure III.1
    illustrates this naming convention relative to the separation vector $\mathbf{h}$.
    In general, the separation vector $\mathbf{h}$ is specified with some direction and
    distance (lag) tolerance. It is not recommended to use expression (III.1)
    with tail and head values corresponding to two different variables. The
    resulting cross "variogram" is not identical with the sample cross semi-
    variogram (see definition (III.2) and the related discussion in section
    III.7). A better measure of cross dependence is the cross covariance
    (see definition (III.3)).

2. *Cross semivariogram:* this measure of cross variability is defined as
    half of the average product of $\mathbf{h}$-increments relative to two different
    attributes:

$$\gamma_{ZY}(\mathbf{h}) = \frac{1}{2N(\mathbf{h})} \sum_{i=1}^{N(\mathbf{h})} (z_i - z_i')(y_i - y_i') \qquad (III.2)$$

    where z_i is the value of attribute z at the tail of pair i and z_i' is the
    corresponding head value; the locations of the two values z_i and z_i' are
    separated by vector $\mathbf{h}$ with specified direction(s) and distance tolerance.
    $(y_i - y_i')$ is the corresponding $\mathbf{h}$-increment of the other attribute y.

---

[1] In fact, it is the covariance (constant minus $\gamma(\mathbf{h})$) which must be positive definite. All
variogram models considered by GSLIB correspond to allowable, positive definite, covariance models. The only exception is the power variogram model, which has no covariance
counterpart and cannot be used with simple kriging.

3. *Covariance:* this measure also called the *non-ergodic* covariance [77] is the traditional covariance commonly used in statistics. The covariance does not implicitly assume that the mean of the tail values is the same as the mean of the head values ([76], p. 59).

$$C(\mathbf{h}) = \frac{1}{N(\mathbf{h})} \sum_{i=1}^{N(\mathbf{h})} x_i y_i - m_{-\mathbf{h}} m_{+\mathbf{h}} \qquad (III.3)$$

where $m_{-\mathbf{h}}$ is the mean of the tail values, i.e., $m_{-\mathbf{h}} = \frac{1}{N(\mathbf{h})} \sum_{1}^{N(\mathbf{h})} x_i$, and $m_{+\mathbf{h}}$ is the mean of the head values, i.e., $m_{+\mathbf{h}} = \frac{1}{N(\mathbf{h})} \sum_{1}^{N(\mathbf{h})} y_i$. If x and y refer to two different attributes, expression (III.3) identifies the sample cross covariance.

4. *Correlogram:* the previous measure (III.3) is standardized by the respective tail and head standard deviations:

$$\rho(\mathbf{h}) = \frac{C(\mathbf{h})}{\sigma_{-\mathbf{h}} \sigma_{+\mathbf{h}}} \qquad (III.4)$$

where $\sigma_{-\mathbf{h}}$ and $\sigma_{+\mathbf{h}}$ are the standard deviation of the tail and head values, i.e.,

$$\sigma_{-\mathbf{h}}^2 = \frac{1}{N(\mathbf{h})} \sum_{i=1}^{N(\mathbf{h})} x_i^2 - m_{-\mathbf{h}}^2, \text{ and } \sigma_{+\mathbf{h}}^2 = \frac{1}{N(\mathbf{h})} \sum_{i=1}^{N(\mathbf{h})} y_i^2 - m_{+\mathbf{h}}^2$$

When x and y refer to two different attributes, expression (III.4) identifies the sample cross correlogram. Moreover, $\rho(\mathbf{h} = 0)$ identifies the (linear) correlation coefficient between the two attributes.

5. *General relative semivariogram:* the semivariogram as defined in (III.1) is standardized by the squared mean of the data used for each lag:

$$\gamma_{GR}(\mathbf{h}) = \frac{\gamma(\mathbf{h})}{\left(\frac{m_{-\mathbf{h}} + m_{+\mathbf{h}}}{2}\right)^2} \qquad (III.5)$$

6. *Pairwise relative semivariogram:* each pair is normalized by the squared average of the tail and head values:

$$\gamma_{PR}(\mathbf{h}) = \frac{1}{2N(\mathbf{h})} \sum_{1}^{N(\mathbf{h})} \frac{(x_i - y_i)^2}{(\frac{x_i + y_i}{2})^2} \qquad (III.6)$$

Practical experience has shown that the general relative (III.5) and pairwise relative (III.6) sample variograms are resistant to data sparsity and outliers when applied to positively skewed sample distributions. They sometimes reveal spatial structure and anisotropy that could not be detected otherwise. Because of the divisors in expressions (III.5) and (III.6), the general and relative pairwise variograms should be limited to strictly positive variables.

7. *Semivariogram of logarithms:* the semivariogram is computed on the natural logarithms of the original variables (provided that they are positive):

$$\gamma_L(\mathbf{h}) = \frac{1}{2N(\mathbf{h})} \sum_1^{N(\mathbf{h})} [\ln(x_i) - \ln(y_i)]^2 \tag{III.7}$$

8. *Semirodogram:* this measure is similar to the traditional variogram; instead of squaring the difference between x_i and y_i, the square root of the absolute difference is taken:

$$\gamma_R(\mathbf{h}) = \frac{1}{2N(\mathbf{h})} \sum_1^{N(\mathbf{h})} \sqrt{|x_i - y_i|} \tag{III.8}$$

9. *Semimadogram:* this measure is similar to the traditional variogram; instead of squaring the difference between x_i and y_i, the absolute difference is taken:

$$\gamma_M(\mathbf{h}) = \frac{1}{2N(\mathbf{h})} \sum_1^{N(\mathbf{h})} |x_i - y_i| \tag{III.9}$$

Madograms and rodograms are particularly useful for establishing large-scale structures (range and anisotropy). They should not be used for modeling the nugget variance of semivariograms.

10. *Indicator semivariogram:* [2] the semivariogram is computed on an internally constructed indicator variable. This requires the specification of a continuous variable and cutoff to create the indicator transform. For the cutoff cut_k and datum value x_i the indicator transform ind_i is defined as:

$$ind_i = \begin{cases} 1, & \text{if } x_i \le cut_k \\ 0, & \text{otherwise} \end{cases} \tag{III.10}$$

where the subscript i refers to a particular location.

Within the GSLIB source code, the variogram type is specified by the integer number in the list given above, i.e., 1 for the traditional semivariogram, 2 for the covariance, etc.

---

[2] An evident alternative is to define the indicator variable outside of the variogram program, then call for variogram type 1 as in (III.1). It is often more convenient to define the indicator variable as 1 if the indicator variable exceeds the cutoff. The indicator variogram is unchanged by this transform, i.e., $\gamma_I = \gamma_J$, with $J = 1 - I$.

Any of the previous experimental distance/correlation measures can be used qualitatively to infer structures of spatial continuity. More than one type of measure can be computed at the same time. For example, modeling of the sample variogram can be facilitated by inferring anisotropy directions and range ratios from the more robust relative variograms or rodogram. In general, the features observed along the abscissa axis (distance $|\mathbf{h}|$) are common to all measures of variability/continuity (perhaps more apparent on some). The actual variogram distance value, however, as read from the ordinate axis, is specific to the variogram type chosen.

Since the actual variogram value depends on the type chosen, the resulting kriging variance should be interpreted only in relative terms. Many excellent discussions on variogram inference, interpretation, and modeling are available in the literature; see [10,27,30,76,77,84,87,94,127,144,145,146]; these discussions will not be reproduced here. Practitioners should expect to spend more time on data analysis and variogram modeling than on all kriging and simulation runs combined.

The MG Case

When a multivariate Gaussian model is being used, as in the MG approach to kriging or simulation, a covariance model for the normal score transforms of the data is required. The corresponding semivariogram model should have a sill of 1.0, which excludes the power model unless used only for small distances $\mathbf{h}$. Any of the three distance measures (III.1), (III.3), or (III.4), may be used since they provide the correct ordinate axis scaling. The other measures, however, can help in detecting nested structures, anisotropy, and evaluating ranges.

III.2 GSLIB Variogram Subroutines

The four variogram subroutines provided with GSLIB differ in their capability to handle 2-D versus 3-D data and regular versus irregular data layout. The first two variogram subroutines **gam2** and **gam3** are for gridded data. The other two variogram subroutines **gamv2** and **gamv3** are for irregularly spaced data in two and three dimensions. **gam2** or **gamv2** can be used for data distributed in 1-D by ignoring the second dimension.

The subroutines can handle many different directions, variables, and variogram types in a single pass. For example, it is possible to consider four different variables, eight directions, all four direct (auto-)variograms, two cross variograms, and the correlogram in one subroutine call. This flexibility has been achieved by making users explicitly state the tail variable, head variable, and variogram type for every "variogram"[3] they want to compute.

---

[3] The generic term "variogram" will be used hereafter to designate any measure of spatial variability/continuity. Whenever appropriate the specific distance measure will be specified, e.g., semivariogram (III.1) or rodogram (III.8).

All directions called for are computed for every variogram.

The output of the "variogram" programs are actually semivariogram or semirodogram values.

Warning: When specifying the tail variable, head variable, and variogram type the user should understand what is being requested. For example, it would be difficult to interpret the cross relative variogram of porosity and permeability. Rather the cross covariance of porosity-permeability should be inferred and modeled through expression (III.3).

Gridded data

When data are on a regular grid, the directions are specified by giving the number of grid nodes that must be shifted to move from a node on the grid to the next nearest node on the grid that lies along the directional vector. Some examples for a cubic grid:

X shift	Y shift	Z shift	
1	0	0	aligned along the X axis
0	1	0	aligned along the Y axis
0	0	1	aligned along the Z axis
1	1	0	horizontal at 45° from Y
1	-1	0	horizontal at 135° from Y.
1	1	1	dipping at -45°, 225° clockwise from Y.

Within the **gam2** and **gam3** subroutines the directions are specified by offsets of the form shown above. No direction or lag tolerances are explicitly specified. In some cases this direction definition is too restrictive and may result in not enough pairs being combined. The solution is either to average multiple directions from **gam2/gam3** with weighting by the number of pairs, or to store the grid as irregularly spaced data and use **gamv2/gamv3** to compute the variograms (at substantially more computational cost).

Irregularly Spaced Data:

The angles and lag distances are entered explicitly for irregularly spaced data. The azimuth angle is measured clockwise from north-south and the dip angle measured in negative degrees down from horizontal (see Figure III.2). Angular half window tolerances are required for both the azimuth and the dip. These tolerances may overlap causing pairs to report to more than one direction. The unit lag distance and the lag tolerance are also required (see Figure III.2). Each pair will report to only one lag even though the lag tolerance may be specified as greater than half the lag distance. The code will require modification if a user wishes to *smooth* the variogram by having pairs report to more than one lag.

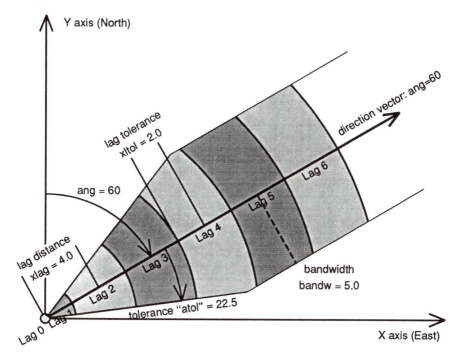

Figure III.2:   Some of the parameters required by **gamv2** and **gamv3**. The azimuth angle is measured clockwise from North; the angle tolerance (half window) is restricted once the deviation from the direction vector exceeds the bandwidth; the shaded areas represent the different lags.

Note that the first lag reported in the output file (*lag0*) corresponds to $|h| \in [0, \epsilon)$. For a cross covariance and cross correlogram this value corresponds to the standard covariance or correlation coefficient of collocated points. The second lag reported to the output file (*lag1*) corresponds to $|h| \in (\epsilon, xlag - xtol)$.

General comments

A number of general comments are useful before documenting each program in more detail:

- The pairwise relative variogram and the variogram of logarithms require that the data be strictly positive. When the pairwise relative variogram is calculated, the sum of each pair is checked and the pair is discarded if that sum is less than a small positive minimum. When computing the variogram of logarithms, both the tail and head value must be greater than a small positive minimum. The user may wish to perform the necessary checks before passing the data to the calculation subroutine.

- There are no default variograms computed. For all subroutines and main programs the user must specify the number of variograms *nvarg* and their corresponding types. Variograms are not automatically computed for the variables in the data file.

- It may be convenient to perform some data transformation outside the subroutines. For example, normal scores, indicator, or logarithm transformations are better performed within a preprocessing program or a customized main program.

- The present allowance for indicator variograms is not especially efficient. For most applications it will not really matter, but the execution time could be considerably reduced with special coding (i.e., using Boolean logic). The current approach is to create a new variable (a 1/0 transform) internally within the main program. A more efficient coding would make use of the particular features of indicator variables, e.g., there must be one 1 and one 0 before anything is added to the experimental variogram, and a prior ordering of the indicator cutoffs would be beneficial.

- The general nature of these variogram programs makes them slightly inefficient. Checking for which type of variogram to compute adds about 3% to the overall execution time. This part of the code could be removed if one is interested in a single variogram type.

III.3   Regularly Spaced 2-D Data

The following files should be located on the diskette(s) before attempting to compile and execute this program/subroutine:

```
                    Parameters for GAM2M
                    ********************

START OF PARAMETERS:
true.dat                            \Data File
1    1    2    3                    \nvar; column numbers...
-1.0e21      1.0e21                 \tmin, tmax (trimming limits)
gam2.var                           \Output File for Variogram
50     50    1                      \nx,    ny,     igrid
1.0   1.0                          \xsiz, ysiz
2   20                             \ndir,nlag
  1  0                             \ixd(i),iyd(i)  i=1,...,ndir
  0  1
4                                  \number of variograms
1    1    1                        \tail, head, variogram type
1    1    3                        \tail, head, variogram type
1    1    6                        \tail, head, variogram type
1    1   10   2.5                  \tail, head, variogram type
```

Figure III.3: An example parameter file for **gam2m**.

gam2m.f an example driver program for **gam2**

gam2.inc an include file with maximum array dimensions

gam2.f the subroutine **gam2**

gam2.par an example parameter file for **gam2m**

The parameters required by the main program **gam2m** are listed below and shown on Figure III.3:

- **datafl:** the input data in a simplified Geo-EAS formatted file. The data are ordered rowwise (x cycles fastest, then y).

- **nvar** and **ivar(1)** ...**ivar(nvar):** the number of variables and their column order in the data file.

- **tmin** and **tmax:** all values, regardless of which variable, strictly less than *tmin* and greater than or equal to *tmax* are ignored.

- **outfl:** the output variograms are written to a single output file named *outfl*. The output file contains the variograms ordered by direction and then variogram type specified in the parameter file (the directions cycle fastest then the variogram number). For each variogram there is a one-line description and then *nlag* lines with the following:

  1. lag number (increasing from 1 to *nlag*).

  2. separation distance for the lag.

  3. the *semivariogram* value (whatever type was specified).

  4. number of pairs for the lag.

  5. mean of the data contributing to the tail.

  6. mean of the data contributing to the head.

  7. the tail and head variances (for the correlogram).

The **vargplt** program documented in section VI.1.5 may be used to create PostScript displays of multiple variograms.

- **nx, ny** and **igrid:** the number of nodes in the x and y direction and the grid number in the input file. One dimensional data may be considered by setting either nx or ny to 1. Often, **gam2** is used to check the variogram reproduction of realizations from a simulation program. In this case there is typically more than one realization or grid in the file. The parameter *igrid* is used to specify which grid is to be used for the variogram computations. Recall that realizations or grids are written one after another; therefore, if *igrid=2* the input file must contain at least $2 \cdot nx \cdot ny$ values and the second set of $nx \cdot ny$ values will be taken as the second grid.

- **xsiz** and **ysiz:** the x and y spacing of the grid nodes.

- **ndir** and **nlag:** the number of directions and lags to consider. The same number of lags are considered for all directions and all directions are considered for all of the *nvarg* variograms specified below.

- **ixd** and **iyd:** these two arrays specify the unit offsets that define each of the *ndir* directions (see section III.2).

- **nvarg:** the number of variograms to compute.

- **ivtail, ivhead** and **ivtype:** for each of the *nvarg* variograms one must specify which variables should be used for the tail and head and which type of variogram is to be computed. For direct variograms the *ivtail* array is identical to the *ivhead* array. Cross variograms are computed by having the tail variable different from the head variable, e.g., if *ivtail(i)* is set to 1, *ivhead(i)* is set to 2, and *ivtype(i)* is set to 2, then distance measure i will be a cross semivariogram between variable 1 and variable 2. Note that *ivtype(i)* should be set to something that makes sense (e.g., types 1,2, or 3); a cross relative variogram would be difficult to interpret. Further, note that for the cross semivariogram (*ivtype=2*) the two variables *ivtail* and *ivhead* are used at both the tail and head locations. The *ivtype* variable corresponds to the integer code in the list given in section III.1.

- **cut:** whenever the *ivtype* is set to 10, i.e., asking for an indicator variogram, then a cutoff must be specified immediately after the *ivtype* parameter on the same line in the input file (see Figure III.3). Note that if *ivtype(i)* is set to 10, then the cutoff applies to variable *ivtail(i)* in the input file (although the *ivhead(i)* variable is not used it must

```
                          Parameters for GAM3M
                          *********************

START OF PARAMETERS:
imagine.dat                             \data file
2    1    2                             \nvar; column numbers...
-1.0e21       1.0e21                    \tmin, tmax (trimming limits)
gam3.var                                \output file for variograms
100     100       50        1           \nx,   ny,    nz,    igrid
1.0    1.0     1.0                      \xsiz, ysiz, zsiz
3   5                                   \ndir,nlag
  1   0   0                             \ixd(i),iyd(i),izd(i),  i=1,ndir
  0   1   0
  0   0   1
4                                       \number of variograms
1    1    1                             \tail, head, variogram type
1    1    3                             \tail, head, variogram type
2    2    1                             \tail, head, variogram type
2    2    3                             \tail, head, variogram type
```

Figure III.4: An example parameter file for **gam3m**.

be present in the file to maintain consistency with the other variogram types).

Regularly spaced data in 1-D can be handled by setting ny to one and iyd to zero.

The maximum size of the input array, the maximum number of variograms, and other maximum dimensioning parameters are specified in the file gam2.inc which is included in both the **gam2** subroutine and the **gam2m** main program.

III.4   Regularly Spaced 3-D Data

The following files should be located on the diskette(s) before attempting to compile and execute this program/subroutine:

gam3m.f an example driver program for **gam3**

gam3.inc an include file with maximum array dimensions

gam3.f the subroutine **gam3**

gam3.par an example parameter file for **gam3m**

The parameters required by the main program **gam3m** are very similar to those required by the **gam2m** program. The parameters required by the main program **gam3m** are listed below and shown on Figure III.4:

- **datafl:** the input data in a simplified Geo-EAS formatted file. The data are ordered level by level and then rowwise (x cycles fastest, then y, and finally z).

- **nvar** and **ivar(1)** ...**ivar(nvar):** the number of variables and their column order in the data file.

- **tmin** and **tmax:** all values, regardless of which variable, strictly less than *tmin* and greater than or equal to *tmax* are ignored.

- **outfl:** the output variograms are written to a single output file named *outfl*. The output file contains the variograms ordered by direction and then variogram type specified in the parameter file (the directions cycle fastest then the variogram number). For each variogram there is a one-line description and then *nlag* lines with the following:

  1. lag number (increasing from 1 to *nlag*).
  2. separation distance for the lag.
  3. the *semivariogram* value (whatever type was specified).
  4. number of pairs for the lag.
  5. mean of the data contributing to the tail.
  6. mean of the data contributing to the head.
  7. the tail and head variances (for the correlogram).

- **nx, ny, nz** and **igrid:** the number of nodes in the x, y, and z directions and the grid number in the input file. Often, **gam3** is used to check the variogram reproduction of realizations from a simulation program. In this case there is typically more than one realization or grid in the file. The parameter *igrid* is used to specify which grid is to be used for the variogram computations. Recall that realizations or grids are written one after another; therefore, if *igrid=2* the input file must contain at least $2 \cdot nx \cdot ny \cdot nz$ values and the second set of $nx \cdot ny \cdot nz$ values will be taken as the second grid.

- **xsiz, ysiz, and zsiz:** the x, y, and z spacing of the grid nodes.

- **ndir** and **nlag:** the number of directions and lags to consider. The same number of lags are considered for all directions and all directions are considered for all of the *nvarg* variograms specified below.

- **ixd, iyd, and izd:** these three arrays specify the unit offsets that define each of the *ndir* directions (see section III.2).

- **nvarg:** the number of variograms to compute.

- **ivtail, ivhead** and **ivtype:** for each of the *nvarg* variograms one must specify which variables should be used for the tail and head and which type of variogram is to be computed. For direct variograms the *ivtail* array is identical to the *ivhead* array. Cross variograms are computed by having the tail variable different from the head variable. The *ivtype* variable corresponds to the integer code in the list given in section III.1.

- **cut:** whenever the *ivtype* is set to 10, i.e., asking for an indicator variogram, then a cutoff must be specified immediately after the *ivtype* parameter on the same line in the input file (see Figure III.4). Note that if *ivtype(i)* is set to 10 then the cutoff applies to variable *ivtail(i)* in the input file (although the *ivhead(i)* variable is not used it must be present in the file to maintain consistency with the other variogram types).

The maximum size of the input array and the maximum number of variograms that can be considered at once are specified in the file gam3.inc which is included in both the **gam3** subroutine and the **gam3m** main program.

III.5   Irregularly Spaced 2-D Data

The following files should be located on the diskette(s) before attempting to compile and execute this program/subroutine:

gamv2m.f an example driver program for **gamv2**

gamv2.inc an include file with maximum array dimensions

gamv2.f the subroutine **gamv2**

gamv2.par an example parameter file for **gamv2m**

Figure III.2 graphically illustrates some of the distance and direction parameters. An example of a parameter file is shown in Figure III.5. The parameters required by **gamv2m** are:

- **datafl:** the input data in a simplified Geo-EAS formatted file.

- **icolx** and **icoly:** the columns for the x and y coordinates.

- **nvar** and **ivar(1)** ...**ivar(nvar):** the number of variables and their column order in the data file.

- **tmin** and **tmax:** all values, regardless of which variable, strictly less than *tmin* and greater than or equal to *tmax* are ignored.

- **outfl:** the output variograms are written to a single output file named *outfl*. The output file contains the variograms ordered by direction and then variogram type specified in the parameter file (the directions cycle fastest then the variogram number). For each variogram there is a one-line description and then *nlag* lines with the following:

  1. lag number (increasing from 1 to *nlag*).
  2. separation distance for the lag.

```
                        Parameters for GAMV2M
                        *********************

START OF PARAMETERS:
cluster.dat                              \Data File in GEOEAS format
1    2                                   \columns for x and y coordinates
1    3    5    6                         \nvar; column numbers...
-1.0e21      1.0e21                      \tmin, tmax (trimming limits)
gamv2.var                               \Output File for Variograms
20                                       \nlag - the number of lags
1.0                                      \xlag - unit separation distance
0.5                                      \xltol- lag tolerance
3                                        \ndir - number of directions
 0.0   90.0    25.0                      \azm(i),atol(i),bandw(i)i=1,ndir
 0.0   22.5    10.0
90.0   22.5    10.0
4                                        \number of variograms
1    1    1                              \tail, head, variogram type
1    1    3                              \tail, head, variogram type
1    1    6                              \tail, head, variogram type
1    1   10 2.5                          \tail, head, variogram type
```

Figure III.5: An example parameter file for **gamv2m**.

3. the *semivariogram* value (whatever type was specified).

4. number of pairs for the lag.

5. mean of the data contributing to the tail.

6. mean of the data contributing to the head.

7. the tail and head variances (for the correlogram).

- **nlag:** the number of lags to compute (same for all directions).

- **xlag:** the unit lag separation distance.

- **xltol:** the lag tolerance. This could be one-half of *xlag* or smaller to allow for data on a pseudo-regular grid. If *xltol* is entered as negative or zero it will be reset to *xlag/2*. Even if *xltol* is greater than *xlag/2* each pair will report to only one lag.

- **ndir:** the number of directions to consider. All these directions are considered for all of the *nvarg* variograms specified below.

- **azm, atol,** and **bandw:** the azimuth angle, the half window azimuth tolerance, and the "bandwidth" or maximum acceptable distance in the direction perpendicular to the direction vector; see Figure III.2. The azimuth is measured in degrees clockwise from north, e.g., $azm = 0$ is north, $azm = 90$ is east, and $azm = 135$ is south-east.

- **nvarg:** the number of variograms to compute.

- **ivtail, ivhead** and **ivtype:** for each of the *nvarg* variograms one must specify which variables should be used for the tail and head and which type of variogram is to be computed. For direct variograms the *ivtail* array is identical to the *ivhead* array. Cross variograms are computed by having the tail variable different from the head variable. The *ivtype* variable corresponds to the integer code in the list given in section III.1.

- **cut:** whenever the *ivtype* is set to 10, i.e., asking for an indicator variogram, then a cutoff must be specified immediately after the *ivtype* parameter on the same line in the input file (see Figure III.5). Note that if *ivtype(i)* is set to 10 then the cutoff applies to variable *ivtail(i)* in the input file (although the *ivhead(i)* variable is not used it must be present in the file to maintain consistency with the other variogram types).

The maximum number of data, the maximum number of variograms, and other maximum dimensioning parameters are specified in the file gamv2.inc which is included in both the **gamv2** subroutine and the **gamv2m** main program.

III.6   Irregularly Spaced 3-D Data

The following files should be located on the diskette(s) before attempting to compile and execute this program/subroutine:

gamv3m.f an example driver program for **gamv3**

gamv3.inc an include file with maximum array dimensions

gamv3.f the subroutine **gamv3**

gamv3.par an example parameter file for **gamv3m**

The **gamv3** subroutine is very similar to the **gamv2** subroutine. The documentation below is largely redundant with the preceding section, but is included for completeness. Figure III.2 graphically illustrates some of the parameters that specify the distance and direction parameters and Figure III.6 shows an example parameter file. The parameters required by **gamv3m** are:

- **datafl:** the input data in a simplified Geo-EAS formatted file.

- **icolx, icoly** and **icolz:** the columns for the x, y, and z coordinates.

- **nvar** and **ivar(1)** ...**ivar(nvar):** the number of variables and their column order in the data file.

- **tmin** and **tmax:** all values, regardless of which variable, strictly less than *tmin* and greater than or equal to *tmax* are ignored.

```
                    Parameters for GAMV3M
                    *********************

START OF PARAMETERS:
imagine.dat                          \data file
1   2   3                            \column for x,y, z coordinates
2   4   5                            \nvar; column numbers...
-1.0e21     1.0e21                    \tmin, tmax (trimming limits)
gamv3.var                            \output file for variograms
20                                   \nlag - the number of lags
1.0                                  \xlag - unit separation distance
0.5                                  \xltol- lag tolerance
4                                    \ndir - number of directions
    0.0  90.0 25.0   0.0  90.0  25.0 \azm,atol,bandwh,dip,dtol,bandwd
    0.0  22.5 25.0   0.0  22.5  25.0
   90.0  22.5 25.0   0.0  22.5  25.0
    0.0  90.0 25.0 -90.0  22.5  25.0
4                                    \number of variograms
1   1   4                            \tail, head, variogram type
1   2   4                            \tail, head, variogram type
2   2   4                            \tail, head, variogram type
1   1   10 2.5                       \tail, head, variogram type
```

Figure III.6: An example parameter file for **gamv3m**.

- **outfl:** the output variograms are written to a single output file named *outfl*. The output file contains the variograms ordered by direction and then variogram type specified in the parameter file (the directions cycle fastest then the variogram number). For each variogram there is a one-line description and then *nlag* lines with the following:

  1. lag number (increasing from 1 to *nlag*).

  2. separation distance for the lag.

  3. the *semivariogram* value (whatever type was specified).

  4. number of pairs for the lag.

  5. mean of the data contributing to the tail.

  6. mean of the data contributing to the head.

  7. the tail and head variances (for the correlogram).

- **nlag:** the number of lags to compute (same for all directions).

- **xlag:** the unit lag separation distance.

- **xltol:** the lag tolerance. This could be one-half of *xlag* or smaller to allow for data on a pseudo-regular grid. If *xltol* is entered as negative or zero it will be reset to *xlag/2*. Even if *xltol* is greater than *xlag/2* each pair will report to only one lag.

- **ndir:** the number of directions to consider. All these directions are considered for all of the *nvarg* variograms specified below.

- **azm, atol, bandwh, dip, dtol,** and **bandwd:** the azimuth angle, the half window azimuth tolerance, the azimuth bandwidth, the dip angle, the half window dip tolerance, and the dip bandwidth. The azimuth is measured in degrees clockwise from north, e.g., $azm = 0$ is north, $azm = 90$ is east, and $azm = 135$ is south-east. The dip angle is measured in negative degrees down from horizontal i.e., $dip = 0$ is horizontal, $dip = -90$ is vertical downward, and $dip = -45$ is dipping down at 45 degrees. *bandwh* is the horizontal "bandwidth" or maximum acceptable horizontal deviation from the direction vector. This parameter is identical to the *bandw* parameter in **gamv2**; see Figure III.2. *bandwd* is the vertical "bandwidth" or maximum acceptable deviation perpendicular to the dip direction in the vertical plane.

- **nvarg:** the number of variograms to compute.

- **ivtail, ivhead** and **ivtype:** for each of the *nvarg* variograms one must specify which variables should be used for the tail and head and which type of variogram is to be computed. For direct variograms the *ivtail* array is identical to the *ivhead* array. Cross variograms are computed by having the tail variable different from the head variable. The *ivtype* variable corresponds to the integer code in the list given in section III.1.

- **cut:** whenever the *ivtype* is set to 10, i.e., asking for an indicator variogram, then a cutoff must be specified immediately after the *ivtype* parameter on the same line in the input file (see Figure III.6). Note that if *ivtype(i)* is set to 10 then the cutoff applies to variable *ivtail(i)* in the input file (although the *ivhead(i)* variable is not used it must be present in the file to maintain consistency with the other variogram types).

The maximum number of data, the maximum number of variograms, and other maximum dimensioning parameters are specified in the file gamv3.inc which is included in both the **gamv3** subroutine and the **gamv3m** main program.

III.7 Application Notes

- A good understanding of the spatial arrangement of the data is essential to make intelligent decisions about variogram computation parameters like lag spacing, directions, and direction tolerances. It is good practice to plot the data locations and contour the data values with any off-the-shelf program prior to choosing the variogram computation parameters; this allows prior detection of data clusters, trends, discontinuities, and other features.

- No special allowance is made for covariance or variogram maps ([76], p. 97). An interested user could take the source code and modify it to consider simultaneously many possible lag distances and directions. The variogram value could then be contoured.

- The coordinate values should be rescaled so that their squared values do not exceed the computer's precision. Similarly, data values may have to be rescaled, particularly when working with different attribute values expressed in widely different unit scales. A common rescaling formula uses the minimum and maximum data values:

$$z' = \frac{z - z_{min}}{z_{max} - z_{min}}$$

- When the data are clustered (e.g., along drillholes or in specific areas) it is common to modify the variogram program to have a smaller lag separation distance $xlag$ for the first few lags. The lags for longer distances are made larger with a corresponding increase in the lag tolerance.

- Certain geometrically complicated situations may require special attention. For example, a 2-D isotropic variogram within the plane of a 2-D dipping tabular layer will require either a prior geometric transformation of coordinates, or some special coding. Variograms along a meandering direction or undulating surface also require specific coding (see [28,93]); **gamv3** does not handle these situations. A worthwhile addition to GSLIB would be a program that would provide 3-D graphic visualization of an ellipsoid defined with all three angles (azimuth, dip, and rake); see Figure II.4 and [57]. Such a program is not included here because it would be machine dependent.

- Care must be taken when attempting to compute omnidirection variograms with a stratigraphic coordinate system. The stratigraphic coordinate must be scaled to the same units as the other directional coordinates.

- It is good practice to run at least two alternate measures of spatial variability/continuity, for example, the semivariogram (III.1) and the correlogram (III.4). The additional computing cost is negligible if they are run simultaneously. If these alternate measures appear similar, everything is fine. If not, their differences must be understood, e.g., by plotting the lag means $m_{-\mathbf{h}}$ and $m_{+\mathbf{h}}$ or the lag variances $\sigma^2_{-\mathbf{h}}$ and $\sigma^2_{+\mathbf{h}}$ versus the distance $|\mathbf{h}|$; see definitions (III.3) and (III.4). Sometimes, the structure read on the semivariogram is an artifact of the preferential clustering of the sample data. These artifacts can be reduced by the measures (III.3 and III.4) that filter the lag means and lag variances; see [77]. Similarly, clustering and outlier values usually affect the rodogram or relative semivariogram measures much less than the semivariogram.

- Within reason, it is justifiable to adopt the measure of spatial continuity that gives the most interpretable and cleanest results. It is easy

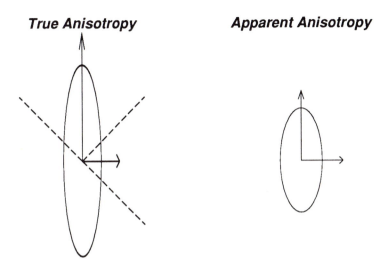

Figure III.7: An illustration of how the apparent anisotropy inferred from experimental variograms, computed with a large angular tolerance, can be considerably less than the true anisotropy. The elliptical outline on the left represents the locus of points that are equally close in terms of variogram distance to the central point. If a 45° tolerance is used to compute variograms the apparent anisotropy is reduced to that shown on the right.

to mask spatial continuity by a poor choice of lag spacing, direction angles, or a poor handling of outlier data values. It is rare to identify spatial continuity that does not exist. There are two notable exceptions to this statement: 1) clustered data may cause certain measures of spatial continuity to show an artificial structure (see above), and 2) the combination of a severe anisotropy and large angular tolerance can artificially increase the range of correlation in the direction of minimum continuity (see below).

- In the presence of few, widely spaced, data (almost always the case in practice) it is common to use a fairly large angular tolerance to obtain enough pairs for a stable "variogram." This is not a problem when the underlying phenomenon is not strongly anisotropic. If the phenomenon is truly anisotropic, however, then directional variograms will always yield an apparent anisotropy that is less than the true anisotropy, i.e., the most continuous direction will appear less continuous and the least continuous direction will appear more continuous by combining pairs that are in different directions. This is illustrated on Figure III.7 where a true 5:1 anisotropy appears as a 2.4:1 anisotropy if the variogram measure is computed with a 45° tolerance. One should use the smallest tolerance possible (in the previous example the apparent anisotropy would be 3.6:1 with a $22.5°$ tolerance) and possibly increase the anisotropy ratios after modeling the experimental variograms.

- Increasing the lag tolerance for the first distance class leads to accepting pairs at a distance significantly greater than the distance interval $(0, \epsilon]$. Considering pairs at relatively long distances, to infer the short scale structure, often leads to experimental semivariograms with too high nugget effects.

- As in any inference endeavor, it is not the properties of the limited sample nor the properties of the elusive RF model which are of prime interest. It is the underlying properties of the (one and only one) physical population which are of interest. In the process of variogram inference and that of variogram modeling (not covered by GSLIB), the practitioner should allow some initiative in cleaning or departing from the data, as long as any such departure is clearly documented. It is the subjective interpretation, possibly based on prior experience or ancillary information, that makes a *good* model; the data, by themselves, are rarely enough.

- When modeling a variogram for the purpose of stochastic simulation, one should try to model with a non-zero range part or all of the apparent nugget variance. Indeed, simulation is most likely to be performed on a grid much denser than the average data spacing. Most likely, at such short scales there exist structures not revealed by the available data.

- When used as a cross variogram (with tail and head values corresponding to different variables Z and Y), the semivariogram measure (III.1) corresponds to the moment:

$$\gamma_{ZY}^{(1)}(\mathbf{h}) = \frac{1}{2}E\left\{[Z(\mathbf{u}) - Y(\mathbf{u}+\mathbf{h})]^2\right\} \tag{III.11}$$

which differs from the traditional cross semivariogram:

$$\gamma_{ZY}^{(2)}(\mathbf{h}) = \frac{1}{2}E\left\{[Z(\mathbf{u}) - Z(\mathbf{u}+\mathbf{h})][Y(\mathbf{u}) - Y(\mathbf{u}+\mathbf{h})]\right\} \tag{III.12}$$

The cross covariance, as required by cokriging systems, is:

$$
\begin{aligned}
C_{ZY}(\mathbf{h}) &= Cov\{Z(\mathbf{u}), Y(\mathbf{u}+\mathbf{h})\} \\
&= E\{Z(\mathbf{u})Y(\mathbf{u}+\mathbf{h})\} - m_Z m_Y
\end{aligned}
$$

with, in general, $C_{ZY}(\mathbf{h}) \neq C_{YZ}(\mathbf{h})$; see "lag effect" in [94], p. 41.

Development of the two alternative definitions (III.11) and (III.12) leads to the two relations:

$$\gamma_{ZY}^{(1)}(\mathbf{h}) = \frac{1}{2}\left[C_Z(0) + C_Y(0) + (m_Z - m_Y)^2 - 2C_{ZY}(\mathbf{h})\right]$$

and

$$\gamma_{ZY}^{(2)}(\mathbf{h}) = C_{ZY}(0) - \frac{1}{2}[C_{ZY}(\mathbf{h}) + C_{YZ}(\mathbf{h})]$$

with m_Z, m_Y and $C_Z(0)$, $C_Y(0)$ being the means and variances of variables Z and Y. If the two variables are standardized then $m_Z = m_Y = 0$ and $C_Z(0) = C_Y(0) = 1$. Introducing the correlogram $\rho_{ZY}(\mathbf{h})$ (equal to $C_{ZY}(\mathbf{h})$ if Z and Y are standardized) the two previous relations become:

$$\gamma_{ZY}^{(1)}(\mathbf{h}) = 1 - \rho_{ZY}(\mathbf{h}) \tag{III.13}$$

and

$$\gamma_{ZY}^{(2)}(\mathbf{h}) = \rho_{ZY}(0) - \frac{1}{2}[\rho_{ZY}(\mathbf{h}) + \rho_{YZ}(\mathbf{h})] \tag{III.14}$$

Remark: Variograms and cross variograms can be used in cokriging systems only under specific constraints on the cokriging weights. It is safe practice to transform sample cross semivariograms of either type (III.11) and (III.12) into the corresponding cross covariance $C_{ZY}(\mathbf{h})$ and then model the covariance/cross covariance matrix with a linear model of coregionalization before using it in the cokriging system.

When modeling spatial cross dependence, there are clear advantages in directly calculating cross covariances, $C_{ZY}(\mathbf{h})$, using the covariance measure (III.3) with tail and head values corresponding to different variables Z and Y; see [77,145]. If cross semivariograms are inferred, then the traditional expressions (III.2) and (III.12) should be preferred.

- Experimental indicator semivariograms will reach their sills (if any) at the non-declustered indicator variances $\hat{F}(z_c)[1 - \hat{F}(z_c)]$, where $\hat{F}(z_c)$ is the sample mean of the corresponding indicator data, not at the declustered variances $F(z_c)[1 - F(z_c)]$ corresponding to the declustered cdf values $F(z_c)$, z_c being the threshold values at which the indicator data are defined; see definition (II.6). Hence, if these experimental indicator semivariograms are to be standardized to a common unit sill, the sample variances $\hat{F}(z_c)[1 - \hat{F}(z_c)]$ should be used. Also, in presence of zonal anisotropy, certain experimental indicator semivariograms may not reach the variance defined on either $F(z_c)$ or $\hat{F}(z_c)$.

- **Variogram modeling** requires interactive and graphical programs not within the scope of GSLIB because they would require a machine-dependent graphics library. The output from GSLIB variogram subroutines could be reformatted to be used in interactive variogram modeling programs such as the one provided in the Geostatistical Toolbox [54], or by any versatile spreadsheet.

The user is warned to exercise caution when using software that automatically fits variograms and cross variograms without user interaction. Good fitting algorithms should require a prior choice of the number of variogram structures and their types and, only then, the parameters (sill, range) would be chosen to match the experimental variogram values.

III.8   Problem Set Two: Variograms

The goal of this second problem set is to experiment with variogram calculation and model fitting techniques. Once again, compiled GSLIB programs are called for. The **vargplt** utility given in Chapter VI may be useful.

The 2-D irregularly spaced data in $\boxed{\text{cluster.dat}}$ and the 2-D grid of reference data in $\boxed{\text{true.dat}}$ will be used for this problem set.

Questions

1. Perform a variogram study on the irregularly spaced data (with **gamv2m**). Try all variogram types and at least three directions (omnidirectional, NS, and EW). Compute and plot (with **vargplt**) the variograms using all 140 data and, then, using only the first 97 data. Comment on each variogram type, the presence or absence of anisotropy, the impact of clustering, etc.

2. Perform a variogram study on the exhaustively sampled grid (with **gam2m**). Just consider a few representative directions - **gam2m** does not compute omnidirectional variograms. Comment on the results and compare to the results of question one.

3. Model one experimental and one exhaustive directional variogram. The program **vmodel** may be used in conjunction with **vargplt** to overlay your model on the experimental variogram points. An interactive variogram modeling program would be helpful (e.g., the program in the Geostatistical Toolbox [54]). Comment on the results.

4. Compute the normal score transform of the data with **nscore** (documented in section VI.2.2), compute the normal score variogram, and model it. Keep this model in preparation for Problem set 6 dealing with Gaussian simulations.

5. Establish the theoretical (given a biGaussian hypothesis) indicator variograms for the three quartile cutoffs using **bigaus** (documented in section VI.2.6). Compute the actual indicator variograms, compare them to the previously obtained theoretical models, and comment on the appropriateness of a multivariate Gaussian RF model; see section V.2.2.

Chapter IV

Kriging

This chapter presents the kriging programs of GSLIB. These programs allow for simple kriging (SK), ordinary kriging (OK), and kriging with various trend models (KT). A cokriging program provides the ability to use secondary variables and an indicator kriging program allows the direct estimation of posterior probability distributions.

Section IV.1 describes the various kriging algorithms and their underlying principles. Section IV.2 presents straightforward 2-D ordinary kriging programs for grid kriging (points or blocks) and cross validation.

Section IV.3 presents more elaborate kriging programs for point/block kriging and cross validation in 2-D and 3-D. These programs allow many alternate kriging algorithms and options such as kriging the trend and filtering certain nested covariance structures.

Section IV.4 presents a 3-D cokriging program that will accept a number of covariates with either simple or ordinary cokriging. The indicator kriging (IK) program is presented in Section IV.5.

Some useful application notes are presented in section IV.6. Finally, a problem set is proposed in section IV.7 that allows the kriging paradigm to be tested and understood. Special problem sets on cokriging and indicator kriging are proposed in sections IV.8 and IV.9, respectively.

IV.1   Kriging with GSLIB

Although kriging was initially introduced to provide estimates for unsampled values, it is being used increasingly to build probabilistic models of uncertainty about these unknown values; see section II.1.4 and Lesson 4 in [82]. In a nutshell, the kriging algorithm provides a minimum error-variance estimate of any unsampled value. Contouring a grid of kriging estimates is the traditional mapping application of kriging. Kriging used as a mapping algorithm is a low-pass filter that tends to smooth out details and extreme values of the original data set.

Since kriging is a minimum error variance estimation algorithm it approximates, and in some cases is identical to the conditional expectation of the variable being estimated. Thus, kriging can be used to estimate a series of posterior conditional probability distributions from which unsmoothed images of the attribute spatial distribution can be drawn. In the multiGaussian (MG) case the conditional distribution is identified by the mean and variance obtained from simple kriging. In the indicator kriging (IK) approach a series of conditional cumulative distribution function (ccdf) values are estimated directly.

The kriging principle, applied both as a mapping algorithm and as a tool to obtain conditional probability distributions, has been presented in numerous papers and textbooks [21,30,59,67,70,76,82,94,105,116,137]. In electrical engineering circles, the kriging algorithm is known as the Wiener filter [19,163]. Only a brief summary and information specific to GSLIB implementations are given in this section.

IV.1.1   Simple Kriging

All versions of kriging are elaborations on the basic linear regression algorithm and corresponding estimator:

$$[Z_{SK}^*(\mathbf{u}) - m(\mathbf{u})] = \sum_{\alpha=1}^{n} \lambda_\alpha(\mathbf{u})\,[Z(\mathbf{u}_\alpha) - m(\mathbf{u}_\alpha)] \qquad (IV.1)$$

where $Z(\mathbf{u})$ is the RV model at location $\mathbf{u}$, the $\mathbf{u}_\alpha$'s are the n data locations, $m(\mathbf{u}) = E\{Z(\mathbf{u})\}$ is the location-dependent expected value of RV $Z(\mathbf{u})$, and $Z_{SK}^*(\mathbf{u})$ is the linear regression estimator, also called the "simple kriging" (SK) estimator.

The SK weights $\lambda_\alpha(\mathbf{u})$ are given by the system (II.13) of normal equations written in their more general non-stationary form as follows:

$$\sum_{\beta=1}^{n} \lambda_\beta(\mathbf{u})C(\mathbf{u}_\beta, \mathbf{u}_\alpha) = C(\mathbf{u}, \mathbf{u}_\alpha), \quad \alpha = 1, \ldots, n \qquad (IV.2)$$

The SK algorithm requires prior knowledge of the $(n + 1)$ means $m(\mathbf{u})$, $m(\mathbf{u}_\alpha), \alpha = 1, \ldots, n$, and the $(n+1)$ by $(n+1)$ covariance matrix $[C(\mathbf{u}_\alpha, \mathbf{u}_\beta),$ $\alpha, \beta = 0, 1, \ldots, n]$ with $\mathbf{u}_0 = \mathbf{u}$. In most practical situations, inference of these means and covariance values requires a prior hypothesis (rather a *decision*) of stationarity of the random function $Z(\mathbf{u})$; see the discussion in section II.1.2. If the RF $Z(\mathbf{u})$ is stationary with constant mean m, and covariance function $C(\mathbf{h}) = C(\mathbf{u}, \mathbf{u} + \mathbf{h}), \forall \mathbf{u}$, the SK estimator reduces to its stationary version:

$$Z_{SK}^*(\mathbf{u}) = \sum_{\alpha=1}^{n} \lambda_\alpha(\mathbf{u})Z(\mathbf{u}_\alpha) + \left[1 - \sum_{\alpha=1}^{n} \lambda_\alpha(\mathbf{u})\right] m \qquad (IV.3)$$

with the traditional stationary SK system:

$$\sum_{\beta=1}^{n} \lambda_\beta(\mathbf{u})C(\mathbf{u}_\beta - \mathbf{u}_\alpha) = C(\mathbf{u} - \mathbf{u}_\alpha), \ \alpha = 1, \ldots, n \qquad \text{(IV.4)}$$

Stationary SK does not adapt to local trends in the data since it relies on the mean value m, assumed known and constant throughout the area. Consequently, SK is rarely used directly for mapping the z-values. Instead, it is the more robust ordinary kriging (OK) algorithm, discussed next, which is used.

According to strict stationary theory, it is SK that should be applied to the normal score transforms in the MG approach. The OK algorithm, however, might be considered if enough data are available to re-estimate locally the normal score mean; see the related discussion in section V.2.3.

IV.1.2 Ordinary Kriging

Ordinary kriging (OK) filters the mean from the SK estimator (IV.3) by requiring that the kriging weights sum to one. This results in the following ordinary kriging (OK) estimator:

$$Z_{OK}^*(\mathbf{u}) = \sum_{\alpha=1}^{n} \nu_\alpha(\mathbf{u})Z(\mathbf{u}_\alpha) \qquad \text{(IV.5)}$$

and the stationary OK system:

$$\begin{cases} \sum_{\beta=1}^{n} \nu_\beta(\mathbf{u})C(\mathbf{u}_\beta - \mathbf{u}_\alpha) + \mu(\mathbf{u}) = C(\mathbf{u} - \mathbf{u}_\alpha), \ \alpha = 1, \ldots, n \\ \sum_{\beta=1}^{n} \nu_\beta(\mathbf{u}) = 1 \end{cases} \qquad \text{(IV.6)}$$

where the $\nu_\alpha(\mathbf{u})$'s are the OK weights and $\mu(\mathbf{u})$ is the Lagrange parameter associated with the constraint $\sum_{\beta=1}^{n} \nu_\beta(\mathbf{u}) = 1$. Comparing expression (IV.4) and (IV.6), note that the SK weights are different from the OK weights.

It can be shown that ordinary kriging amounts to re-estimating, at *each* new location $\mathbf{u}$, the mean m as used in the SK expression; see [97,116]. Since ordinary kriging is most often applied within moving search neighborhoods, i.e., using different data sets for different locations $\mathbf{u}$, the implicit re-estimated mean denoted $m^*(\mathbf{u})$ depends on the location $\mathbf{u}$. Thus, the OK estimator (IV.5) is, in fact, a simple kriging of type (IV.3) where the constant mean value m is replaced by the location-dependent estimate $m^*(\mathbf{u})$:

$$\begin{aligned} Z_{OK}^*(\mathbf{u}) &= \sum_{\alpha=1}^{n} \nu_\alpha(\mathbf{u})Z(\mathbf{u}_\alpha) \qquad \text{(IV.7)} \\ &\equiv \sum_{\alpha=1}^{n} \lambda_\alpha(\mathbf{u})Z(\mathbf{u}_\alpha) + \left[1 - \sum_{\alpha=1}^{n} \lambda_\alpha(\mathbf{u})\right] m^*(\mathbf{u}) \end{aligned}$$

Hence, ordinary kriging as applied within moving data neighborhoods is already a non-stationary algorithm, in the sense that it corresponds to a non-stationary RF model with varying mean but stationary covariance. This ability to rescale locally the RF model $Z(\mathbf{u})$ to a different mean value $m^*(\mathbf{u})$ explains the extreme robustness of the OK algorithm. Ordinary kriging has been and will remain the anchor algorithm of geostatistics.

The subroutine **okb2d** provides a basic 2-D ordinary kriging program stripped of fancy data classification and search strategies. This program could be easily modified to handle simple kriging: remove the condition (IV.6) on the weights and the corresponding Lagrange parameter $\mu(\mathbf{u})$.

The more complex **ktb3d** subroutine allows SK, OK, and kriging with various trend models in either 2-D or 3-D.

IV.1.3   Kriging with a Trend Model

The term "universal kriging" has been traditionally used to denote what is, in fact, kriging with a prior trend model [59,73,94,97,126,162]. The terminology "kriging with a trend model" (KT) is more appropriate since the underlying RF model is the sum of a trend component plus a residual:

$$Z(\mathbf{u}) = m(\mathbf{u}) + R(\mathbf{u}) \tag{IV.8}$$

The trend component, defined as $m(\mathbf{u}) = E\{Z(\mathbf{u})\}$, is usually modeled as a smoothly varying deterministic function of the coordinates vector $\mathbf{u}$ whose unknown parameters are fitted from the data:

$$m(\mathbf{u}) = \sum_{l=0}^{L} a_l f_l(\mathbf{u}) \tag{IV.9}$$

The $f_l(\mathbf{u})$'s are known functions of the location coordinates and the a_l's are unknown parameters. The trend value $m(\mathbf{u})$ is itself unknown since the parameters a_l are unknown.

The residual component $R(\mathbf{u})$ is usually modeled as a stationary RF with zero mean and covariance $C_R(\mathbf{h})$.

Kriging with the trend model (IV.9) results in the so-called "universal" kriging estimator and system of equations. This system, which is actually a system of constrained normal equations [59,105], would be better named the KT system. The KT estimator is written:

$$Z_{KT}^*(\mathbf{u}) = \sum_{\alpha=1}^{n} \xi_\alpha(\mathbf{u}) Z(\mathbf{u}_\alpha) \tag{IV.10}$$

and the KT system is:

$$\begin{cases} \sum_{\beta=1}^{n} \xi_\beta(\mathbf{u}) C_R(\mathbf{u}_\beta - \mathbf{u}_\alpha) + \sum_{l=0}^{L} \mu_l(\mathbf{u}) f_l(\mathbf{u}_\alpha) = C_R(\mathbf{u} - \mathbf{u}_\alpha), \; \alpha = 1, \dots, n \\ \sum_{\beta=1}^{n} \xi_\beta(\mathbf{u}) f_l(\mathbf{u}_\beta) = f_l(\mathbf{u}), \; l = 0, \dots, L \end{cases}$$

$$\tag{IV.11}$$

where the $\xi_\beta(\mathbf{u})$'s are the KT weights and the $\mu_l(\mathbf{u})$'s are the $(L+1)$ Lagrange parameters associated to the $(L+1)$ constraints on the weights.

Remarks:

Ideally, the functions $f_l(\mathbf{u})$ that define the trend should be specified by the physics of the problem. For example, a sine function $f_l(\mathbf{u})$ with period ω_l would be considered if one knows that this periodic component contributes to the spatial or temporal variability of $z(\mathbf{u})$; the amplitude of the periodic component, i.e., the parameter a_l, is then implicitly estimated from the z-data through the KT system.

In the absence of any information about the shape of the trend, the dichotomization (IV.8) of the z-data into trend and residual components is somewhat arbitrary: what is regarded as stochastic fluctuations ($R(\mathbf{u})$) at large scale may later be modeled as a trend if additional data allow focusing on the smaller scale variability. In the absence of a physical interpretation, the trend is usually modeled as a low order (≤ 2) polynomial of the coordinates $\mathbf{u}$, e.g., with $\mathbf{u} = (x, y)$:

- a linear trend in 1-D: $m(\mathbf{u}) = a_0 + a_1 x$

- a linear trend in 2-D limited to the 45° direction: $m(\mathbf{u}) = a_0 + a_1(x+y)$

- a quadratic trend in 2-D: $m(\mathbf{u}) = a_0 + a_1 x + a_2 y + a_3 x^2 + a_4 y^2 + a_5 xy$

By convention, $f_0(\mathbf{u}) = 1$, $\forall \mathbf{u}$. Hence the case $L = 0$ corresponds to ordinary kriging with a constant but unknown mean: $m(\mathbf{u}) = a_0$.

Trend models using higher order polynomials ($n > 2$) or arbitrary non-monotonic functions of the coordinates $\mathbf{u}$ are better replaced by a random function component with a large-range variogram; see section IV.1.6.

When only z-data are available, the residual covariance $C_R(\mathbf{h})$ is inferred from linear combinations of z-data that filter the trend $m(\mathbf{u})$. For example, differences of order 1 such as $z(\mathbf{u} + \mathbf{h})$ - $z(\mathbf{u})$ would filter any trend of order zero $m(\mathbf{u}) = a_0$; differences of order 2 such as $z(\mathbf{u} + 2\mathbf{h}) - 2z(\mathbf{u} + \mathbf{h}) + z(\mathbf{u})$ would filter any trend of order 1 such as $m(\mathbf{u}) = a_0 + a_1 \mathbf{u}$; see the related discussion in section II.1.3.

In most practical situations it is possible to locate sub-areas or directions along which the trend can be ignored, in which case $Z(\mathbf{u}) \simeq R(\mathbf{u})$, and the residual covariance can be directly inferred from the local z-data covariance.

Exactitude of Kriging

If the location $\mathbf{u}$ to be estimated coincides with a datum location $\mathbf{u}_\alpha$, the normal system (SK,OK, or KT) returns the datum value for the estimate. Thus, kriging is an exact interpolator in the sense that it honors the (hard) data values at their locations.

IV.1.4   Kriging the Trend

Rather than estimating the sum $Z(\mathbf{u}) = m(\mathbf{u}) + R(\mathbf{u})$ one could estimate only the trend component $m(\mathbf{u})$. Starting directly from the original z-data the KT system (IV.11) is easily modified to yield a KT estimate for $m(\mathbf{u})$,

$$m_{KT}^*(\mathbf{u}) = \sum_{\alpha=1}^{n} \zeta_\alpha(\mathbf{u}) Z(\mathbf{u}_\alpha) \qquad\qquad (\text{IV.12})$$

and the KT system:

$$\begin{cases} \sum_{\beta=1}^{n} \zeta_\beta(\mathbf{u}) C_R(\mathbf{u}_\beta - \mathbf{u}_\alpha) + \sum_{l=0}^{L} \delta_l(\mathbf{u}) f_l(\mathbf{u}_\alpha) = 0, & \alpha = 1, \ldots, n \\ \sum_{\beta=1}^{n} \zeta_\beta(\mathbf{u}) f_l(\mathbf{u}_\beta) = f_l(\mathbf{u}), & l = 0, \ldots, L \end{cases}$$

$$(\text{IV.13})$$

This algorithm identifies the least squares fit of the trend model (IV.9) when the residual model $R(\mathbf{u})$ is assumed to have no correlation: $C_R(\mathbf{h}) = 0, \forall \mathbf{h} \neq 0$; see [33], p. 405.

The program ktb3d allows, as an option, kriging the trend as in expression (IV.12) and system (IV.13).

The direct KT estimation of the trend component can also be interpreted as a low-pass filter that removes the random (high frequency) component $R(\mathbf{u})$. The same principle underlies the algorithm of "factorial kriging" (see section IV.1.6 and [138]) and that of the Wiener-Kalman filter [19].

The traditional notation (IV.9) for the trend does not reflect the general practice of kriging with moving data neighborhoods. Because the data used for estimation change from one location $\mathbf{u}$ to another, the resulting implicit estimates of the parameters a_l's are different. Hence the following notation for the trend is more appropriate:

$$m(\mathbf{u}) = \sum_{l=0}^{L} a_l(\mathbf{u}) f_l(\mathbf{u}) \qquad\qquad (\text{IV.14})$$

Ordinary kriging (OK) corresponds to the case $L = 0$ and $m(\mathbf{u}) = a_0(\mathbf{u})$ which is not fundamentally different from the more general expression (IV.14). In the case of OK, the varying trend $m(\mathbf{u})$ is specified as a single unknown value $a_0(\mathbf{u})$, whereas in the case of KT the trend is split (often arbitrarily) into $(L + 1)$ components $f_l(\mathbf{u})$.

When the trend functions $f_l(\mathbf{u})$ are not based on physical considerations (often the case in practice), and in interpolation conditions, it can be shown [97] that the choice of the trend model does not change the estimated values $z_{KT}^*(\mathbf{u})$ or $m_{KT}^*(\mathbf{u})$. When working with moving neighborhoods the important aspect is the residual covariance $C_R(\mathbf{h})$, not the choice of the trend model. The trend model, however, is important in extrapolation conditions, i.e., when the data locations $\mathbf{u}_\alpha$ do not surround within the covariance range the location $\mathbf{u}$ being estimated. Extrapolating a constant yields significantly different results for either $z_{KT}^*(\mathbf{u})$ or $m_{KT}^*(\mathbf{u})$ than extrapolating either a line or a parabola (non-constant trend).

The practitioner is warned against overzealous modeling of the trend and the unnecessary usage of "universal kriging" (KT) or intrinsic random functions of order k (IRF-k): in most interpolation situations the simpler and well-proven OK algorithm within moving search neighborhoods will suffice. In extrapolation situations, almost by definition, the sample data alone cannot justify the trend model chosen.

IV.1.5 Kriging with an External Drift

Kriging with an external drift variable [111] is an extension of KT. The trend model is limited to two terms $m(\mathbf{u}) = a_0 + a_1 f_1(\mathbf{u})$, with the term $f_1(\mathbf{u})$ set equal to a secondary (external) variable. The *smooth* variability of the second variable is deemed related to that of the primary variable $Z(\mathbf{u})$ being estimated.

Let $y(\mathbf{u})$ be the secondary variable; the trend model is then:

$$E\{Z(\mathbf{u})\} = m(\mathbf{u}) = a_0 + a_1 y(\mathbf{u}) \qquad (\text{IV.15})$$

$y(\mathbf{u})$ is assumed to reflect the spatial trends of the z-variability up to a linear rescaling of units (corresponding to the two parameters a_0 and a_1).

The estimate of the z-variable and the corresponding system of equations are identical to the KT estimate (IV.10) and system (IV.11) with $L = 1$, and $f_1(\mathbf{u}) = y(\mathbf{u})$, i.e.:

$$Z_{KT}^* = \sum_{\alpha=1}^{n} \xi_\alpha(\mathbf{u}) Z(\mathbf{u}_\alpha) \ , \ \text{ and}$$

$$\begin{cases} \sum_{\beta=1}^{n} \xi_\beta(\mathbf{u}) C_R(\mathbf{u}_\beta - \mathbf{u}_\alpha) + \mu_0(\mathbf{u}) + \mu_1(\mathbf{u}) y(\mathbf{u}_\alpha) = & C_R(\mathbf{u} - \mathbf{u}_\alpha), \\ & \alpha = 1, \ldots, n \\ \sum_{\beta=1}^{n} \xi_\beta(\mathbf{u}) = 1 \\ \sum_{\beta=1}^{n} \xi_\beta(\mathbf{u}) y(\mathbf{u}_\beta) = y(\mathbf{u}) \end{cases}$$

$$(\text{IV.16})$$

Kriging with an external drift is a simple and efficient algorithm to incorporate a secondary variable in the estimation of the primary variable $z(\mathbf{u})$. The fundamental (hypothesis) relation (IV.15) must make physical sense. For example, if the secondary variable $y(\mathbf{u})$ represents the travel time to a seismic reflective horizon, assuming a constant velocity, the depth $z(\mathbf{u})$ of that horizon should be (in average) proportional to the travel time $y(\mathbf{u})$. Hence a relation of type (IV.15) makes sense. But, if the primary variable is permeability, it is not clear that the general trends of the spatial variability of permeability is revealed by the variability of seismic data. More demanding alternatives to kriging with an external drift are cokriging or soft kriging; see later sections IV.1.7 and IV.1.12.

Two conditions must be met before applying the external drift algorithm: (1) the external variable must vary smoothly in space, otherwise the resulting

KT system (IV.16) may be unstable; and (2) the external variable must be known at *all* locations $\mathbf{u}_\alpha$ of the primary data values and at all locations $\mathbf{u}$ to be estimated.

Note that the residual covariance rather than the covariance of the original variable $Z(\mathbf{u})$ must be used in the KT system. Both covariances are equal in areas or along directions where the trend $m(\mathbf{u})$ can be ignored (can be set to zero). Note also that the cross covariance between variables $Z(\mathbf{u})$ and $Y(\mathbf{u})$ plays no role in system (IV.16); this is different from cokriging.

Kriging with an external drift yields maps whose trends reflect the y-variability. This is a result of the decision (IV.15). It is not proof that those trends necessarily pertain to the z-variable.

An option of program **ktb3d** allows kriging with an external drift.

IV.1.6 Factorial Kriging

Rather than splitting the RF model $Z(\mathbf{u})$ into a deterministic trend $m(\mathbf{u})$ plus a stochastic component $R(\mathbf{u})$, one could consider a model consisting of two or more *independent* stochastic components (also called "factors" in relation to factor analysis):

$$Z(\mathbf{u}) = Z_0(\mathbf{u}) + Z_1(\mathbf{u}) + \ldots + Z_K(\mathbf{u}) \qquad \text{(IV.17)}$$

The Z-covariance is then the sum of the $(K+1)$ component covariances:

$$C_Z(\mathbf{h}) = \sum_{k=0}^{K} C_k(\mathbf{h}) \qquad \text{(IV.18)}$$

For example, the $(K+1)$ RF's $Z_k(\mathbf{u})$ can be modeled from the $(K+1)$ nested covariance structures $C_k(\mathbf{h})$ used to model the overall Z-sample covariance.

Filtering

A kriging estimator of the partial sum of any number of the previous $(K+1)$ components $Z_k(\mathbf{u})$ can be obtained by filtering out the covariance contribution of the non-selected components [19,107,138,160]. For example, starting from the OK system (IV.6), filtering of the first k_0 component leads to the estimator:

$$\left[\sum_{k=k_0}^{K} Z_k(\mathbf{u}) \right]_{OK}^{*} = \sum_{\alpha=1}^{n} d_\alpha(\mathbf{u}) Z(\mathbf{u}_\alpha) \qquad \text{(IV.19)}$$

with the kriging system:

$$\begin{cases} \sum_{\beta=1}^{n} d_\beta(\mathbf{u}) C_Z(\mathbf{u}_\beta - \mathbf{u}_\alpha) + \mu_0(\mathbf{u}) = \sum_{k=k_0}^{K} C_k(\mathbf{u} - \mathbf{u}_\alpha), \alpha = 1, \ldots, n \\ \qquad\qquad\qquad\qquad = C_Z(\mathbf{u} - \mathbf{u}_\alpha) - \sum_{k=0}^{k_0-1} C_k(\mathbf{u} - \mathbf{u}_\alpha) \\ \\ \sum_{\beta=1}^{n} d_\beta(\mathbf{u}) = 1 \end{cases}$$

$$\text{(IV.20)}$$

The last equation of system (IV.20) ensures unbiasedness of the estimator (IV.19) in the case $E\{Z_k(\mathbf{u})\} = 0, k = 0, \ldots, k_0 - 1$. Other unbiasedness conditions should be considered otherwise. Alternatively, no unbiasedness condition are needed if all factors have a mean of zero.

Kriging the trend (IV.12) and filtering the high frequency components (corresponding to, say $k \leq k_0$) are two examples of low-pass filters: in one case, a deterministic trend is fitted to the z-data; in the other case, a smooth covariance-type interpolator (the right hand side covariance of system (IV.20)) is fitted to the same z-data. Note that in both cases the resulting z-map need not honor the z-data at their locations. Note that strict nugget effect filtering only changes the kriging estimate at the data locations.

Both programs **okb2d** and **ktb3d** can be easily modified to handle any such type of filtering.

Warning

Be aware that the result of any factorial kriging depends heavily on the (usually arbitrary) decomposition (IV.17). If that decomposition is based solely on the z-sample covariance, then it is an artifact of the model (IV.18) rather than a proof of physical significance of the "factors" $Z_k(\mathbf{u})$.

IV.1.7 Cokriging

The term "kriging" is traditionally reserved for linear regression using data on the same attribute as that being estimated. For example, an unsampled porosity value $z(u)$ is estimated from neighboring porosity sample values defined on the same volume support.

The term "cokriging" is reserved for linear regression that also uses data defined on different attributes. For example, the porosity value $z(\mathbf{u})$ may be estimated from a combination of porosity samples and related acoustic data values.

In the case of a single secondary variable (y), the ordinary cokriging estimator of $Z(\mathbf{u})$ is written:

$$Z_{COK}^*(\mathbf{u}) = \sum_{\alpha_1=1}^{n_1} \lambda_{\alpha_1}(\mathbf{u}) Z(\mathbf{u}_{\alpha_1}) + \sum_{\alpha_2=1}^{n_2} \lambda'_{\alpha_2}(\mathbf{u}) Y(\mathbf{u}'_{\alpha_2}) \qquad \text{(IV.21)}$$

where the λ_{α_1}'s are the weights applied to the n_1 z-samples and the λ'_{α_2}'s are the weights applied to the n_2 y-samples.

Kriging requires a model for the Z-covariance. Cokriging requires a *joint* model for the matrix of covariance functions including the Z-covariance $C_Z(\mathbf{h})$, the Y-covariance $C_Y(\mathbf{h})$, the cross $Z - Y$ covariance $C_{ZY}(\mathbf{h}) = Cov\{Z(\mathbf{u}), Y(\mathbf{u}+\mathbf{h})\}$, and the cross $Y - Z$ covariance $C_{YZ}(\mathbf{h})$.

The covariance matrix requires K^2 covariance functions when K different variables are considered in a cokriging exercise. The inference becomes extremely demanding in terms of data and the subsequent joint modeling is particularly tedious. This is the main reason why cokriging has not been extensively used in practice. Algorithms such as kriging with an external drift (see section IV.1.5) and collocated cokriging (see hereafter) have been developed to shortcut the tedious inference and modeling process required by cokriging.

Another reason that cokriging is not used extensively in practice is the screen effect of the better correlated data (usually the z-samples) over the data less correlated with the z-unknown (the y-samples). Unless the primary variable, that which is being estimated, is undersampled with respect to the secondary variable, the weights given to the secondary data tend to be small, and the reduction in estimation variance brought by cokriging is not worth the additional inference and modeling effort.

Other than tedious inference and matrix notations, cokriging is the same as kriging. Cokriging with trend models could be developed, and a cokriging that filters specific components of the spatial variability of either Z or Y can be designed. These notation-heavy developments will not be given here: the reader is referred to [43,94,123,124,160].

GSLIB gives only one cokriging program, cokb3d, a 3-D program based on okb2d rather than ktb3d. This simple program does not do justice to the full theoretical potential of cokriging. The three most commonly applied types of cokriging, however, are included:

1. **traditional ordinary cokriging:** the sum of the weights applied to the primary variable is set to one, and the sum of the weights applied to any other variable is set to zero. In the case of two variables as in expression (IV.21), these two conditions are:

$$\sum_{\alpha_1} \lambda_{\alpha_1}(\mathbf{u}) = 1 \text{ and } \sum_{\alpha_2} \lambda'_{\alpha_2}(\mathbf{u}) = 0 \qquad (IV.22)$$

The problem with this traditional formalism is that the second condition (IV.22) tends to limit severely the influence of the secondary variable(s).

2. **standardized ordinary cokriging:** often, a better approach (see [76], p. 400-416) consists of creating new secondary variables with the same mean as the primary variable. Then, all the weights are constrained to sum to one.

In this case expression (IV.21) could be rewritten as:

$$Z^*_{COK}(\mathbf{u}) = \sum_{\alpha_1=1}^{n_1} \lambda_{\alpha_1}(\mathbf{u})Z(\mathbf{u}_{\alpha_1}) + \sum_{\alpha_2=1}^{n_2} \lambda'_{\alpha_2}(\mathbf{u}) \left[Y(\mathbf{u}'_{\alpha_2}) + m_Z - m_Y\right]$$

$$(IV.23)$$

with the single condition: $\sum_{\alpha_1=1}^{n_1} \lambda_{\alpha_1}(\mathbf{u}) + \sum_{\alpha_2=1}^{n_2} \lambda'_{\alpha_2}(\mathbf{u}) = 1$ where $m_Z = E\{Z(\mathbf{u})\}$ and $m_Y = E\{Y(\mathbf{u})\}$ are the stationary means of Z and Y.

3. **simple cokriging:** there is no constraint on the weights. Just like simple kriging (IV.1), this version of cokriging requires working on data residuals or, equivalently, on variables whose means have all been standardized to zero. This is the case when applying simple cokriging in an MG approach (the normal score transforms of each variable have a stationary mean of zero).

Except when using traditional ordinary cokriging, covariance measures should be inferred, modeled, and used in the cokriging system rather than variograms or cross variograms.

Collocated Cokriging:

A reduced form of cokriging consists of retaining only the collocated secondary variable $y(\mathbf{u})$, provided that it is available at the location $\mathbf{u}$ being estimated [45,98]. The cokriging estimator (IV.21) is written:

$$Z^*_{COK}(\mathbf{u}) = \sum_{\alpha_1}^{n_1} \lambda_{\alpha_1}(\mathbf{u}) Z(\mathbf{u}_{\alpha_1}) + \lambda' Y(\mathbf{u}) \qquad (IV.24)$$

The corresponding cokriging system requires knowledge of only the Z-covariance $C_Z(\mathbf{h})$ and the $Z - Y$ cross covariance $C_{ZY}(\mathbf{h})$. The latter can be approximated through a Markov-type hypothesis; see section IV.1.12 and relation IV.51, and [98]:

$$C_{ZY}(\mathbf{h}) = B \cdot C_Z(\mathbf{h}), \quad \forall \mathbf{h} \qquad (IV.25)$$

where $B = \sqrt{C_Y(0)/C_Z(0)} \cdot \rho_{ZY}(0)$, $C_Z(0)$, $C_Y(0)$ are the variances of Z and Y, and $\rho_{ZY}(0)$ is the linear coefficient of correlation of collocated $z - y$ data.

GSLIB does not provide any specific program for collocated cokriging, however, such a program can be easily designed starting from either **cokb3d** or **ktb3d** with the external drift option; see later sections IV.3 and IV.4.

IV.1.8 Non-Linear Kriging

All non-linear kriging algorithms are actually linear kriging (SK or OK) applied to specific non-linear transforms of the original data. The non-linear transform used specifies the non-linear kriging algorithm considered. Examples are:

- **lognormal kriging,** or kriging applied to logarithms of the data; see [86,130,133].

- **multiGaussian kriging,** or kriging applied to the normal score transforms of the data; see [94,158]. MultiGaussian kriging is actually a generalization of lognormal kriging.

- **indicator kriging,** or kriging of indicator transforms of the data; see section IV.1.9 and [88].

- **disjunctive kriging,** or kriging of specific polynomial transforms of the data [117].

For example, consider the log transform $y(\mathbf{u}) = \ln z(\mathbf{u})$ of the original strictly positive variable $z(\mathbf{u})$. Simple or ordinary kriging of the log data yields an estimate $y^*(\mathbf{u})$ for $\ln z(\mathbf{u})$. Unfortunately, a "good" estimate of $\ln z(\mathbf{u})$ need not lead straightforwardly to a good estimate of $z(\mathbf{u})$; in particular the antilog back-transform $e^{y^*(\mathbf{u})}$ is a biased estimator of $Z(\mathbf{u})$. The unbiased back-transform of the simple lognormal kriging estimate $y^*(\mathbf{u})$ is actually

$$z^*(\mathbf{u}) = \exp\left[y^*(\mathbf{u}) + \frac{\sigma^2_{SK}(\mathbf{u})}{2} \right]$$

where $\sigma^2_{SK}(\mathbf{u})$ is the simple lognormal kriging variance. The exponentiation involved in that back-transform is particularly delicate since it exponentiates any error in the process of estimating either the lognormal estimate $y^*(\mathbf{u})$ or its SK variance $\sigma^2_{SK}(\mathbf{u})$, [86]. This extreme sensitivity to the back-transform explains why direct non-linear kriging of unsampled values has fallen into disuse. Instead, non-linear kriging algorithms, essentially the MG and IK approaches, have been developed for stochastic simulations.

In the MG approach, the ccdf mean and variance obtained by SK applied to the normal score transform data are *not* back-transformed. Instead, it is the simulated normal score values, as drawn from the normal ccdf, which are back-transformed. That back-transform is the straight inverse normal score transform without any need for a bias correction.

Similarly, the IK estimates of the ccdf values are *not* back-transformed. Instead, realizations of the original z-attribute are drawn directly from the ccdf (see sections IV.1.9 and V.3.2).

MultiGaussian kriging is straightforward: it consists of SK or OK of the normal score data using the covariance modeled from the sample normal score covariance. Such kriging does not require any specific program (**okb2d** or **ktb3d** could be used); multiGaussian kriging has been coded in the corresponding Gaussian simulation program **sgsim** (see section V.6.3). The normal score transform itself and the corresponding straight back-transform are presented in sections V.2.1, VI.2.2, and VI.2.3 (programs **nscore** and **backtr**).

As for indicator kriging (IK), its theoretical background is equally straightforward [82,88]; however, its implementation is delicate enough to justify the next section IV.1.9.

IV.1.9   Indicator Kriging

Indicator kriging of a continuous variable is *not* aimed at estimating the indicator transform $i(\mathbf{u}; z_k)$ set to 1 if $z(\mathbf{u}) \leq z_k$, and to 0 if not. Indicator kriging provides a least-squares estimate of the conditional cumulative distribution function (ccdf) at cutoff z_k:

$$
\begin{aligned}
[i(\mathbf{u}; z_k)]^* &= E\left\{I(\mathbf{u}; z_k|(n))\right\}^* \\
&= Prob^*\left\{Z(\mathbf{u}) \le z_k|(n)\right\}
\end{aligned}
\tag{IV.26}
$$

where (n) represents the conditioning information available in the neighborhood of location $\mathbf{u}$.

The IK process is repeated for a series of K cutoff values $z_k, k = 1, \ldots, K$, which discretize the interval of variability of the *continuous* attribute z. The ccdf, built from assembling the K indicator kriging estimates of type (IV.26), represents a probabilistic model for the uncertainty about the unsampled value $z(\mathbf{u})$.

If $z(\mathbf{u})$ is itself a binary categorical variable, e.g., set to 1 if a specific rock type prevails at $\mathbf{u}$, to 0 if not, then there is no need for any prior indicator transform. The direct kriging of $z(\mathbf{u})$ provides an estimate for the probability that $z(\mathbf{u})$ be one, i.e., for that rock type to prevail at location $\mathbf{u}$.

If $z(\mathbf{u})$ is a continuous variable, then the correct selection of the cutoff values z_k at which indicator kriging takes place is essential: too many cutoff values and the inference and computation becomes needlessly tedious and expensive; too few, and the details of the distribution are lost.

Simple IK

The stationary mean of the binary indicator RF $I(\mathbf{u}; z)$ is the cumulative distribution function (cdf) of the RF $Z(\mathbf{u})$ itself; indeed:

$$
\begin{aligned}
E\left\{I(\mathbf{u}; z)\right\} &= 1 \cdot Prob\{Z(\mathbf{u}) \le z\} + 0 \cdot Prob\{Z(\mathbf{u}) > z\} \\
&= Prob\left\{Z(\mathbf{u}) \le z\right\} = F(z)
\end{aligned}
\tag{IV.27}
$$

The SK estimate of the indicator transform $i(\mathbf{u}; z)$ is thus written, according to expression (IV.3):

$$
[i(\mathbf{u}; z)]^*_{SK} = [Prob\left\{Z(\mathbf{u}) \le z|(n)\right\}]^*_{SK}
\tag{IV.28}
$$

$$
= \sum_{\alpha=1}^{n} \lambda_\alpha(\mathbf{u}; z)i(\mathbf{u}_\alpha; z) + \left[1 - \sum_{\alpha=1}^{n} \lambda_\alpha(\mathbf{u}; z)\right] F(z)
$$

where the $\lambda_\alpha(\mathbf{u}; z)$'s are the SK weights corresponding to cutoff z. These weights are given by a SK system of type (IV.4):

$$
\sum_{\beta=1}^{n} \lambda_\beta(\mathbf{u}; z)C_I(\mathbf{u}_\beta - \mathbf{u}_\alpha; z) = C_I(\mathbf{u} - \mathbf{u}_\alpha; z), \ \alpha = 1, \ldots, n
\tag{IV.29}
$$

where $C_I(\mathbf{h}; z) = Cov\{I(\mathbf{u}; z), I(\mathbf{u} + \mathbf{h}; z)\}$ is the indicator covariance at cutoff z. If K cutoff values z_k are retained, simple IK requires K indicator covariances $C_I(\mathbf{h}; z_k)$ in addition to the K cdf values $F(z_k)$. If the z-data are preferentially clustered, the sample cdf values should be declustered before being used (see program **declus** in section VI.2.1).

Ordinary IK

Just like any simple kriging, simple IK is dependent on the stationarity decision and on the cdf values $F(z)$ interpreted as mean indicator values. When data are abundant, ordinary indicator kriging within moving data neighborhoods may be considered; this amounts to re-estimating locally the prior cdf values $F(z)$; see section IV.1.2.

Both simple and ordinary kriging are implemented in the GSLIB program ik3d.

Median IK

In indicator kriging the K cutoff values z_k are usually chosen so that the corresponding indicator covariances $C_I(\mathbf{h}; z_k)$ are different one from another. There are cases, however, when the sample indicator covariances/variograms appear proportional to each other, i.e., the sample indicator correlograms are all similar in shape. The corresponding continuous RF model $Z(\mathbf{u})$ is the so-called "mosaic"[1] model [89] such that:

$$\rho_Z(\mathbf{h}) = \rho_I(\mathbf{h}; z_k) = \rho_I(\mathbf{h}; z_k, z_{k'}), \ \forall z_k, z_{k'} \qquad \text{(IV.30)}$$

where $\rho_Z(\mathbf{h})$ and $\rho_I(\mathbf{h}; z_k, z_{k'})$ are the correlograms and cross correlograms of the continuous RF $Z(\mathbf{u})$ and its indicator transforms.

Then, the single correlogram function is better estimated either directly from the sample Z-correlogram or from the sample indicator correlogram at the median cutoff $z_k = M$, such that $F(M) = 0.5$. Indeed, at the median cutoff, the indicator data are evenly distributed as 0 and 1 values with, by definition, no outlier values.

Indicator kriging under the model (IV.30) is called "median indicator kriging" [88]. It is a particularly simple and fast procedure since it calls for a single easy-to-infer median indicator variogram that is used for all K cutoffs. Moreover, if the indicator data configuration is the same for all cutoffs,[2] one single IK system needs to be solved with the resulting weights being used for all cutoffs. For example, in the case of simple IK,

$$[i(\mathbf{u}; z_k)]^*_{SK} = \sum_{\alpha=1}^{n} \lambda_\alpha(\mathbf{u}) i(\mathbf{u}_\alpha; z_k) + \left[1 - \sum_{\alpha=1}^{n} \lambda_\alpha(\mathbf{u})\right] F(z_k) \qquad \text{(IV.31)}$$

where the $\lambda_\alpha(\mathbf{u})$'s are the SK weights common to all cutoffs z_k and given by the single SK system:

$$\sum_{\beta=1}^{n} \lambda_\beta(\mathbf{u}) C(\mathbf{u}_\beta - \mathbf{u}_\alpha) = C(\mathbf{u} - \mathbf{u}_\alpha), \ \alpha = 1, \ldots, n \qquad \text{(IV.32)}$$

---

[1]The term "mosaic model" is unfortunate, because the realizations of this continuous RF model do not appear as pavings of subareas (mosaic pieces) with constant z-values.

[2]Unless inequality constraint-type data $z(\mathbf{u}_\alpha) \in (a_\alpha, b_\alpha]$ are considered, the indicator data configuration is the same for all cutoffs z_k's as long as the same data locations $\mathbf{u}_\alpha$'s are retained for all cutoffs. For more details see section IV.1.12.

The covariance $C(\mathbf{h})$ is modeled from either the z-sample covariance or, better, the sample median indicator covariance.

Note that the weights $\lambda_\alpha(\mathbf{u})$ are also the SK weights of the simple kriging estimate of $z(\mathbf{u})$ using the $z(\mathbf{u}_\alpha)$-data.

Median indicator kriging can be performed with program ik3d.

Using Inequality Data

In the IK expression (IV.28), the indicator data $i(\mathbf{u}_\alpha; z)$ originate from data $z(\mathbf{u}_\alpha)$ that are deemed perfectly known; thus, the indicator data $i(\mathbf{u}_\alpha; z)$ are "hard" in the sense that they are valued either 0 or 1 and are available at any cutoff value z.

There are applications where some of the z-information takes the form of inequalities such as:

$$z(\mathbf{u}_\alpha) \in (a_\alpha, b_\alpha] \qquad (IV.33)$$

or $z(\mathbf{u}_\alpha) \leq b_\alpha$ equivalent to $z(\mathbf{u}_\alpha) \in (-\infty, b_\alpha]$, or $z(\mathbf{u}_\alpha) > a_\alpha$ equivalent to $z(\mathbf{u}_\alpha) \in (a_\alpha, +\infty]$. The indicator data corresponding to the constraint interval (IV.33) are available only outside that interval:

$$i(\mathbf{u}_\alpha; z) = \begin{cases} 0 \text{ for } z \leq a_\alpha \\ \text{undefined for } z \in (a_\alpha, b_\alpha] \\ 1 \text{ for } z > b_\alpha \end{cases} \qquad (IV.34)$$

Except for the undefined (missing) indicator data in the interval $(a_\alpha, b_\alpha]$, the IK algorithm applies identically; yet the constraint interval information (IV.33) is honored by the resulting ccdf; see hereafter and [81]. The program ik3d allows simple and ordinary indicator kriging in 2-D or 3-D with, possibly, constraint intervals of the type (IV.33).

Exactitude of IK

If the location $\mathbf{u}$ to be estimated coincides with a datum location $\mathbf{u}_\alpha$, whether a hard datum or a constraint interval of type (IV.33), the exactitude of kriging entails that the ccdf returned is either a zero variance cdf identifying the class to which the datum value $z(\mathbf{u}_\alpha)$ belongs or a cdf honoring the constraint interval up to the interval amplitude, in the sense that:

$$\begin{aligned} [i(\mathbf{u}_\alpha; z)]^* &= Prob^*\{Z(\mathbf{u}_\alpha) \leq z | (n)\} \\ &= \begin{cases} 0, & \text{if } z \leq a_\alpha \\ 1, & \text{if } z > b_\alpha \end{cases} , \text{if } z(\mathbf{u}_\alpha) \in (a_\alpha, b_\alpha] \end{aligned}$$

This is true whether simple, ordinary, or median IK is used.

Programs ik3d and postik

Indicator kriging is primarily used to generate conditional probabilities within the stochastic simulation programs sisim and mbsim. GSLIB offers one independent IK program, ik3d, for simple and ordinary indicator kriging of cumulative indicators defined on continuous variables (see definition (II.6)). IK of categorical variables can be done by directly applying **okb2d** or **ktb3d** to the categorical data.

The IK algorithm itself does not ensure that the resulting probability estimates, ccdf's for continuous variables, discrete probabilities for categorical variables, verify the order relations for legitimate probabilities, [3] i.e., for ccdf's of continuous variables $z(\mathbf{u})$:

$$Prob\{Z(\mathbf{u}) \leq z|(n)\} = F(\mathbf{u}; z|(n)) \in [0, 1] \qquad \text{(IV.35)}$$

$$\text{and}: \ F(\mathbf{u}; z_{k'}|(n)) \geq F(\mathbf{u}; z_k|(n)), \ \forall \ z_{k'} > z_k$$

For conditional probabilities of an exhaustive set of mutually exclusive categorical variables $I_k(\mathbf{u}), k = 1, \ldots, K$:

$$Prob\{I_k(\mathbf{u}) = 1|(n)\} = F(\mathbf{u}; k|(n)) \in [0, 1] \qquad \text{(IV.36)}$$

$$\text{and}: \ \sum_{k=1}^{K} F(\mathbf{u}; k|(n)) = 1$$

The indicator $i_k(\mathbf{u})$ is set to 1 if category k prevails at location $\mathbf{u}$, to zero if not.

Regardless of the estimation algorithm used it is imperative to correct order relation deviations. **ik3d** performs these corrections and provides a detailed report of the number and magnitude of corrections at each cutoff.

The program **postik** (section VI.2.10) performs within-class interpolation, see section V.1.6, to provide any required quantile value or probabilities of exceeding any given threshold value. **postik** also returns the mean of the ccdf, called the "E-type" estimate of $z(\mathbf{u})$, and defined as:

$$[z(\mathbf{u})]_E^* = \int_{-\infty}^{+\infty} z \, dF(\mathbf{u}; z|(n)) \qquad \text{(IV.37)}$$

$$\approx \sum_{k=1}^{K+1} z_k' \left[F(\mathbf{u}; z_k|(n)) - F(\mathbf{u}; z_{k-1}|(n)) \right]$$

where $z_k, k = 1, \ldots, K$ are the K cutoffs retained, and $z_0 = z_{min}$, $z_{K+1} = z_{max}$ are the minimum and maximum of the z-range, to be entered as input parameters. The conditional mean value z_k' within each class, $(z_{k-1}, z_k]$, is obtained by the interpolation procedure specified as input to **postik** (see section V.1.6).

---

[3] Order relation problems are not specific to IK. Most methods that attempt to estimate a ccdf, including IK and DK, will incur such problems. The exceptions to this rule are methods that rely on a prior multivariate distribution model to determine the ccdf exactly; for example, the MG approach assumes that all ccdf's are Gaussian, exactly determined by a mean and variance.

Exactitude of the E-type Estimate

Because the ccdf returned by IK honors both hard z-data and constraint intervals of the type (IV.33), the corresponding E-type estimate also honors that information. More precisely, at a datum location $\mathbf{u}_\alpha$, $[z(\mathbf{u}_\alpha)]_E^* = z(\mathbf{u}_\alpha)$, if the z-datum is hard, and $[z(\mathbf{u}_\alpha)]_E^* \in (a_\alpha, b_\alpha]$, if the information at $\mathbf{u}_\alpha$ is the constraint interval $z(\mathbf{u}_\alpha) \in (a_\alpha, b_\alpha]$. In practice, the exactitude of the E-type estimate is limited by the finite discretization into K cutoff values z_k. For example, in the case of a hard z-datum, the estimate is: $[z(\mathbf{u}_\alpha)]_E^* \in (z_{k-1}, z_k]$, with z_k being the upper bound of the interval containing the datum value $z(\mathbf{u}_\alpha)$. Thus, the E-type estimate attached to IK provides a straightforward solution to the difficult problem of constrained interpolation; as opposed to the quadratic programming solution which would limit the estimate $z^*(\mathbf{u}_\alpha)$ to either bound a_α or b_α if the constraint interval is active [47,108].

The IK solution is particularly fast if median IK is used. However, if median IK is used with constraint intervals of type (IV.33), at any given location the data configuration may change from one cutoff z_k to another; consequently, one may have to solve a different IK system for each cutoff.

Correcting for Order Relation Problems

The flexibility of the IK approach is obtained at the cost of order relation problems. IK-derived conditional probabilities may not verify the order relations conditions (IV.35) or (IV.36). In any particular study one would expect to meet at least one order relation deviation for up to one-half or two-thirds of the IK-derived ccdf's. Fortunately, the average magnitude of the probability corrections is usually on the order of 0.01, much smaller than shown on Figure IV.1. Program ik3d provides statistics of the order relation problems encountered.

In the case of categorical probabilities, the first constraint (IV.36) is easily met by resetting the estimated value $F^*(\mathbf{u}; z|(n))$ to the nearest bound, 0 or 1, if originally valued outside the interval $[0, 1]$. This resetting corresponds exactly to the solution provided by quadratic programming [108].

The second constraint (IV.36) is tougher because it involves K separate krigings. One solution consists of kriging only $(K - 1)$ probabilities leaving aside one category k_o, chosen with a large enough prior probability p_{k_o}, so that:

$$F^*(\mathbf{u}; k_0|(n)) = 1 - \sum_{k \neq k_o} F^*(\mathbf{u}; k|(n))$$

Another solution, applied after the first constraint (IV.36) has been met, is to restandardize each estimated probability $F^*(\mathbf{u}; k|(n)) \in [0, 1]$ by the sum $\sum_k F^*(\mathbf{u}; k|(n)) < 1$.

Correcting for order relations of continuous variable ccdf's is more delicate, because of the ordering of the cumulative indicators.

Figure IV.1 shows an example with the following order relation problems:

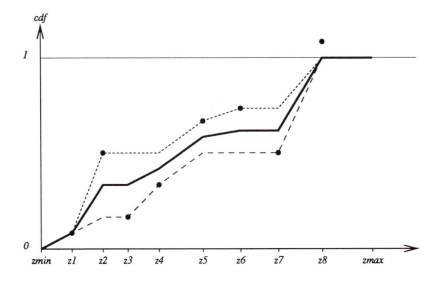

Figure IV.1: Order relation problems and their correction. The dots are the ccdf values returned by IK. The corrected ccdf is obtained by averaging the forward and downward corrections.

$$F(\mathbf{u}; z_3|(n)) < F(\mathbf{u}; z_2|(n))$$

$$F(\mathbf{u}; z_7|(n)) < F(\mathbf{u}; z_6|(n))$$

$$F(\mathbf{u}; z_8|(n)) > 1$$

There are two sources of order relation problems:

1. Negative indicator kriging weights. One solution is to constrain the IK system to deliver only non-negative weights, [12]. One would have to forfeit, however, the sometimes beneficial properties of having a non-convex kriging estimate [85].

2. Lack of data in some classes; see hereafter.

Practice has shown that the majority of order relation problems are due to a lack of data, more precisely, to cases when IK is attempted at a cutoff z_k which is the upper bound of a class $(z_{k-1}, z_k]$ that contains no z-data. In such case the indicator data set is the same for both cutoffs z_{k-1} and z_k and yet the corresponding indicator variogram models are likely different; therefore, the resulting ccdf values will likely be different with a good chance for order relation problems.

There are many implementations, specific to each data set considered, that will reduce the occurrence of order relation problems. Examples include:

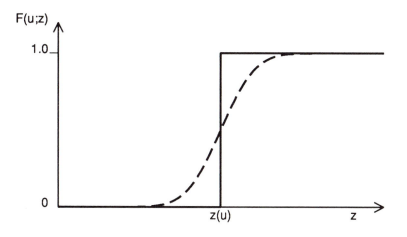

Figure IV.2: The binary step function $i(\mathbf{u}; z)$ (solid line) is replaced by the continuous kernel cdf $F(\mathbf{u}; z)$ (dashed line).

- Smooth the original discrete distribution of z-data into a more continuous distribution by replacing each set of binary indicator data $\{i(\mathbf{u}_\alpha; z_k),$ $k = 1, \ldots, K\}$ with a prior parametric kernel cdf $F(\mathbf{u}_\alpha; z)$ that varies continuously in $[0, 1]$; see [104,141]. The IK procedure is unchanged; the indicator data $i(\mathbf{u}_\alpha; z_k)$ in expression (IV.28) are simply replaced by the kernel values $F(\mathbf{u}_\alpha; z_k)$; see Figure IV.2. The price to pay for getting a more continuous ccdf is the loss of exactitude due to the spread of the kernel function $F(\mathbf{u}_\alpha; z)$ around the hard datum value $z(\mathbf{u}_\alpha)$. The indicator kriging subroutine ik3d could be used with a pre-processor that generates such kernel cdf values.

- Define a continuum in the variability of the indicator variogram parameters with increasing (or decreasing) cutoff values. There should not be sudden large changes in these variogram parameters from one cutoff to the next. It is good practice to model all indicator variograms with the same nested sum of basic structures, e.g.,

$$\gamma_I(\mathbf{h}; z_k) = C_0(z_k) + C_1(z_k)Sph(|\mathbf{h}|/a(z_k)) \qquad \text{(IV.38)}$$

with, e.g., $Sph(\cdot)$ being a spherical variogram model with unit sill and range. The variability of each parameter (e.g., C_0, C_1, a) is plotted versus the cutoff value z_k to ensure smoothly changing parameter values. If the decision of stationarity is reasonable, then one would expect smoothly changing parameters. Moreover, these variogram parameter plots allow interpolation or extrapolation beyond the cutoff values z_k initially retained. An extreme case would be to retain the same indicator variogram for all cutoffs. Choosing the robust median indicator variogram would lead to the "median IK" approach presented in relation

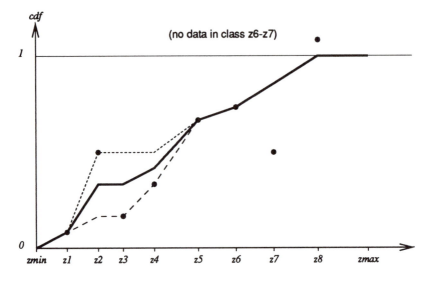

Figure IV.3: Order relation problems and their correction ignoring the class $(z_6, z_7]$ that did not contain any z-data.

(IV.31). Median IK drastically reduces the number of order relation deviations at the expense of less flexibility: only one indicator correlogram is available to model the bivariate cdf of the RF $Z(\mathbf{u})$.

- Identify the cutoff values z_k to the z-data values retained in the neighborhood of the location $\mathbf{u}$ being estimated. The corresponding indicator variogram parameters are then interpolated from the parameters of the available indicator variogram models; see relation (IV.38) and [40].

- Retain only those prior cutoff values z_k such that the class $(z_{k-1}, z_k]$ has at least one datum; see Figure IV.3.

Such implementation procedures reduce, but do not eliminate, order relation problems. A final correction of the IK-returned ccdf values is necessary. The following correction algorithm, implemented in the program **ik3d**, considers the average of an upward and downward correction:

1. Upward correction resulting in the upper line of Figure IV.1:

    - Start with the lowest cutoff z_1.
    - If the IK-returned ccdf value $F(\mathbf{u}_\alpha; z_1|(n))$ is not within $[0, 1]$, reset it to the closest bound.
    - Proceed to the next cutoff z_2. If the IK-returned ccdf value $F(\mathbf{u}_\alpha; z_2|(n))$ is not within $[F(\mathbf{u}_\alpha; z_1|(n)), 1]$, reset it to the closest bound.

- Loop through all remaining cutoffs z_k, $k = 3, \ldots, K$.

2. Downward correction resulting in the lower line of Figure IV.1:

  - start with the largest cutoff z_K.
  - if the IK-returned ccdf value $F(\mathbf{u}_\alpha; z_K|(n))$ is not within $[0, 1]$ reset it to the closest bound.
  - proceed to the next lower cutoff z_{K-1}. If the IK-returned ccdf value $F(\mathbf{u}_\alpha; z_{K-1}|(n))$ is not within $[0, F(\mathbf{u}_\alpha; z_K|(n))]$ reset it to the closest bound.
  - loop downward through all remaining cutoffs z_k, $k = K - 2, \ldots, 1$.

3. Average the two sets of corrected ccdf's resulting in the thick middle line of Figure IV.1.

As mentioned above, one option is to ignore ccdf values for cutoffs that are upper bounds of classes with no z-data. Consider the example of Figure IV.1 and suppose there are no z-data in class $(z_6, z_7]$, yet all other classes contain at least one z-datum. The ccdf value $F(\mathbf{u}_\alpha; z_7|(n))$ is ignored and the correction algorithm given above is applied to the remaining cutoff values. This correction is illustrated on Figure IV.3. Note that this latter correction is implemented within the indicator simulation programs, but not within **ik3d**.

Order relations correction entails departure from the reproduction of the indicator covariance models as ensured from the theory of sequential indicator simulation ([82], p. 35). They may also entail departure from honoring constraint intervals if the correction is done within that interval. Order relation problems represent the most severe drawback of the indicator approach. They are the price to pay for attributing more than a single sample covariance to the implicit RF model.

IV.1.10 Indicator Cokriging

Indicator kriging does not make full use of the information contained in the original z-data. In the case of a continuous variable, IK considers only whether the z-data values exceed the cutoff value z. In the case of several categorical variables, e.g., defining the presence or absence of several rock types, IK retains only information about one rock type at a time.

An alternative is to consider all K indicator data $i(\mathbf{u}_\alpha; z_k)$, $k = 1, \ldots, K$ for the estimation of each ccdf value. The corresponding estimate would be a co-indicator kriging (coIK) estimate defined as:

$$[i(\mathbf{u}; z_{k_0})]^*_{coIK} = \sum_{k=1}^{K} \sum_{\alpha=1}^{n} \lambda_{\alpha,k}(\mathbf{u}; z_{k_0}) \cdot i(\mathbf{u}; z_k) \qquad (\text{IV.39})$$

The corresponding cokriging system would call for a matrix of K^2 direct and cross indicator covariances of the type

$$C_I(\mathbf{h}; z_k, z_{k'}) = Cov\{I(\mathbf{u}; z_k), I(\mathbf{u} + \mathbf{h}; z_{k'})\}$$

The direct inference and joint modeling of the K^2 covariances is not practical for large K. One solution is to call for an a-priori bivariate distribution model; another is to work on linear transforms of the indicator variables which are less cross correlated; see next section IV.1.11.

Adopting an a-priori bivariate distribution model amounts to forfeiting actual data-based inference of some or all of the indicator (cross)covariances; see relation (II.7). The most widely used bivariate distribution model is the bivariate Gaussian model after normal scores transform of the original variable: $Z(\mathbf{u}) \to Y(\mathbf{u}) = \varphi(Z(\mathbf{u}))$; see [9] and [110]. All indicator (cross)covariances are then determined from the normal score transform Y-covariance by relations like (V.20). Then, one is better off calling for the complete multivariate Gaussian model; see discussions on the MG approach in sections IV.1.8 and V.2.

A slight generalization of the bivariate Gaussian model is offered by the (bivariate) isofactorial models used in disjunctive kriging (DK) [117]. The generalization is obtained by a non-linear rescaling of the original variable $Z(\mathbf{u})$ by transforms $\psi(\cdot)$ different from the normal score transform $\varphi(\cdot)$. Besides the problem of which function $\psi(\cdot)$ to retain, the bivariate isofactorial distribution adopted for the new transform $Y(\mathbf{u}) = \psi(Z(\mathbf{u}))$ is not fundamentally different from the bivariate Gaussian model; see relation (V.22) and the related discussion. Then, one may be better off using the most congenial and parsimonious[4] of all RF models, the multivariate Gaussian model.

General experience with cokriging, see section IV.1.7 and [152,168], indicates that cokriging improves little from kriging if primary and secondary variable(s) are equally sampled, which is generally the case with indicator data defined at various cutoff values. In addition, when working with continuous variables, the corresponding cumulative indicator data do carry substantial information from one cutoff to the next one; in which case, the loss of information associated with using IK instead of coIK is not as large as it appears. Last, many more order relation problems occur in the practice of coIK.

Lighter alternatives to coIK that use more of the original information than IK are probability kriging [75,89,149,150] and indicator principal component kriging; see next section IV.1.11.

IV.1.11   Indicator Principal Component Kriging (IPCK)

IPCK [152,153] consists of kriging the principal components of the indicator covariance matrix $[C_I(\mathbf{h}_0; z_k, z_{k'}), k, k' = 1, \ldots, K]$ defined at some separation vector $\mathbf{h}_0$, usually taken as zero or a very small distance.

The K principal components $Y_k(\mathbf{u}), k = 1, \ldots, K$, are linear combinations

---

[4]Another extremely parsimonious RF model is that corresponding to relations (IV.30) and median indicator kriging. The median IK model, like the multivariate Gaussian model, is fully characterized by a single covariance, that of $Z(\mathbf{u})$, and the marginal cdf of $Z(\mathbf{u})$.

of the original K indicator variables:

$$Y_k(\mathbf{u}) = \sum_{k'=1}^{K} a_{k,k'} I(\mathbf{u}; z_{k'}) \qquad \text{(IV.40)}$$

$$\text{i.e., } \mathbf{Y} = \mathbf{A}^T \cdot \mathbf{I}$$

The $(K \times K)$ matrix $\mathbf{A}^T = [a_{k,k'}]$ is the transpose of the orthogonal matrix $\mathbf{A}$ originating from the spectral decomposition of the covariance matrix (e.g., see [148], p. 226):

$$[C_I(\mathbf{h}_0; z_k, z_{k'})] = \mathbf{A}\mathbf{\Lambda}\mathbf{A}^T \qquad \text{(IV.41)}$$

$\mathbf{\Lambda}$ is a diagonal matrix with the K eigenvalues of the indicator covariance matrix.

The K principal components $Y_k(\mathbf{u})$ are, by construction, orthogonal at the separation vector $\mathbf{h}_0$, i.e.,

$$Cov\{Y_k(\mathbf{u}), Y_{k'}(\mathbf{u} + \mathbf{h}_0)\} = 0, \ \forall k \neq k'$$

IPCK requires that this orthogonality extend to all other separation vectors $\mathbf{h}$ [153], i.e.,

$$Cov\{Y_k(\mathbf{u}), Y_{k'}(\mathbf{u} + \mathbf{h})\} \simeq 0, \ \forall \mathbf{h}, \forall k \neq k' \qquad \text{(IV.42)}$$

Then, cokriging of the K principal component $y_k(\mathbf{u})$ values reduces to their individual and separate kriging at considerably less inference and computation cost.

Moreover, when the original variable $Z(\mathbf{u})$ is continuous, it has been observed [152] that the K principal components $Y_k(\mathbf{u})$ are also ranked in decreasing order of spatial correlation; the higher order principal components tend to show a pure nugget effect behavior:

$$Cov\{Y_k(\mathbf{u}), Y_k(\mathbf{u} + \mathbf{h})\} \simeq 0, \text{ for } \mathbf{h} \neq 0, \text{ for large } k \text{ (close to } K) \qquad \text{(IV.43)}$$

Therefore, the direct kriging of the higher order principal components $y_k(\mathbf{u})$ reduces to a mere averaging of the $y_k(\mathbf{u}_\alpha)$-data within the moving search neighborhood. Thus, the original cokriging of K indicator values $i(\mathbf{u}, z_k)$ has been reduced to K' direct krigings of the first K' indicator principal components, with $K' < K$.

Starting from the kriged indicator principal components $y_k^*(\mathbf{u}; z_k)$, $k = 1$, $\dots, K$ a linear back-transform provides the corresponding unbiased approximate cokriging estimates for the indicator values (recall that these estimates are in fact ccdf models):

$$[i(\mathbf{u}; z_k)]^*_{IPCK} = \sum_{k'=1}^{K} a_{k',k} \cdot y_{k'}^*(\mathbf{u}) \qquad \text{(IV.44)}$$

$$\text{i.e., } \mathbf{I}^* = \mathbf{A} \cdot \mathbf{Y}^*$$

The limited experience presently available with IPCK has shown that the results are not significantly better than straight IK (IV.28) when applied to *continuous* Z-variables. The CPU time, however, may be reduced considerably. One reason is that cumulative indicator data are correlated from one cutoff to the next; thus they already carry part of the information that would be brought by coIK or IPCK.

The main application of IPCK is for evaluating ccdf's of several ($K > 1$) categorical variables, an application that would typically require indicator cokriging [154]. One drawback of working with principal components $y_k(\mathbf{u})$ rather than the original categorical indicators is that, during the variogram modeling phase, one cannot capitalize on prior soft structural information about continuity or anisotropy of the spatial distribution of the corresponding categories, e.g., facies. Another drawback is the smearing of specific facies anisotropy over the various indicator principal components resulting in a smearing of that anisotropy over the various facies ccdf models (IV.44).

Implementation of IPCK requires an orthogonal matrix decomposition of the type (IV.41) providing the loading coefficients $a_{k,k'}$ for both the principal component transform (IV.40) and back transform (IV.44). The program ipcprep given in section VI.2.9 performs that decomposition. Otherwise, IPCK does not require any special coding; the kriging of the indicator principal components could be done with the GSLIB programs okb2d or ktb3d. A specific program ipcsim is provided in section V.7.4 for indicator principal component simulation.

IV.1.12   Soft Kriging: The Markov Bayes Model

The major advantage of the indicator kriging approach to generating posterior conditional distributions (ccdf's) is its ability to account for soft data. As long as the soft or fuzzy data can be coded into prior local probability values, indicator kriging can be used to integrate that information into a posterior probability value [6,81,99].

The prior information can take one of the following forms:

- local hard indicator data $i(\mathbf{u}_\alpha; z)$ originating from local hard data $z(\mathbf{u}_\alpha)$:

$$i(\mathbf{u}_\alpha; z) = 1 \text{ if } z(\mathbf{u}_\alpha) \leq z, = 0 \text{ if not} \qquad (IV.45)$$

  or $i_k(\mathbf{u}_\alpha) = 1$ if $\mathbf{u}_\alpha \in$ category $k, = 0$ if not

- local hard indicator data $j(\mathbf{u}_\alpha; z)$ originating from ancillary information that provides hard inequality constraints on the local value $z(\mathbf{u}_\alpha)$. If $z(\mathbf{u}_\alpha) \in (a_\alpha, b_\alpha]$, then:

$$j(\mathbf{u}_\alpha; z) = \begin{cases} 0 \text{ if } z \leq a_\alpha \\ \text{undefined (missing) if } z \in (a_\alpha, b_\alpha] \\ 1 \text{ if } z > b_\alpha \end{cases} \qquad (IV.46)$$

- local soft indicator data $y(\mathbf{u}_\alpha; z)$ originating from ancillary information providing prior (pre-posterior) probabilities about the value $z(\mathbf{u}_\alpha)$:

$$y(\mathbf{u}_\alpha; z) \;=\; Prob\{Z(\mathbf{u}_\alpha) \leq z | \text{ local information}\} \qquad \text{(IV.47)}$$
$$\in \;\; [0, 1]$$
$$\neq \;\; F(z) : \text{ global prior as defined hereafter}$$

usually $E\{Y(\mathbf{u}; z)\} = F(z)$.

- *global* prior information common to all locations $\mathbf{u}$ within the stationary area A:

$$F(z) = Prob\{Z(\mathbf{u}) \leq z\}, \forall\, \mathbf{u} \in A \qquad \text{(IV.48)}$$

At any location $\mathbf{u} \in A$, prior information about the value $z(\mathbf{u})$ is characterized by any one of the four previous types of prior information. The IK process consists of a Bayesian updating of the local prior cdf into a posterior cdf using information supplied by neighboring local prior cdf's [14,99,168]:

$$[Prob\{Z(\mathbf{u}) \leq z | (n + n')\}]^*_{IK} = \qquad \text{(IV.49)}$$

$$\lambda_0 F(z) + \sum_{\alpha=1}^{n} \lambda_\alpha(\mathbf{u}; z) i(\mathbf{u}_\alpha; z) + \sum_{\alpha'=1}^{n'} \nu_{\alpha'}(\mathbf{u}; z) y(\mathbf{u}'_\alpha; z)$$

The $\lambda_\alpha(\mathbf{u}; z)$'s are the weights attached to the n neighboring hard indicator data of type (IV.45) *or* (IV.46), the $\nu_{\alpha'}(\mathbf{u}; z)$'s are the weights attached to the n' neighboring soft indicator data of type (IV.47), and λ_0 is the weight attributed to the global prior cdf. To ensure unbiasedness, λ_0 is usually set to:

$$\lambda_0 = 1 - \sum_{\alpha=1}^{n} \lambda_\alpha(\mathbf{u}; z) - \sum_{\alpha'=1}^{n'} \nu_{\alpha'}(\mathbf{u}; z)$$

The ccdf estimate (IV.49) can be seen as an indicator cokriging that pools information of different types: the hard i **or** j indicator data and the soft y-prior probabilities. When the soft information is not present or is ignored ($n' = 0$), expression (IV.49) reverts to the simple IK expression (IV.28).

Note that if different local sources of information are available at any one single location $\mathbf{u}_\alpha$, different prior probabilities of type (IV.47), $y_1(\mathbf{u}_\alpha; z) \neq y_2(\mathbf{u}_\alpha; z)$, may be derived and all used in an extended expression of type (IV.49). Care should be taken that the corresponding cokriging covariance matrices are positive definite to ensure a unique solution.

Updating

If the spatial distribution of the soft y-data is modeled by the covariance $C_I(\mathbf{h}; z)$ of the hard indicator data, i.e., if:

$$C_Y(\mathbf{h}; z) \;=\; Cov\{Y(\mathbf{u}; z), Y(\mathbf{u} + \mathbf{h}; z)\} \qquad \text{(IV.50)}$$
$$= C_{IY}(\mathbf{h}; z) \;=\; Cov\{I(\mathbf{u}; z), Y(\mathbf{u} + \mathbf{h}; z)\}$$
$$= C_I(\mathbf{h}; z) \;=\; Cov\{I(\mathbf{u}; z), I(\mathbf{u} + \mathbf{h}; z)\}$$

then there is no updating of the prior probability values $y(\mathbf{u}'_\alpha; z)$ at their locations $\mathbf{u}'_\alpha$, i.e.,

$$[Prob\{Z(\mathbf{u}'_\alpha) \leq z|(n + n')\}]^*_{IK} \equiv y(\mathbf{u}'_\alpha; z), \; \forall \, z$$

Most often, the soft z-data originate from information related to, but different from, the hard data $z(\mathbf{u}_\alpha)$. Thus the soft y-indicator spatial distribution is likely different from that of the hard i-indicator data, therefore:

$$C_Y(\mathbf{h}; z) \neq C_{IY}(\mathbf{h}; z) \neq C_I(\mathbf{h}; z)$$

Then, the indicator *cokriging* (IV.49) amounts to a full updating of all prior cdf's that are not already hard.

At the location of a constraint interval $j(\mathbf{u}_\alpha; z)$ of type (IV.46), indicator kriging *or* cokriging amounts to in-filling the interval $(a_\alpha, b_\alpha]$ with spatially interpolated ccdf values. Thus, if simulation is performed at that location, a z-attribute value would be drawn necessarily from within the interval. Were quadratic programming used to impose the same interval, the solution in the case of an active constraint would be limited to either bound a_α or b_α [47,108].

Covariance Inference

With enough data one could infer directly and model the matrix of covariance functions (one for each cutoff z): $[C_Y(\mathbf{h}; z) \neq C_{IY}(\mathbf{h}; z) \neq C_I(\mathbf{h}; z)]$.

An alternative to this tedious exercise is provided by the Markov-Bayes[5] model [99,168], whereby:

$$
\begin{aligned}
C_{IY}(\mathbf{h}; z) &= B(z)C_I(\mathbf{h}; z), \; \forall \, \mathbf{h} & \text{(IV.51)}\\
C_Y(\mathbf{h}; z) &= B^2(z)C_I(\mathbf{h}; z), \; \forall \, \mathbf{h} > 0 \\
&= |B(z)|C_I(\mathbf{h}; z), \; \mathbf{h} = 0.
\end{aligned}
$$

The coefficients $B(z)$ are obtained from calibration of the soft y-data to the hard z-data; more precisely:

$$B(z) = m^{(1)}(z) - m^{(0)}(z) \in [-1, +1] \qquad \text{(IV.52)}$$

with:

$$
\begin{aligned}
m^{(1)}(z) &= E\{Y(\mathbf{u}; z)|I(\mathbf{u}; z) = 1\} \\
m^{(0)}(z) &= E\{Y(\mathbf{u}; z)|I(\mathbf{u}; z) = 0\}
\end{aligned}
$$

Consider a calibration data set $\{y(\mathbf{u}_\alpha; z), i(\mathbf{u}_\alpha; z), \alpha = 1, \ldots, n\}$ where the soft probabilities $y(\mathbf{u}_\alpha; z)$ valued in $[0, 1]$ are compared to the actual hard values $i(\mathbf{u}_\alpha; z)$ valued 0 or 1. $m^{(1)}(z)$ is the mean of the y-values corresponding to $i=1$; the best situation is when $m^{(1)}(z) = 1$, that is when all

---

[5]The name "Markov-Bayes" model (for the ccdf) relates to the constitutive Markov approximation (IV.53) and to the Bayesian process of updating prior cdf-type data.

y-values exactly predict the outcome $i=1$. Similarly, $m^{(0)}(z)$ is the mean of the y-values corresponding to $i=0$, best being when $m^{(0)}(z) = 0$.

The parameter $B(z)$ measures how well the soft y-data separate the two actual cases $i = 1$ and $i = 0$. The best case is when $B(z) = \pm 1$, and the worst case is when $B(z) = 0$, that is $m^{(1)}(z) = m^{(0)}(z)$.

The case $B(z) = -1$ corresponds to soft data predictably wrong and is best handled by correcting the wrong probabilities $y(\mathbf{u}_\alpha; z)$ into $1 - y(\mathbf{u}_\alpha; z)$.

When $B(z) = 1$, the soft prior probability data $y(\mathbf{u}'_\alpha; z)$ in expression (IV.49) are treated as hard indicator data; in particular, they are not updated. Conversely, when $B(z) = 0$, the soft data $y(\mathbf{u}'_\alpha; z)$ are ignored, i.e., their weights in expression (IV.49) become zero.

In many practical situations the soft y-data are much more numerous than the hard i-data. Also, the y-values tend to be uniformly distributed, with no clustering, over the study area. In such case, it can be shown that the cokriging system is insensitive to the soft autocovariance $C_Y(\mathbf{h}; z)$. When validating the Markov-derived model, the first relation (IV.51) is by far the most critical.

An approximation which does not call for the soft autocovariance $C_Y(\mathbf{h}; z)$ consists of retaining only one single soft y-datum, that nearest to the location $\mathbf{u}$ being estimated; see the collocated cokriging option in section IV.1.7 and [98].

The model (IV.51) is analytically derived [99] from the Markov-type hypothesis (rather, an approximation) stating that: *"hard information $I(\mathbf{u}; z)$ screens any soft collocated information $Y(\mathbf{u}; z)$,"* i.e.,

$$Prob\{Z(\mathbf{u}') \le z | i(\mathbf{u}; z), y(\mathbf{u}; z)\} \tag{IV.53}$$
$$= Prob\{Z(\mathbf{u}') \le z | i(\mathbf{u}; z)\}, \ \forall \, y(\mathbf{u}; z), \ \forall \, \mathbf{u}, \mathbf{u}', z$$

Indicator cokriging using the Markov-Bayes model (IV.51) is implemented within the simulation program mbsim. Indicator kriging, using the simpler model (IV.50) corresponding to $B(z)=1$, can be performed with the program ik3d. A full indicator cokriging using any model for the coregionalization of hard and soft indicator data can be implemented through the general cokriging program cokb3d.

A very efficient Markov-Bayes model is obtained by considering that all indicator covariances $C_I(\mathbf{h}; z)$ are proportional to each other; see relation (IV.30).

Soft Data Inference

Consider the case of a primary continuous variable $z(\mathbf{u})$ informed by a related secondary variable $v(\mathbf{u})$. The series of hard indicator data valued 0 or 1, $i(\mathbf{u}_\alpha; z_k), k = 1, \ldots, K$ are derived from each hard datum value $z(\mathbf{u}_\alpha)$.

The soft indicator data series, $y(\mathbf{u}'_\alpha, z_k) \in [0, 1], k = 1, \ldots, K$, corresponding to the secondary variable value $v(\mathbf{u}'_\alpha)$, can be obtained from a calibration scattergram of z-values versus collocated v-values; see Figure IV.4. The

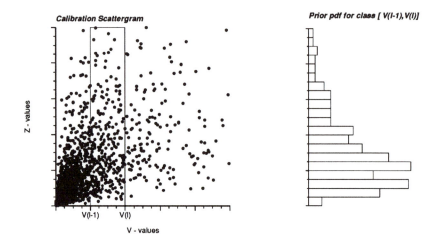

Figure IV.4: Inference of the soft prior probabilities from a calibration scattergram. The prior z-probability pdf at a location $\mathbf{u}'_\alpha$ where the secondary variable is $v(\mathbf{u}'_\alpha) \in (v_{l-1}, v_l]$ is identified to the calibration conditional pdf, drawn in the right of the figure.

range of v-values is discretized into L classes $(v_{l-1}, v_l], l = 1, \ldots, L$. For class $(v_{l-1}, v_l]$, the y-prior probability cdf can be modeled from the cumulative histogram of primary data values $z(\mathbf{u}'_\alpha)$ such that the collocated secondary data values $v(\mathbf{u}'_\alpha)$ fall into class $(v_{l-1}, v_l]$:

$$y(\mathbf{u}'_\alpha; z) = Prob\{Z(\mathbf{u}'_\alpha) \leq z | v(\mathbf{u}'_\alpha) \in (v_{l-1}, v_l]\} \qquad (IV.54)$$

Note that the secondary variable $v(\mathbf{u})$ need not be continuous. The classes $(v_{l-1}, v_l]$ can be replaced by categories of v-values, e.g., if the information v relates to different lithofacies or mineralizations.

The calibration scattergram that provides the prior y-probability values may be borrowed from a better sampled field different from the field under study. That calibration scattergram may be based on data other than those used to calibrate the covariance parameters $B(z)$ involved in relation (IV.51) (be aware that consistency problems may arise [168]).

In presence of sparse z-v samples, the fluctuations of the sample calibration scattergram may be smoothed using bivariate kernel functions [141] or, more simply, the resulting prior cdf's (IV.54) may be smoothed.

IV.1.13   Block Kriging

The linearity of the kriging algorithm allows direct estimation of *linear* averages of the attribute $z(\mathbf{u})$. For example, consider the estimation of the block average defined as:

$$z_V(\mathbf{u}) = \frac{1}{|V|} \int_{V(\mathbf{u})} z(\mathbf{u}')d\mathbf{u}' \simeq \frac{1}{N} \sum_{j=1}^{N} z(\mathbf{u}_j') \qquad \text{(IV.55)}$$

where $V(\mathbf{u})$ is a block of measure $|V|$ centered at $\mathbf{u}$, and the $\mathbf{u}_j'$'s are N points discretizing the volume $V(\mathbf{u})$.

One could estimate the N point values $z(\mathbf{u}_j')$ and average the point estimates into an estimate for the block value $z_V(\mathbf{u})$. If kriging is performed with the *same* data for all N point estimates, the N point kriging systems of type (IV.4) or (IV.6) can be averaged into a single "block kriging" system (see [94], p. 306). The righthand side point-to-point covariance values $C(\mathbf{u} - \mathbf{u}_\alpha)$ are replaced by point-to-block covariance values defined as:

$$\overline{C}(V(\mathbf{u}), \mathbf{u}_\alpha) \simeq \frac{1}{N} \sum_{j=1}^{N} C(\mathbf{u}_j' - \mathbf{u}_\alpha) \qquad \text{(IV.56)}$$

The programs **okb2d** and **ktb3d** allow both point and block kriging. The user can adjust the level of discretization N.

Warning:

Block kriging is a source of common misinterpretations when applied to non-linear transforms of the original variable $z(\mathbf{u})$.

For example, the average of the log transforms $y(\mathbf{u}') = \ln z(\mathbf{u}')$ is not the log transform of the average of the $z(\mathbf{u}')$'s:

$$y_V(\mathbf{u}) = \frac{1}{|V|} \int_{V(\mathbf{u})} \ln z(\mathbf{u}')d\mathbf{u}' \neq \ln z_V(\mathbf{u})$$

Therefore, the antilog of the block estimate $y_V^*(\mathbf{u})$ is *not* a kriging estimate of $z_V(\mathbf{u})$; see the related discussion in section IV.1.8.

Similarly, when working with the indicator transform:

$$\frac{1}{|V|} \int_{V(\mathbf{u})} i(\mathbf{u}'; z)d\mathbf{u}' \neq i_V(\mathbf{u}; z) = \left\{ \begin{array}{ll} 1, & \text{if } z_V(\mathbf{u}) \leq z \\ 0, & \text{if not} \end{array} \right. \qquad \text{(IV.57)}$$

Therefore, the average $\frac{1}{N} \sum_{j=1}^{N} [i(\mathbf{u}_j'; z)]_{IK}^*$ as obtained from block IK should *not* be used to model the ccdf of the average $z_V(\mathbf{u})$ of the original z-variable. Rather, $\frac{1}{N} \sum_{j=1}^{N} [i(\mathbf{u}_j'; z)]_{IK}^*$ is an estimate of the proportion of point values $z(\mathbf{u}_j')$ within $V(\mathbf{u})$ that are less than the cutoff z.

Inference of the block ccdf

Approximation of the conditional distribution of the block average $Z_V(\mathbf{u})$ could be done by correcting the variance of the point ccdf at the central location **u** of the block; several such correction algorithms are available; see [76], p. 468, [110] and [129].

Instead, we strongly recommend that the block ccdf be approached through small-scale stochastic simulations [61,74]. In this approach, many realizations of the small-scale distributions of point values $z(\mathbf{u}'_j)$ within the block $V(\mathbf{u})$ are generated and the point simulated values are averaged into simulated block values. The proportion of the simulated block values no greater than the cutoff z provides an estimate of the block indicator $I_V(\mathbf{u}; z)$ conditional expectation, i.e., an estimate of the block ccdf. The advantage of this approach is that it allows sensitivity analysis of the impact of the variogram model(s) at short scale.

IV.1.14   Cross Validation

There are so many interdependent subjective decisions in a geostatistical study that it is a good practice to validate the entire geostatistical model and kriging plan prior to any production run. The exercise of cross-validation is analogous to a dress rehearsal: it is intended to detect what could go wrong, but it does not ensure that the show will be successful.

The geostatistical model is validated by re-estimating the known values under implementation conditions as close as possible to those of the forthcoming production runs. These implementation conditions include the variogram model(s), the type of kriging, and the search strategy. These re-estimation techniques are discussed in most practical statistics and geostatistics books [32,48,76,84,122,156].

In cross-validation actual data are dropped one at a time and re-estimated from some of the remaining neighboring data.[6] Each datum is replaced in the data set once it has been re-estimated.

The term jackknife applies to resampling without replacement, i.e., when one set of data values is re-estimated from another non-overlapping data set [48].

The subroutines **xvok2d** and **xvkt3d**, based on **okb2d** and **ktb3d** respectively, allow both cross-validation and the jackknife.

Analysing the re-estimation scores

Consider a total of K estimation procedures or implementation variants, with the goal being to select the "best" procedure. Let $z(\mathbf{u}_j), j = 1, \ldots, N$ be the original data marked for re-estimation and let $z^{*(k)}(\mathbf{u}_j), j = 1, \ldots, N$ be their re-estimated values by procedure number k, with $k = 1, \ldots, K$.

The following criteria are suggested for analysing the re-estimation scores:

---

[6] The cross-validation should be as difficult as the actual estimation of unsampled values. For example, if data are aligned along drillholes or lines then all data originating from the same hole or line should be ignored when re-estimating a datum value in order to approach the sampling density available in actual estimation.

- The distribution of the N errors $\{z^{*(k)}(\mathbf{u}_j) - z(\mathbf{u}_j), j = 1, \ldots, N\}$ should be symmetric, centered on a zero mean, with a minimum spread.

- The plot of the error $z^{*(k)}(\mathbf{u}_j) - z(\mathbf{u}_j)$ versus the estimated value $z^{*(k)}(\mathbf{u}_j)$ should be centered around the zero error line, a property called "conditional unbiasedness" ([76], p. 264; [94], p. 458). Moreover, the plot should have an equal spread, i.e., the error variance should not depend on whether the true value is low or high, a property known as homoscedasticity of the error variance.

- The N errors for a given procedure k should be independent one from another. This could be checked by contouring the error values: the contour map should not show any general trend. Alternatively, the variogram of the error values could be calculated: it should show a pure nugget effect.

Warning

It is worth recalling that the kriging variance is but a ranking index of data configurations. Since the kriging variance does not depend on the data values, it should not be used to select a variogram model or a kriging implementation; nor should it be used as the sole criterion to determine the location of additional samples [32,85].

IV.2 2-D Ordinary Kriging Subroutines

The following files should be located on the diskette(s) before attempting to compile and execute the **okb2dm** program:

| okb2dm.f | an example driver program for **okb2d**

| okb2d.inc | an include file with maximum array dimensions

| okb2d.f | the subroutine **okb2d** and other required routines

| okb2d.par | an example parameter file for **okb2dm**

This is a straightforward 2-D ordinary kriging subroutine that can be used as is or as a basis for custom kriging programs. The code is not cluttered with extra features such as super block searching. This makes it easy to understand, but inefficient when dealing with lots of data.

The parameters for the main program **okb2dm** are shown on Figure IV.5 and documented below. The input variables are passed in the argument list to the subroutine **okb2d**:

- **datafl:** the input data in a simplified Geo-EAS formatted file.

```
                    Parameters for OKB2D
                    ********************

START OF PARAMETERS:
cluster.dat                             \data file
1   2   3                               \columns for x,y, and variable
-1.0e21   1.0e21                        \data trimming limits
okb2d.out                               \output File of Kriged Results
3                                       \debugging level: 0,1,2,3
okb2d.dbg                               \output file for debugging
5    5.0   10.0                         \nx,xmn,xsiz
5    5.0   10.0                         \ny,ymn,ysiz
1    1                                  \x and y block discretization
4   16                                  \min and max data for kriging
20.0                                    \maximum search radius
1   0.2                                 \nst, nugget effect
1   10.0    0.8    0.00    1.00         \it,aa,cc,angl,anis: structure 1
```

Figure IV.5: An example parameter file for okb2d.

- **icolx, icoly,** and **icolvr:** the columns for the x and y coordinates, and the variable to be kriged.

- **tmin** and **tmax:** all values strictly less than *tmin* and greater than or equal to *tmax* are ignored.

- **outfl:** the output grid is written to this file. The output file will contain both the kriging estimate and the kriging variance for all points/blocks. The output grid cycles fastest on x and then y.

- **idbg:** an integer debugging level between 0 and 3. The higher the debugging level the more output. Normally level 0 or 1 should be chosen. If there are suspected problems, or if you would like to see the actual kriging matrices, level 2 or 3 can be chosen. It is advisable to restrict the actual number of points being estimated when the debugging level is high (the debugging file can become extremely large).

- **dbgfl:** the debugging output is written to this file.

- **nx, xmn, xsiz:** definition of the grid system (x axis).

- **ny, ymn, ysiz:** definition of the grid system (y axis).

- **nxdis** and **nydis:** the number of discretization points for a block. If both *nxdis* and *nydis* are set to 1 then point kriging is performed.

- **ndmin** and **ndmax:** the minimum and maximum number of data points to use for kriging a block.

- **radius:** the maximum isotropic search radius.

- **nst** and **c0:** the number of variogram structures and the isotropic nugget constant. The nugget constant does not count as a structure.

```
                    Parameters for XVOK2D
                    *********************

START OF PARAMETERS:
cluster.dat                     \data file
1    2    3                     \columns for x,y, and variable
0                               \1=jackknife, 0=cross validate
nodata.dat                      \jackknife data file
1    2    0                     \x,y, and variable(0 if cross v)
-1.0e21   1.0e21                \data trimming limits
xvok2d.out                      \output file of kriged results
1                               \Debugging level: 0,1,2,3
xvok2d.dbg                      \output file for debugging
4   16                          \min and max data for kriging
20.0                            \maximum search radius
1 0.20                          \nst, nugget effect
1 10.0 0.80 0.00 1.00           \it,aa,cc,ang1,anis: structure 1
```

Figure IV.6: An example parameter file for **xvok2d**.

- For each of the *nst* nested structures one must define *it*, the type of structure; *aa*, the *a* parameter; *cc*, the *c* parameter; *ang*1 and *anis*, the azimuth and geometric/zonal anisotropy parameters. A detailed description of these parameters is given in section II.3.

Cross Validation Program

The following files should be located on the diskette(s) before attempting to compile and execute this cross validation program **xvok2d**:

$\boxed{\text{xvok2dm.f}}$ an example driver program for **xvok2d**

$\boxed{\text{xvok2d.inc}}$ an include file with maximum array dimensions

$\boxed{\text{xvok2d.f}}$ the subroutine **xvok2d** and other required routines

$\boxed{\text{xvok2d.par}}$ an example parameter file for **xvok2d**

This program, based on **okb2d**, serves to cross validate or jackknife a kriging plan. Prior to a **okb2d** kriging run it is good practice to check the results of a cross-validation. Obvious problems can be identified with either the input parameters (e.g., search or variogram specification) or data (e.g., anomalous values). It is not considered sound practice to arbitrarily change the variogram parameters to achieve a good cross-validation.

The parameters required by **xvok2d** are very similar to those required by **okb2d**. The reader can refer to the description given above for documentation of most parameters. The differences are shown on Figure IV.6 and given below:

1. **ijack:** an indicator specifying whether performing a jackknife (*ijack=1*) or cross validation (*ijack=0*).

2. **jackfl:** a data file in a simplified Geo-EAS format which contains the locations that will be submitted to jackknife. The column number of the x coordinate, y coordinate, and variable must be specified. If the column for the variable is negative or zero then all locations in *jackfl* will be estimated but no cross validation scores will be reported; this allows an irregular grid network to be estimated.

3. The parameters in **okb2d** that define the grid are not required.

4. The output is a file containing the x location, y location, the true value, the estimated value, and the kriging variance for all locations that were estimated.

IV.3   Flexible 3-D Kriging Subroutines

The following files should be located on the diskette(s) before attempting to compile and execute this program/subroutine:

| ktb3dm.f | an example driver program for **ktb3d**

| kt3d.inc | an include file with maximum array dimensions

| kt3d.f | some subroutines common to **ktb3d** and **xvkt3d**

| ktb3d.f | the subroutine **ktb3d**

| ktb3d.par | an example parameter file for **ktb3dm**

The program **ktb3d** provides a fairly advanced 3-D kriging program for points or blocks by simple kriging (SK), ordinary kriging (OK), or kriging with a polynomial trend model (KT) with up to nine monomial terms. The program works in 2-D and is faster than **okb2d** if there are many data. One of the features that makes this program fairly fast is the super block search; see section II.4.

The program is set up so that a novice programmer can make changes to the form of the polynomial drift. The external drift concept has been incorporated, adding an additional unbiasedness constraint to the ordinary kriging system. When using an external drift, it is necessary to know the value of the drift variable at all data locations and all the locations that will be estimated (i.e., all grid nodes); see section IV.1.5.

The input parameters are shown in Figure IV.7. They are:

- **datafl:** the input data in a simplified Geo-EAS formatted file If an external drift variable is considered *datafl* should include an additional column for the values of that variable at the primary data locations.

```
                        Parameters for KTB3D
                        ********************
START OF PARAMETERS:
cluster.dat                      \data file
1    2    0    3                  \column for x,y,z and variable
-1.0e21    1.0e21                 \data trimming limits
ktb3d.out                        \output file of kriged results
1                                \debugging level: 0,1,2,3
ktb3d.dbg                        \output file for debugging
5    5.0    10.0                 \nx,xmn,xsiz
5    5.0    10.0                 \ny,ymn,ysiz
1    5.0    10.0                 \nz,zmn,zsiz
1    1    1                      \x,y and z block discretization
4    16                          \min, max data for kriging
0                                \max per octant (0-> not used)
20.0                             \maximum search radius
0.0   0.0   0.0   1.0   1.0      \search: ang1,2,3,anis1,2
0    2.302                       \1=use sk with mean, 0=ok+drift
0 0 0 0 0 0 0 0 0                \drift: x,y,z,xx,yy,zz,xy,xz,zy
0                                \0, variable; 1, estimate trend
0                                \1, then consider external drift
5                                \column number in original data
extdrift.dat                     \Gridded file with drift variabl
4                                \column number in gridded file
1    0.2                         \nst, nugget effect
1    10.0    0.8                 \it,aa,cc:          structure 1
0.0   0.0   0.0   1.0   1.0      \ang1,ang2,ang3,anis1,anis2:
```

Figure IV.7: An example parameter file for ktb3d.

- **icolx, icoly, icolz** and **icolvr:** the columns for the $x, y,$ and z coordinates, and the variable to be estimated.

- **tmin** and **tmax:** all values strictly less than *tmin* and greater than or equal to *tmax* are ignored.

- **outfl:** the output grid is written to this file. The output contains the estimate and the kriging variance for every point/block on the grid, cycling fastest on x then y, and finally z. Unestimated points are flagged with a large negative number (-999.). The parameter UNEST, in the source code, can be changed if a different number is preferred.

- **idbg:** an integer debugging level between 0 and 3. The higher the debugging level the more output. The normal levels are 0 and 1 which summarize the results. Levels 2 and 3 provide all the kriging matrices and data used for the estimation of every point/block. It is suggested not to use a high debugging level with a large grid.

- **dbgfl:** the debugging output is written to this file.

- **nx, xmn, xsiz:** definition of the grid system (x axis).

- **ny, ymn, ysiz:** definition of the grid system (y axis).

- **nz, zmn, zsiz:** definition of the grid system (z axis).

- **nxdis, nydis** and **nzdis:** the number of discretization points for a block. If $nxdis, nydis$ and $nzdis$ are all set to 1 then point kriging is performed.

- **ndmin** and **ndmax:** the minimum and maximum number of data points to use for kriging a block.

- **noct:** the maximum number to retain from an octant (an octant search is not used if $noct=0$).

- **radius:** the maximum search radius after accounting for the following anisotropy:

- **sang1, sang2, sang3, sanis1,** and **sanis2:** the parameters that describe the search anisotropy. See the discussion on anisotropy specification in Figure II.4.

- **isk** and **skmean:** if isk is set to 1 then simple kriging will be performed. Note that if doing SK the stationary mean is required ($skmean$) and power variogram models ($it = 4$) are not allowed.

- **idrif(i),i=1...9:** indicators for those drift terms to be included in the trend model. $idrif(i)$ is set to 1 if the drift term number i should be included, and is set to zero if not. The nine drift terms correspond to the following:

  i = 1 linear drift in x

  i = 2 linear drift in y

  i = 3 linear drift in z

  i = 4 quadratic drift in x

  i = 5 quadratic drift in y

  i = 6 quadratic drift in z

  i = 7 cross quadratic drift in xy

  i = 8 cross quadratic drift in xz

  i = 9 cross quadratic drift in yz

- **itrend:** indicator of whether to estimate the trend ($itrend = 1$) or the variable ($itrend = 0$). The trend may be kriged with ordinary kriging (all $idrif(i)$ values set to 0) or with any combination of trend kriging (some $idrif(i)$ terms set to 1).

- **iext:** if set to 1, kriging with an external drift is implemented. If $iext$ is anything but 1 then an external drift will *not* be considered.

- **iextv:** [7] the column number, in the original data file (*datafl*), for the external drift variable at each primary data location. This column should be full valued (i.e., no missing values unless the primary variable is also missing). This external variable is used only if *iext* = 1.

- **extfl:** a file for the gridded external drift variable. The external drift variable is needed at all grid locations to be estimated. The origin of the grid network, the number of nodes, and the spacing of the grid nodes should be exactly the same as the grid being kriged in **ktb3d**. This variable is used only if *iext* = 1.

- **iextfl:** the column number in *extfl* for the gridded external drift variable. This variable is used only if *iext* = 1.

- **nst** and **c0:** the number of variogram structures and the nugget constant. The nugget constant does not count as a structure.

- For each of the *nst* nested structures one must define *it* the type of structure, *aa* the *a* parameter, *cc* the *c* parameter, the angles, and the geometric/zonal anisotropy parameters: *ang*1, *ang*2, *ang*3, *anis*1, and *anis*2. A detailed description of these parameters is given in section II.3.

The maximum dimensioning parameters are specified in the file $\boxed{\text{kt3d.inc}}$. All memory is allocated in common blocks declared in this file. This file is included in all functions and subroutines except for the low level routines which are independent of the application. The advantage of this practice is that modifications can be made easily in one place without altering multiple source code files and long argument lists. One disadvantage is that a programmer must avoid all variable names declared in the include file (a global parameter could accidentally be changed). Another potential disadvantage is that the program logic can become obscure because no indication has to be given about the variables that are changed in each subroutine. Our intention is to provide adequate in-line documentation.

Cross Validation Program

The following files should be located on the diskette(s) before attempting to compile and execute the cross-validation program **xvkt3d**:

$\boxed{\text{xvkt3dm.f}}$ an example driver program for **xvkt3d**

$\boxed{\text{kt3d.inc}}$ an include file with maximum array dimensions

$\boxed{\text{kt3d.f}}$ some subroutines common to **ktb3d** and **xvkt3d**

---

[7]The external drift variable values are found in two different data files: *datafl* for those values at the locations of the primary variable data values, and *extfl* for those values at the grid node locations to be estimated.

```
                         Parameters for XVKT3D
                         *********************

START OF PARAMETERS:
cluster.dat                             \data file
1    2    0    3                        \column for x,y,z and variable
0                                       \1=jackknife, 0=cross validate
xvkt3d.dat                              \Jackknife Data File
1    2    0    0                        \column for x,y,z and variable
-1.0e21   1.0e21                        \data trimming limits
xvkt3d.out                              \output file of kriged results
1                                       \Debugging level: 0,1,2,3
xvkt3d.dbg                              \output file for debugging
10   2.5    5.0                         \nx,xmn,xsiz
10   2.5    5.0                         \ny,ymn,ysiz
1    2.5    5.0                         \nz,zmn,zsiz
1    12                                 \min, max data for kriging
0                                       \max per octant (0-> not used)
20.0                                    \maximum search radius
0.0  0.0  0.0  1.0  1.0                 \search: ang1,2,3,anis1,2
0     0.0                               \1=use sk with mean, 0=ok+drift
0 0 0 0 0 0 0 0 0                       \drift: x,y,z,xx,yy,zz,xy,xz,zy
0                                       \1, then consider external drift
5                                       \column number in original data
5                                       \column number in jackknife file
1    0.20                               \nst, nugget effect
1    10.0   0.8                         \it,aa,cc:        structure 1
0.0  0.0   0.0  1.0 1.0                 \ang1,ang2,ang3,anis1,anis2:
```

Figure IV.8: An example parameter file for **xvkt3d**.

| xvkt3d.f | the subroutine **xvkt3d**

| xvkt3d.par | an example parameter file for **xvkt3dm**

This program, based on **ktb3d**, serves to cross validate or jackknife a kriging plan. Prior to a **ktb3d** kriging run it is good practice to check the results of a cross-validation. Obvious problems can be identified with either the input parameters (e.g., search or variogram specification) or data (e.g., anomalous values). It is not considered sound practice to change arbitrarily the variogram or drift parameters to achieve a good cross-validation.

The parameters required by **xvkt3d** are very similar to those required by **ktb3d**. The reader can refer to the description given above for most parameters. The notable differences are shown on Figure IV.8 and described below:

1. **ijack:** an indicator specifying whether performing a jackknife (*ijack=1*) or cross validation (*ijack=0*).

2. **jackfl:** a data file in a simplified Geo-EAS format which contains the locations that will be submitted to jackknife. The column number of the x, y, z coordinates, and of the variable must be specified. If the column for the variable is negative or zero then all locations in *jackfl* will be estimated but no cross validation scores will be reported; this allows an irregular grid network to be estimated.

3. The parameters in **ktb3d** that define the grid are still required because they are used to establish the limits of the super block search network.

4. The output file will contain the x location, y location, z location, true value, estimated value, and kriging variance for all locations.

5. For generality, this program also accepts the *itrend* = 1 option. Be aware that poor cross-validation results can be obtained when comparing the estimated trend to the measured values which include short scale fluctuations.

IV.4 Cokriging Subroutine

The following files should be located on the diskette(s) before attempting to compile and execute the cokriging subroutine:

$\boxed{\text{cokb3dm.f}}$ example driver program

$\boxed{\text{cokb3d.inc}}$ include file with array dimensions

$\boxed{\text{cokb3d.f}}$ cokriging and other required subroutines

$\boxed{\text{cokb3d.par}}$ example parameter file

This cokriging subroutine is based on **okb2d**. In particular, it does not have the super block search of **ktb3d**. The construction of the kriging matrix using the linear model of coregionalization ([76], p. 390) and setting the various unbiasedness conditions are the essential additions.

The parameters required by **cokb3d** are illustrated on Figure IV.9 and documented below:

- **datafl:** the input data in a simplified Geo-EAS formatted file.

- **nvar:** the number of variables (primary plus all secondary). For example, *nvar=2* if there is only one secondary variable.

- **icolx, icoly,** and **icolvr():** the columns for the x and y coordinates, the primary variable to be kriged, and all secondary variables.

- **tmin** and **tmax:** all values (for all variables) strictly less than *tmin* and greater than or equal to *tmax* are ignored.

- **ktype:** the kriging type must be specified: 0 = simple cokriging, 1 = standardized ordinary cokriging with re-centered variables and a single unbiasedness constraint; see relation (IV.23), 2 = traditional ordinary cokriging.

```
                        Parameters for COKB3D
                        *********************

START OF PARAMETERS:
somedata.dat                          \data file
3                                     \number of variables primary+oth
1    2    3    4    5    6             \columns for x,y,z and variables
-1.0e21    1.0e21                      \data trimming limits
1                                     \ktype (0=SK,1=OK,2=OK-trad)
0.00  0.00  0.00  0.00                \mean(i),i=1,nvar
cokb3d.out                            \output file for kriging results
1                                     \debugging level: 0,1,2,3
cokb3d.dbg                            \output file for debugging
50    2.5    5.0                      \nx,xmn,xsiz
50    2.5    5.0                      \ny,ymn,ysiz
10    2.5    5.0                      \nz,zmn,zsiz
1    1    1                           \x, y, and z block discret
1    8    8                           \min, max primary, max secondary
60.0   10.0                           \maximum search radius: prim,sec
1    1                                \semivariogram for "i" and "j"
1    11.0                             \   number of structures, nugget
1    60.0   39.0                      \   it,aa,cc:        structure 1
0.0  0.0    0.0   1.0 1.0             \   ang1,ang2,ang3,anis1,anis2:
1    2                                \semivariogram for "i" and "j"
1    0.0                              \   number of structures, nugget
1    60.0   14.5                      \   it,aa,cc:        structure 1
0.0  0.0    0.0   1.0 1.0             \   ang1,ang2,ang3,anis1,anis2:
1    3                                \semivariogram for "i" and "j"
1    0.0                              \   number of structures, nugget
1    60.0   5.0                       \   it,aa,cc:        structure 1
0.0  0.0    0.0   1.0 1.0             \   ang1,ang2,ang3,anis1,anis2:
2    2                                \semivariogram for "i" and "j"
1    9.0                              \   number of structures, nugget
1    60.0   15.0                      \   it,aa,cc:        structure 1
0.0  0.0    0.0   1.0 1.0             \   ang1,ang2,ang3,anis1,anis2:
2    3                                \semivariogram for "i" and "j"
1    0.0                              \   number of structures, nugget
1    60.0   3.8                       \   it,aa,cc:        structure 1
0.0  0.0    0.0   1.0 1.0             \   ang1,ang2,ang3,anis1,anis2:
3    3                                \semivariogram for "i" and "j"
1    1.1                              \   number of structures, nugget
1    60.0   1.8                       \   it,aa,cc:        structure 1
0.0  0.0    0.0   1.0 1.0             \   ang1,ang2,ang3,anis1,anis2:
```

Figure IV.9: An example parameter file for **cokb3d**.

- **mean():** the mean of the primary and all secondary variables are required input if either simple cokriging or standardized ordinary cokriging are used.

- **outfl:** the output grid is written to this file. The output file will contain both the kriging estimate and the kriging variance for all points/blocks. The output grid cycles fastest on x then y then z.

- **idbg:** an integer debugging level between 0 and 3. The higher the debugging level the more output. Normally level 0 or 1 should be chosen.

- **dbgfl:** the debugging output is written to this file.

- **nx, xmn, xsiz:** definition of the grid system (x axis).

- **ny, ymn, ysiz:** definition of the grid system (y axis).

- **nz, zmn, zsiz:** definition of the grid system (z axis).

- **nxdis, nydis** and **nzdis:** the number of discretization points for a block. If $nxdis, nydis$ and $nzdis$ are set to 1 then point cokriging is performed.

- **ndmin, ndmaxp** and **ndmaxs:** the minimum and maximum number of primary data, and the maximum number of secondary data (regardless of which secondary variable) to use for kriging a block.

- **radiusp** and **radiuss:** the maximum isotropic search radius for the primary and the secondary data.

The direct and cross variograms may be specified in any order; they are specified according to the variable *number*. Variable "1" is the primary (regardless of its column ordering in the input data files) and the secondary variables are numbered from "2" depending on their ordered specification in *icolvr()*. It is unnecessary to specify the j to i cross variogram if the i to j cross variogram has been specified; the cross covariance is assumed to be symmetric unless the user has explicitly entered a non-symmetric cross-variogram. For each i to j variogram the following are required:

- **nst** and **c0:** the number of variogram structures and the isotropic nugget constant. The nugget constant does not count as a structure.

- For each of the *nst* nested structures one must define *it*, the type of structure; *aa*, the a parameter; *cc*, the c parameter; *ang.* and *anis.*, the angle and geometric/zonal anisotropy parameters. A detailed description of these parameters is given in section II.3.

The semivariogram matrix corresponding to the input of Figure IV.9) is; see [94], p. 259:

$$
\begin{bmatrix}
\gamma_Z(\mathbf{h}) & \gamma_{ZU}(\mathbf{h}) & \gamma_{ZV}(\mathbf{h}) \\
\gamma_{UZ}(\mathbf{h}) & \gamma_U(\mathbf{h}) & \gamma_{UV}(\mathbf{h}) \\
\gamma_{VZ}(\mathbf{h}) & \gamma_{VU}(\mathbf{h}) & \gamma_V(\mathbf{h})
\end{bmatrix}
=
\begin{bmatrix}
11.0 & 0.0 & 0.0 \\
0.0 & 9.0 & 0.0 \\
0.0 & 0.0 & 1.1
\end{bmatrix} \gamma_0(\mathbf{h})
$$

$$
+
\begin{bmatrix}
39.0 & 14.5 & 5.0 \\
14.5 & 15.0 & 3.8 \\
5.0 & 3.8 & 1.8
\end{bmatrix} \gamma_1(\mathbf{h})
$$

where $\gamma_0(\mathbf{h})$ is a nugget effect model and $\gamma_1(\mathbf{h})$ is a spherical structure of sill 1 and range 60. The program checks for a positive definite variogram model. The user must tell the program, by keying in a *yes*, to continue if the model is found not to be positive definite.

IV.5 Indicator Kriging Subroutine

The following files should be located on the diskette(s) before attempting to compile and execute the IK program:

ik3dm.f an example driver program for **ik3d**

ik3d.inc an include file with maximum array dimensions

ik3d.f the indicator kriging and other required subroutines

ik3d.par an example parameter file for **ik3dm**

The program **ik3d** considers ordinary or simple indicator kriging, performs all order relation corrections, and, optionally, allows the direct input of already transformed indicator data, including possibly soft data and constraint intervals. As opposed to program **mbsim**, soft data of type (IV.46) are not updated.

The parameters required by **ik3d** are shown in Figure IV.10 and are described below:

- **datafl:** the input data in a simplified Geo-EAS formatted file.

- **icolx, icoly, icolz, and icolvr:** the columns for the x, y, and z coordinates, and the variable to be estimated.

- **directik:** already transformed indicator values are read from this file. A title line, number of variables, and identification lines should be present (just like in any of the simplified Geo-EAS files used in GSLIB). The program then expects the x, y, and z coordinates and *ncut* values per line. Missing values are identified as less than *tmin* which would correspond to a constraint interval. Otherwise, the soft cdf data should steadily increase from 0 to 1.

```
                    Parameters for IK3D
                    ********************
START OF PARAMETERS:
cluster.dat                         \data file
1    2    0    3                    \column for x,y,z and variable
direct.ik                           \direct indicator input (soft)
-1.0e21   1.0e21                    \data trimming limits
ik3d.out                            \output file of kriging results
2                                   \debugging level: 0,1,2,3
ik3d.dbg                            \output file for debugging
10   2.5    5.0                     \nx,xmn,xsiz
10   2.5    5.0                     \ny,ymn,ysiz
1    0.0    5.0                     \nz,zmn,zsiz
1    16                             \min, max data for kriging
20.0                                \maximum search radius
0.0  0.0  0.0  1.0  1.0             \search: ang1,2,3,anis1,2
0                                   \max per octant (0-> not used)
0    2.5                            \0=full IK, 1=Med IK (cutoff)
1                                   \0=SK, 1=OK
5                                   \number cutoffs
0.5  0.12   1    0.15               \cutoff, global cdf, nst, nugget
     1   10.0   0.85                \          it, aa, cc
     0.0  0.0  0.0  1.0  1.0        \          ang1,ang2,ang3,anis1,2
1.0  0.29   1    0.10               \cutoff, global cdf, nst, nugget
     1   10.0   0.90                \          it, aa, cc
     0.0  0.0  0.0  1.0  1.0        \          ang1,ang2,ang3,anis1,2
2.5  0.50   1    0.10               \cutoff, global cdf, nst, nugget
     1   10.0   0.90                \          it, aa, cc
     0.0  0.0  0.0  1.0  1.0        \          ang1,ang2,ang3,anis1,2
5.0  0.74   1    0.10               \cutoff, global cdf, nst, nugget
     1   10.0   0.90                \          it, aa, cc
     0.0  0.0  0.0  1.0  1.0        \          ang1,ang2,ang3,anis1,2
10.0 0.88   1    0.15               \cutoff, global cdf, nst, nugget
     1   10.0   0.85                \          it, aa, cc
     0.0  0.0  0.0  1.0  1.0        \          ang1,ang2,ang3,anis1,2
```

Figure IV.10: An example parameter file for **ik3dm**.

- **tmin** and **tmax:** all values strictly less than *tmin* and greater than or equal to *tmax* are ignored.

- **outfl:** the output grid is written to this file. The output file will contain the order relation corrected conditional distributions for all nodes in the grid. The ccdf values will be written to the same line or "record"; there are *ncut* variables in the output file. Unestimated nodes are identified by all ccdf values equal to -9.9999.

- **idbg:** an integer debugging level between 0 and 3. The higher the debugging level the more output.

- **dbgfl:** the debugging output is written to this file.

- **nx, xmn, xsiz:** definition of the grid system (x axis).

- **ny, ymn, ysiz:** definition of the grid system (y axis).

- **nz, zmn, zsiz:** definition of the grid system (z axis).

- **ndmin** and **ndmax:** the minimum and maximum number of data points to use for kriging a block.

- **radius:** the maximum isotropic search radius.

- **ang1, ang2, ang3, anis1,** and **anis2:** the geometric anisotropy parameters for an anisotropic search. A detailed description of these parameters is given in section II.3.

- **noct:** the maximum number of samples per octant (octant search is not used if *noct*=0).

- **mik** and **mikcut:** If *mik* = 0, then a full IK is performed. If *mik* = 1, then a median IK approximation is used where the kriging weights from the cutoff closest to *mikcut* are used for all cutoffs.

- **ktype:** simple kriging is performed if *ktype=0*, ordinary kriging is performed if *ktype=1*.

- **ncut:** the number of cutoffs.

The following information is required for each of the *ncut* cutoffs:

- **cut, cdf, nst,** and **c0:** the cutoff value, the global cdf corresponding to the cutoff value, the number of variogram structures and the nugget constant, for the corresponding indicator variogram.

- For each of the *nst* nested structures one must define the type of structure *it*, the relative contribution c, the a parameter, the geometric anisotropy parameters $ang1, ang2, ang3, anis1$, and $anis2$. A detailed description of these parameters is given in section II.3. Gaussian and power variogram models ($it = 3, 4$) are not allowed with indicator variables.

IV.6   Application Notes

- Details of the grid definition are given in section II.2.3, the parameter files and program execution are described in II.2.4, and the variogram specification is described in II.3.

- In general, there are two different objectives when using kriging as a mapping algorithm: 1) to show a smooth picture of global trends, and 2) to provide local estimates to be further processed (e.g., selection of high grade areas, flow simulation, etc.). In the first case a large number of samples, hence, a large search neighborhood should be used for each kriging in order to provide a smooth continuous surface. In the second case one may have to limit the number of samples within the local search neighborhood. This minimizes smoothing if the local data mean is known to fluctuate considerably and also speeds up the matrix inversion.

- It is not good practice to limit systematically the search radius to the largest variogram range. Although a datum location $\mathbf{u}_\alpha$ may be beyond the range, i.e., if $C(\mathbf{u} - \mathbf{u}_\alpha) = 0$, the datum value $z(\mathbf{u}_\alpha)$ still provides information about the unknown mean value $m(\mathbf{u})$ at the location $\mathbf{u}$ being estimated.

- When using quadrant and octant searches, the maximum number of samples should be set larger than normal. The reason is that quadrant and octant searches cause artifacts when the data are spaced regularly and aligned with the centers of the blocks being estimated, i.e., from one point/block to the next there can be dramatic changes in the data configuration. The artifacts take the form of artificial short scale fluctuations in the estimated trend or mean surface. This problem is less significant when more samples are used. The user can check for this problem by mapping the trend with program **ktb3d**.

- When using KT it is good practice to contour or post the estimated trend with color or gray scale on some representative 2-D views. This may help to obtain a better understanding of the data spatial distribution and it will also indicate potential problems with erratic estimation and extrapolation. Problems are more likely to occur with higher order drift terms. The trend is estimated instead of the actual variable by setting the parameter *itrend* = 1 in program **ktb3d**. Recall the warning given in section IV.1.4 against overzealous modeling of the trend. In most interpolation situations, OK with a moving data neighborhood will suffice [97].

- The KT estimate (IV.10) or (IV.12) does not depend on a factor multiplying the residual covariance $C_R(\mathbf{h})$, but the resulting KT (kriging) variances do. Thus, if KT is used only for its estimated values, the scaling of the residual covariance is unimportant. But if the kriging variance is to be used, e.g., as in lognormal or multiGaussian simulation (see section V.2.3), then the scaling of the residual covariance is important. In such cases, even if the KT or OK estimate is retained it is the smaller SK (kriging) variance that should be used; see [86,130] and the discussion "SK or OK?" in section V.2.3.

• When using OK, KT, or any kriging system with unbiasedness-type conditions on the weights, to avoid matrix instability the units of the unbiasedness constraints should be scaled to that of the covariance. This rescaling is done automatically in all GSLIB kriging programs.

  Whenever the KT matrix is singular, reverting to OK often allows a diagnostic of the source of the problem.

• There are two ways to handle an exhaustively sampled secondary variable: kriging with an external drift and cokriging. Kriging with an external drift only uses the secondary variable to inform the shape of the primary variable mean surface whereas cokriging uses the secondary variable to its full potential (subject to the limitations of the unbiasedness constraint and the linear model of coregionalization). Checking hypothesis IV.15 should always be attempted before applying the external drift formalism.

  A better but more demanding way to account for secondary information is to use it in the conditioning of posterior conditional distributions; see the discussion of the Markov-Bayes algorithm in section IV.1.12. A program is given for simulation with the Markov-Bayes algorithm (see program mbsim) but not estimation. It would be straightforward to merge the Markov-Bayes construction of the conditional distribution (taken from mbsim) and the indicator kriging algorithm (taken from ik3d).

• If the data are on a regular grid then it is strongly recommended to customize the search procedure to save CPU time; the exact location of any datum is then exactly specified by its grid index location.

• The kriging algorithm, particularly in its SK version, is a non-convex interpolation algorithm. It allows for negative weights and for the estimate to be valued outside the data interval. This feature has merits whenever the variable being interpolated is very continuous, or in extrapolation cases. It may create unacceptable estimated values such as negative grades or percentages greater than 100%. The solution is to reset the estimated value to the nearest bound of the physical interval [108] rather than imposing constraints on the kriging weights [12]. A more demanding solution is to consider the physical interval as a constraint interval applicable everywhere (see relation IV.33), and to retain the E-type estimate (IV.37) possibly using median IK to save CPU time.

- Negative IK weights create order relation deviations. Therefore, for the purpose of IK one might consider implementing the conditions that all IK weights be positive and add up to one. This option paired with median IK would eliminate all order relation problems. Note that the CPU advantage of median IK is lost if constraint intervals are considered.

- Fixing the IK cutoffs for all locations **u** may be inappropriate depending on which z-data values are present in the neighborhood of **u**. For example, if there are no z-data values in the interval $(z_{k-1}, z_k]$, separating indicator krigings at both cutoff values z_{k-1} and z_k is artificial. It would be better to take for cutoff values the z-data values actually falling in the neighborhood of **u**: this solution could be easily implemented [36]. Note that the variogram parameters (nugget constant, sills, and ranges) would have to be modeled with smoothly changing parameters so that the variogram at any specified cutoff could be derived; see relation (IV.38).

IV.7   Problem Set Three: Kriging

The goals of this problem set are to experiment with kriging in order to judge the relative importance of variogram parameters, data configuration, screening, support effect, and cross validation. The GSLIB programs **okb2d**, **ktb3d**, **xvok2d**, and **xvkt3d** are called for. This problem set is quite extensive and important since, as mentioned in section II.1.4, kriging is used extensively in stochastic simulation techniques in addition to mapping.

The data for this problem set are based on the now familiar 2-D irregularly spaced data in cluster.dat. A subset of 29 samples in data.dat is used for some of the problems and three other small data sets parta.dat, partb.dat, and partc.dat, are used to illuminate various aspects of kriging.

Part A: Variogram Parameters

When performing a kriging study, it is good practice to select one or two representative locations and have a close look at the kriging solution including the kriging matrix, the kriging weights, and how the estimate compares to nearby data.

The original clustered data are shown in the upper half of Figure IV.11. An enlargement of the area around the point $x = 41$ and $y = 29$ is shown in the lower half. This part of the problem set consists of estimating point $x = 41, y = 29$ using OK with different variogram parameters. The goal is to gain an intuitive understanding of the relative importance of the various variogram parameters. Note, in most cases a standardized semivariogram is used such that the nugget constant C_0 plus the variance contributions of all nested structures sum to 1.0. Further, if you are unsure of your chosen variogram parameters then use the program **vmodel** in conjunction with **vargplt** to plot the variogram in a number of directions.

A data file containing the local data needed, $\boxed{\text{parta.dat}}$, is located in the **data** subdirectory on the diskettes. Start with these data to answer the following questions:

1. Consider an isotropic spherical variogram with a range of 10 plus a nugget effect. Vary the relative nugget constant from 0.0 (i.e., $C_0 = 0.0$ and $C = 1.0$) to 1.0 (i.e., $C_0 = 1.0$ and $C = 0.0$) in increments of 0.2. Prepare a table showing the kriging weight applied to each datum, the kriging estimate, and the kriging variance. Comment on the results.

2. Consider an isotropic spherical variogram with a variance contribution $C = 0.8$ and nugget constant $C_0 = 0.2$. Vary the isotropic range from 1.0 to 21.0 in increments of 2.0. Prepare a table showing the kriging weight applied to each datum, the kriging estimate, and the kriging variance. Comment on the results.

3. Consider an isotropic spherical variogram with a range of 12.0. Vary the nugget constant and variance contribution values without changing their relative proportions (i.e., keep $C_0/(C_0 + C) = 0.2$). Try $C_0 = 0.2$ ($C = 0.8$), $C_0 = 0.8$ ($C = 3.2$), and $C_0 = 20.0$ ($C = 80.0$). Comment on the results.

4. Consider alternative isotropic variogram models with a variance contribution $C = 1.0$ and nugget constant $C_0 = 0.0$. Keep the effective range constant at 15 (the relation between the effective range and the a parameter is given in section II.3). Consider the spherical, exponential, and Gaussian variogram models. Prepare a table showing the kriging weight applied to each datum, the kriging estimate, and the kriging variance. Use the programs **vmodel** and **vargplt** to plot the variogram models if you are unfamiliar with their shapes. Comment on the results.

5. Consider a spherical variogram with a range of 10.0, a variance contribution $C = 0.8$, and nugget constant $C_0 = 0.2$. Consider a geometric anisotropy with the direction of major continuity at 75 degrees clockwise from the y axis and anisotropy ratios of 0.33, 1.0, and 3.0. Prepare a table showing the kriging weight applied to each datum, the kriging estimate, and the kriging variance. Comment on the results.

Part B: Data Configuration

This part aims at developing intuition about the influence of the data configuration. Consider the situation shown on Figure IV.12 with the angle *ang* varying between $0°$ and $90°$.

A data file containing these data $\boxed{\text{partb.dat}}$ is located in the **data** subdirectory. Start with this data, an isotropic spherical variogram with a range

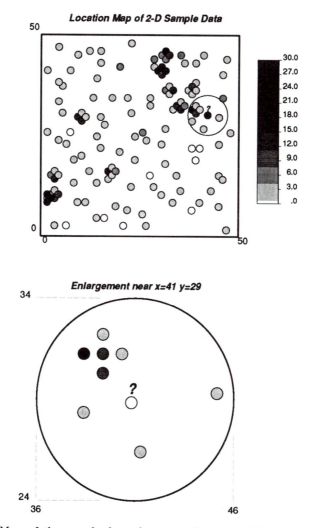

Figure IV.11: Map of the sample data shown on the original data posting and enlarged around the point being estimated.

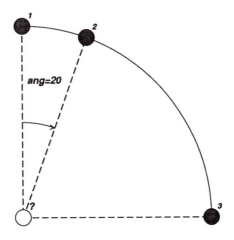

Figure IV.12: Three data points aligned on an arc of radius 1.0 from the point (?) to be estimated.

of 2.0, a variance contribution of $C = 0.8$, and nugget constant $C_0 = 0.2$, and use **okb2d** to answer the following:

1. How does the weight applied to data point 2 vary with *ang*? Vary *ang* in 5 or 10 degree increments (you may want to customize the kriging program to recompute automatically the coordinates of the second data point). Plot appropriate graphs and comment on the results.

2. Repeat question 1 with $C_0 = 0.9$ and $C = 0.1$.

Part C: Screen Effect

This part will reaffirm the intuition developed in part A about the influence of data screening. Consider the situation shown on Figure IV.13 with the angle *ang* varying between $0°$ and $90°$. The first data point is fixed at 1.0 unit away while the second is at 2.0 units away from the point being estimated.

A data file containing these two data $\boxed{\text{partc.dat}}$ is located in the **data** subdirectory. Start with this file, an isotropic spherical variogram with a range of 5.0, a variance contribution of $C = 1.0$, and nugget constant $C_0 = 0.0$, and use **okb2d** to answer the following:

1. How does the weight applied to point 2 vary with *ang*? Vary *ang* in 5 or 10 degree increments (you may want to customize the kriging program to recompute automatically the coordinates of the second data point). Plot appropriate graphs and comment on the results.

2. Repeat question 1 with $C_0 = 0.9$ and $C = 0.1$.

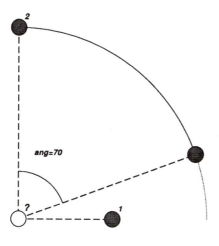

Figure IV.13: The first data point is fixed at one unit away and the second describes a circle two units away from the point being estimated.

Part D: Support Effect

Consider the 29 samples of example in Part A, except this time estimate a block 2.0 miles by 2.0 miles centered at $x = 41$ and $y = 29$. See Figure IV.14 for an illustration of the area being estimated.

Start with the same files as part A for **okb2d**. Consider a spherical variogram with a range of 10.0, a variance contribution of $C = 0.9$, and nugget constant $C_0 = 0.1$.

1. Change the number of discretization points from 1 by 1 to 4 by 4 and 10 by 10. Note the effect on the kriging weights, the estimate, and the kriging variance when estimating a block rather than a point. Comment on the results.

2. Repeat question 1 with $C_0 = 0.9$ and $C = 0.1$.

Part E: Cross Validation

Consider the data set $\boxed{\text{data.dat}}$ and the cross validation programs **xvok2d** and **xvkt3d**.

1. Cross validate the 29 data with **xvok2d**. Consider the exhaustive variogram model built in problem Set 2. Perform the following:

   • Plot scatterplots of the true values versus the estimates and comment on conditional unbiasedness and homoscedasticity of error variance.

   • Create a data file that contains the error (estimate minus true), the true value, the estimate, and the estimation variance.

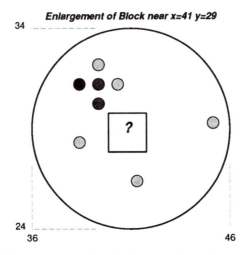

Figure IV.14: Estimation of a block centered at point $x = 41$, $y = 29$.

- Plot a histogram of the errors and comment.

- Plot a map of the 29 errors (contour or gray scale) and comment.

- Plot a scatterplot of the error absolute value versus the corresponding kriging variance and comment.

2. Justify your choice of search radius, minimum and maximum number of data, and other kriging parameters. Back up your justification with additional cross validation runs if necessary.

3. Try **xvkt3d** with simple kriging and the declustered mean you established in problem Set 1. Compare the results to ordinary kriging and comment.

4. Try **xvkt3d** with linear drift terms in both the x and y directions. Compare the results to ordinary kriging and comment.

Part F: Kriging on a Grid

Consider the complete 50 by 50 square mile area informed by $\boxed{\text{data.dat}}$. For this part use the program **ktb3d** and answer the following:

1. Plot a gray scale map of the true values in $\boxed{\text{true.dat}}$ using **gscale**.

2. Perform kriging on a one mile by one mile point grid by simple kriging (with your declustered mean from problem Set 1). Note the mean and variance of the kriging estimates, plot a gray scale map, and comment.

3. Study the effect of the maximum number of data by considering **ndmax** at 4, 8, and 24. Note the variance of the kriging estimates and comment on the look of the gray scale map.

4. Study the effect of the search radius by setting the isotropic radius at 5, 10, and 20. Keep **ndmax** constant at 24. Again, note the mean and variance of the kriging estimates and the look of the gray scale map.

5. Repeat question 2 with ordinary kriging. Comment on the results. Compare the mean of the kriging estimates to the declustered mean established in problem Set 1. How could you determine the declustering weight that should be applied to each datum from such kriging?

6. Repeat question 2 using kriging with a trend model (linear drift terms in both the x and y directions). Comment on the results.

7. Repeat question 2 with ordinary block kriging (grid cell is discretized 4 by 4). Comment on the results.

8. Estimate the mean/trend surface (by resetting the *itrend* flag to 1) with the KT parameter file used earlier. Note that you will have an estimate of the point trend at the node locations. Comment on the results. Compare these results to those obtained from block kriging.

IV.8 Problem Set Four: Cokriging

The goal of this problem set is to experiment with cokriging and kriging with an external drift for mapping purposes. The GSLIB programs **gamv2**, **vmodel**, **vargplt**, are needed to compute and model the matrix of covariances; programs **ktb3d** and **cokb3d** are required.

The 2-D irregularly spaced data in ⌐data.dat⌐ is considered as the sample data set for the primary variable. An exhaustive secondary variable data set is given in ⌐ydata.dat⌐ in the **data** subdirectory.

The 2-D estimation exercise consists of estimating the primary variable, denoted $z(\mathbf{u})$, at all 2500 nodes using:

1. a relatively sparse sample of 29 primary data in file ⌐data.dat⌐ and duplicated in ⌐ydata.dat⌐. These 29 data are preferentially located in high z-concentration zones.

2. in addition, a much larger sample of 2500 secondary data, denoted $y(\mathbf{u})$, in file ⌐ydata.dat⌐.

Since the secondary information is available at all nodes to be estimated and at all nodes where primary information is available, it can be interpreted as an external drift for the primary variable $z(\mathbf{u})$; see section IV.1.5.

All of the estimates can be checked versus the actual z-values in ⌐true.dat⌐.

Part A: Cokriging

1. Calculate the 3 sample covariances and cross covariances, i.e., $C_Z(\mathbf{h})$, $C_Y(\mathbf{h})$, $C_{ZY}(\mathbf{h})$, and model them with a legitimate model of coregionalization; see [76], p. 390.

2. Use program cokb3d to perform cokriging of the primary variable at all 2500 nodes of the reference grid. Document which type of cokriging (which unbiasedness conditions) you have retained. You may want to consider all 3 types of cokriging coded in cokb3d; see section IV.1.7.

3. **Validation:** Compare the cokriging estimates to the reference true values given in $\boxed{\text{true.dat}}$. Plot the histogram of the errors (estimate minus true) and the scattergram of the errors versus the true values. Comment on the conditional unbiasedness and homoscedasticity of the error distribution. Plot a gray (or color) scale map of the cokriging estimates and error values. Compare to the corresponding map of true values and direct kriging estimates. Comment on any improvement brought by cokriging.

Part B: Kriging with an External Drift

1. Use program ktb3d with the external drift option[8] to estimate the primary variable at all 2500 nodes of the reference grid.

2. **Validation:** Compare the resulting estimates to the previous cokriging estimates and to the reference true values. Use any or all of the previously proposed comparative graphs and plots. Comment on the respective performance of kriging, cokriging, and kriging with an external drift.

IV.9   Problem Set Five: Indicator Kriging

The goal of this problem set is to provide an introduction to indicator kriging as used to build probabilistic models of uncertainty (ccdf's) at unsampled locations. From these models, optimal estimates (possibly non least-squares) can be derived together with various non-parametric probability intervals.

The GSLIB programs gamv2, ik3d, and postik are called for. The 140 irregularly spaced data in $\boxed{\text{cluster.dat}}$ are considered as the sample data. The reference data set is in $\boxed{\text{true.dat}}$.

The following are typical steps in an IK study. A detailed step-by-step description of the practice of IK is contained in [84].

---

[8] This example is a favorable case for using the external drift concept in that the secondary y-data were generated as spatial averages of z-values (although taken at a different level from an original 3-D data set). Compare Figures A.26 and A.27.

1. **Determination of a representative histogram:** Retain the declustered (weighted) cdf based on all 140 z-data; see problem Set 1. Determine the nine decile cutoffs $z_k, k = 1, \ldots, 9$ from the declustered cdf $F(z)$.

2. **Indicator variogram modeling:** Use **gamv2** to compute the corresponding nine indicator variograms (once again the 140 clustered data in ⎢cluster.dat⎢ are used). Standardize these variograms by their respective (non-declustered) data variances and model them. Use the same nested structures at all cutoffs with the same total unit sill. The relative sills and range parameters may vary; see the example of relation IV.38. The graph of these parameters versus the cutoff value z_k should be reasonably smooth.

3. **Indicator kriging:** Using **ik3d**, perform an indicator kriging at each of the 2500 nodes of the reference grid. The program returns the $K = 9$ ccdf values at each node with order relation problems corrected.

   Program **postik** provides the E-type estimate, i.e., the mean of the calculated ccdf, any conditional p-quantile, and the probability of (not) exceeding any given threshold value.

4. **E-type estimate:** Retrieve from **postik** the 2500 E-type estimates. Plot the corresponding gray scale map and histogram of errors ($z^* - z$). Compare to the corresponding figures obtained through ordinary kriging and cokriging (problem sets three and four).

5. **Quantile and probability maps:** Retrieve from **postik** the 2500 conditional 0.1 quantile and probability of exceeding the high threshold value $z_9 = 5.39$. The corresponding isopleth maps indicate zones that are most certainly high z-valued. Compare the M-type estimate map (0.5 quantile) to the E-type estimate and comment.

6. **Median indicator kriging:** Assume a median IK model of type (IV.30), i.e., model all experimental indicator correlograms at all cutoffs by the same model better inferred at the median cutoff. Compute the E-type estimate and probability maps and compare.

7. **Soft structural information:** Beyond the 140 sample data, soft structural information indicates that high concentration z-values tend to be aligned NW-SE whereas median-to-low concentration values are more isotropic in their spatial distribution. Modify correspondingly the various indicator variogram models to account for that structural information, and repeat the runs of questions 3 to 5. Comment on the flexibility provided by multiple IK with multiple indicator variograms.

Chapter V

Simulation

The collection of stochastic simulation programs included with GSLIB is presented in this chapter. Section V.1 presents the principles of stochastic simulation including the important concept of sequential simulation.

Section V.2 presents the Gaussian-related algorithms which are the algorithms of choice for continuous variables. Section V.3 presents the indicator simulation algorithms which are better suited to categorical variables, cases where the Gaussian RF model is inappropriate, or where accounting for additional soft/fuzzy data is important. Sections V.4 and V.5 present Boolean and simulated annealing algorithms respectively.

Sections V.6 and V.7 give detailed descriptions of the Gaussian and Indicator simulation programs in GSLIB. Different programs for many different situations are available for each of these major approaches. Elementary programs for Boolean simulations and simulations based on simulated annealing are presented in Section V.8 and V.9.

Some useful application notes are presented in section V.10. Finally, a problem set is proposed in section V.11 which allows the simulation programs to be tested and understood.

V.1 Principles of Stochastic Simulation

Consider the distribution over a field A of one (or more) attribute(s) $z(\mathbf{u}), \mathbf{u} \in A$. Stochastic simulation is the process of building alternative, equally probable, high resolution models of the spatial distribution of $z(\mathbf{u})$; each realization is denoted with the superscript l: $\{z^{(l)}(\mathbf{u}), \mathbf{u} \in A\}$. The simulation is said to be "conditional" if the resulting realizations honor the hard data values at their locations:

$$z^{(l)}(\mathbf{u}_\alpha) = z(\mathbf{u}_\alpha), \ \forall \, l \tag{V.1}$$

The variable $z(\mathbf{u})$ can be categorical, e.g., indicating the presence or absence of a particular rock type, or it can be continuous such as porosity in a reservoir or concentrations over a contaminated site.

Simulation differs from kriging or any interpolation algorithm, in two major aspects:

1. In most interpolation algorithms the goal is to provide a "best" local estimate $z^*(\mathbf{u})$ of each unsampled value $z(\mathbf{u})$ taken one at a time without specific regard to the resulting spatial statistics of the estimates $z^*(\mathbf{u}), \mathbf{u} \in A$. In simulation, the resulting global features (texture) and statistics of the simulated values $z^{(l)}(\mathbf{u}), \mathbf{u} \in A$, take precedence over local accuracy.

2. For a given set of local data and conditioning statistics, kriging used as an interpolation algorithm provides a single numerical model $\{z^*_K(\mathbf{u}), \mathbf{u} \in A\}$ which is "best" in some local accuracy sense. Whereas simulation provides many alternative numerical models $\{z^{(l)}(\mathbf{u}), \mathbf{u} \in A\}, l = 1, \dots, L$, each of which is a "good" representation of the reality in some global sense; see Figure V.1. The differences among these L alternative models or realizations provides a measure of joint spatial uncertainty.

Different simulation algorithms impart different global statistics and spatial features on each realization. For example, simulated categorical values can be made to honor specific geometrical patterns as in Boolean-type simulations (see section V.4); or, the covariance of simulated continuous values can be made to honor a prior covariance model as in Gaussian-related simulations; see section V.2. A hybrid approach could be considered to generate simulated numerical models that reflect widely different types of features. For example, one may start with a Boolean process or categorical indicator simulation to generate the geometric architecture of the lithofacies, following with a Gaussian algorithm to simulate the distribution of continuous petrophysical properties within each separate lithofacies, then a simulated annealing process could be used to modify locally the petrophysical properties to match, say, well test data [36].

No single simulation algorithm is flexible enough to allow the reproduction of the wide variety of features and statistics encountered in practice. It is the responsibility of each user to select the set of appropriate algorithms and customize them if necessary. Presently, no single paper or book provides an exhaustive list or classification of available simulation algorithms. See the bibliography and particularly the reference lists of [15,18,25,39,46,66,91,106, 135,155] for a start.

The GSLIB software offers code for simulation algorithms frequently used in mining, hydrogeology, petroleum applications, and for which the authors have a working experience. Notable omisions are:

- Frequency domain simulations [18,64,109]. These algorithms can be seen as duals of Gaussian algorithms where the covariance is reproduced through sampling of its discretized Fourier transform (spectrum).

- Random fractal simulation [49,68,69,159] which can be seen as particular implementations of frequency domain simulations.

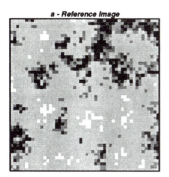

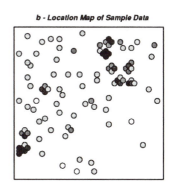

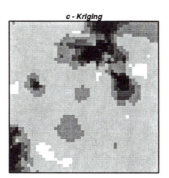

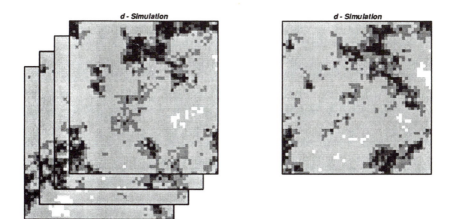

Figure V.1: Kriging versus Stochastic Simulation: The reference image (a) has been sparsely sampled (b). The kriging estimates based on the sparse sample are given in a gray scale map in (c); note the smoothing effect more pronounced in areas of sparser sampling. The same sample was used to generate multiple stochastic images (shown in (d)); note the difference between the various stochastic images and the better reproduction of the reference spatial variability.

- Simulation of marked point processes [128,135,147]. These processes can be seen as generalizations of the widely used Boolean process; see the discussion in section V.4.

- Deterministic simulation of physical processes, e.g., of sedimentation, erosion, transport, redeposition, etc. [20,155]. Although such simulations can result in alternative images, their emphasis is on the genesis of the images and not on the reproduction of local data. Conversely, random function-based stochastic simulations tend to bypass the actual genetic process and rely on global statistics that are deemed representative of the *end* phenomenon actually observed.

The following five subsections provide a brief discussion of paradigms and implementation procedures that are common to many stochastic simulation algorithms.

V.1.1   Reproduction of Major Heterogeneities

Major large scale heterogeneities are usually the most consequential features of the numerical models generated. For example, facies boundaries are often critical for flow performance in a reservoir or for slope stability in an open pit mine. Similarly, boundaries between different types of vegetation may condition the spatial distribution of a particular insect. Those heterogeneities should be given priority in the simulation process. A continuous RF model $Z(\mathbf{u})$ cannot reproduce severe heterogeneities or discontinuities, as found when crossing a physical boundary such as that of a lithotype.

If the phenomenon under study is a mixture of different physical and/or statistical populations, the geometry of the mixture should be modeled and simulated first; then, the attribute(s) within each homogeneous population can be simulated.[1] For example, the spatial architecture of major lithofacies in an oil reservoir, or mineralizations in a mining deposit, should be represented first, e.g., through prior rock type / facies zoning or the simulation of categorical-type variables. Then, the distribution of petrophysical properties or mineral grades within each facies or mineralization can be simulated [7,29,95,154,169]. A two-step approach to simulation is not only more consistent with the underlying physical phenomenon (e.g., geology), it also avoids stretching the stationarity/homogeneity hypothesis[2] underlying most continuous RF models.

Similarly, reproduction of patterns of spatial continuity of extreme values should be given priority especially in situations where connected patterns of extreme values condition the results obtained from processing the simulated

---

[1]There are cases when that sequence should be reversed, e.g., when one population is of quasi-zero measure (volume) as in the case of fracture lines in 2-D, or planes in 3-D. Such fractures and the corresponding displacements can be simulated afterwards and superimposed on the prior continuous simulation.

[2]The degree of spatial correlation of the attribute $z(\mathbf{u})$ is usually quite different within a facies than across facies, unless there has been important diagenetic remobilization.

numerical models [91,92]. For example, the reproduction of the geometry of fractures intersecting multiple facies may be more important than reproduction of the facies geometry.

V.1.2 Joint Simulation of Several Variables

In many applications reproduction of the spatial dependence of several variables is critical. For example, a realistic model for a porphyry copper-gold deposit must account for the spatial cross-correlation of the copper and gold grades. An oil reservoir model typically calls for joint knowledge of porosity, permeability, and fluid saturations.

Most random function-based simulation algorithms can be generalized, at least in theory, to joint simulation of several variables by considering a vectorial random function $\mathbf{Z}(\mathbf{u}) = \{Z_1(\mathbf{u}), Z_2(\mathbf{u}), \ldots, Z_K(\mathbf{u})\}$; see [23,61,94, 106,123,152]. The problem resides in the inference and modeling of the cross-covariance matrix

$$\boldsymbol{\Sigma} = [C_{k,k'}(\mathbf{h}) = Cov\{Z_k(\mathbf{u}), Z_{k'}(\mathbf{u}+\mathbf{h})\}, k, k' = 1, \ldots, K]$$

Because of implementation problems, joint simulation of several variables is rarely implemented to the full extent allowed by theory. Rather, various approximations are considered; the two most common:

1. Replace the simulation of the K dependent variables $Z_k(\mathbf{u})$ by that of K *independent* factors $Y_k(\mathbf{u})$ from which the original Z-variables can be reconstituted:

$$\text{if}: \mathbf{Y} = \varphi(\mathbf{Z}), \text{ then}: \mathbf{z}^{(l)} = \varphi^{-1}(\mathbf{y}^{(l)}) \qquad \text{(V.2)}$$

with: $\mathbf{y}^{(l)} = [y_1^{(l)}(\mathbf{u}), \ldots, y_K^{(l)}(\mathbf{u}), \mathbf{u} \in A]$ being the set of simulated independent factors $y_k^{(l)}(\mathbf{u})$.

$\mathbf{z}^{(l)} = [z_1^{(l)}(\mathbf{u}), \ldots, z_K^{(l)}(\mathbf{u}), \mathbf{u} \in A]$ is the set of resulting simulated values $z_k^{(l)}(\mathbf{u})$ whose interdependence is guaranteed by the common inverse transform φ^{-1}.

Except for specific cases where physics-based non-linear factors $\mathbf{y}$ are evident and particularly cases involving constraints on the sums of variables (as in chemical analyses), the vast majority of applications consider only linear factors based on an orthogonal decomposition of some covariance matrix $\boldsymbol{\Sigma}$ of the original K variables $Z_k(\mathbf{u})$; see section IV.1.11 and [17,152]. The decomposition of the covariance matrix,

$$\boldsymbol{\Sigma}(\mathbf{h}_o) = [Cov\{Z_k(\mathbf{u}), Z_{k'}(\mathbf{u}+\mathbf{h}_o)\}, k, k' = 1, \ldots, K]$$

corresponds to a single separation vector $\mathbf{h}_o$. The Z-cross correlations are reproduced only for that specific separation vector. The zero distance, $\mathbf{h}_o = 0$, is often chosen for convenience, the argument being that cross-correlation is maximum at the zero distance $\mathbf{h}_o = 0$.

2. The second approximation is much simpler. The most important or better auto-correlated variable, hereafter called the primary variable $Z_1(\mathbf{u})$, is simulated first; then, all other covariates $Z_k(\mathbf{u}), k \neq 1$, are simulated by drawing from specific conditional distributions. For example, typically in reservoir modeling [8]:

- The porosity $\Phi(\mathbf{u}) = Z_1(\mathbf{u})$ within a given lithofacies is simulated first since its spatial variability is reasonably smooth and easy to infer.

- The horizontal permeability $K_h(\mathbf{u}) = Z_2(\mathbf{u})$ is then simulated from the conditional distribution $Prob\{K_h(\mathbf{u}) \leq z_2 | \Phi(\mathbf{u}) = z_1^{(l)}\}$ of $K_h(\mathbf{u})$ given the simulated porosity value $z_1^{(l)}$ at the same location. Such conditional distributions are directly inferred from the sample scattergram of K_h versus Φ corresponding to the facies prevailing at $\mathbf{u}$.

- Next, given the simulated value for horizontal permeability $K_h(\mathbf{u}) = z_2^{(l)}(\mathbf{u})$, the simulated value for vertical permeability is drawn from the sample scattergram of K_h versus K_v, again specific to the facies prevailing at $\mathbf{u}$.

The cross-correlation between K_v and Φ is thus indirectly approximated. Similarly, the spatial auto-correlations of the secondary variables $K_h(\mathbf{u})$ and $K_v(\mathbf{u})$ are indirectly reproduced through that of the primary variable $\Phi(\mathbf{u})$. This approach assumes that conditioning by a selected collocated covariate screens the influence of any other data. For the previous example:

$$Prob\{K_h(\mathbf{u}) \leq z_2 | \text{information in the neighborhood of } \mathbf{u}\} \qquad (V.3)$$
$$\approx Prob\{K_h(\mathbf{u}) \leq z_2 | \Phi(\mathbf{u}) = z_1^{(l)}\}$$

A Late Remark:

At the time of this book going to press, another approach for joint conditional simulation of several spatially distributed variables has been suggested as a PhD research topic at Stanford (Almeida). This approach builds on the concept of collocated cokriging with a Markov model for all cross-covariances, as presented in section IV.1.7. The following is addressed to readers already familiar with the sequential Gaussian (sGs) algorithm (section V.2.1 to V.2.3).

Consider the normal score transforms $Y_k(\mathbf{u}), k = 1, \ldots, K$ of K dependent (original) variables $Z_k(\mathbf{u})$. Conditional simulation of the first variable $Y_1(\mathbf{u})$ is performed using the sGs algorithm. Then, conditional simulation of the second variable $Y_2(\mathbf{u})$ is performed with sGs considering, at any location $\mathbf{u}$, the neighboring y_2-data *and* only the collocated previously simulated value $y_1(\mathbf{u})$. Conditioning to this y_1-datum actually calls for a cokriging system but much simplified if a Markov model of type (IV.25) is adopted for the cross-correlogram between $Y_1(\mathbf{u})$ and $Y_2(\mathbf{u})$. Next, conditional simulation of the third variable $Y_3(\mathbf{u})$ is performed again using sGs and considering, in addition to the y_3-data, the collocated previously simulated value(s) $y_1(\mathbf{u})$ and/or $y_2(\mathbf{u})$ .... Backtransformation of the K normal score simulated fields $y_k(\mathbf{u})$ provides the required K *jointly simulated* fields $z_k(\mathbf{u})$.

Note that both this approach and kriging with an external drift (section IV.1.5) call for the secondary variable to be available at all locations being simulated. A full cokriging approach would be necessary if the secondary data are irregularly located throughout the field of interest.

V.1.3 The Sequential Simulation Approach

Approximation (V.3) allows drawing the value of a variable $Z(\mathbf{u})$ from its conditional distribution given the value of the most related covariate at the same location $\mathbf{u}$. The sequential simulation principle is a generalization of that idea: the conditioning is extended to include all data available within a neighborhood of $\mathbf{u}$, including the original data and all previously simulated values [78,82,135,136].

Consider the joint distribution of N random variables Z_i with N very large. The N RV's Z_i may represent the same attribute at the N nodes of a dense grid discretizing the field A, or they can represent N different attributes measured at the same location, or they could represent a combination of K different attributes defined at the N' nodes of a grid with $N = KN'$.

Next, consider the conditioning of these N RV's by a set of n data of *any* type symbolized by the notation $|(n)$. The corresponding N-variate ccdf is denoted:

$$F_{(N)}(z_1, \ldots, z_N|(n)) = Prob\{Z_i \leq z_i, i = 1, \ldots, N|(n)\} \qquad (\text{V.4})$$

Expression (V.4) is completely general with no intrinsic limitations; some or all of the variables Z_i could be categorical.

Successive application of the conditional probability relation shows that drawing an N-variate sample from the ccdf (V.4) can be done in N successive steps, each involving a univariate ccdf with increasing levels of conditioning:

- draw a value $z_1^{(l)}$ from the univariate ccdf of Z_1 given the original data (n). The value $z_1^{(l)}$ is now considered as a conditioning datum for all subsequent drawings; thus, the information set (n) is updated to $(n+1) = (n) \cup \{Z_1 = z_1^{(l)}\}$.

- draw a value $z_2^{(l)}$ from the univariate ccdf of Z_2 given the updated data set $(n+1)$, then update the information set to $(n+2) = (n+1) \cup \{Z_2 = z_2^{(l)}\}$.

- sequentially consider all N RV's Z_i.

The set $\{z_i^{(l)}, i = 1, \ldots, N\}$ represents a simulated joint realization of the N dependent RV's Z_i. If another realization is needed, $\{z_i^{(l')}, i = 1, \ldots, N\}$, the entire sequential drawing process is repeated.

This sequential simulation procedure requires the determination of N univariate ccdf's, more precisely:

$$Prob\{Z_1 \leq z_1 | (n)\} \qquad\qquad (V.5)$$
$$Prob\{Z_2 \leq z_2 | (n+1)\}$$
$$Prob\{Z_3 \leq z_3 | (n+2)\}$$
$$\cdots$$
$$Prob\{Z_N \leq z_N | (n+N-1)\}$$

The sequential simulation principle is independent of the algorithm or model used to establish the sequence (V.5) of univariate ccdf's. In the program **sgsim**, all ccdf's (V.5) are assumed Gaussian and their means and variances are given by a series of N simple kriging systems; see sections IV.1.1 and V.6.3. In the program **sisim**, the ccdf's are obtained directly by indicator kriging; see sections IV.1.9 and V.3.2.

Implementation Considerations

- Strict application of the sequential simulation principle calls for the determination of more and more complex ccdfs, in the sense that the size of the conditioning data set increases from (n) to $(n + N - 1)$. In practice, the argument is that the closer[3] data screen the influence of more remote data; therefore, only the closest data are retained to condition any of the N ccdfs (V.5). Since the number of previously simulated values may become overwhelming as i progresses from 1 to $N \gg n$, one may want to give special attention to the original data (n) even if they are more remote.

- The neighborhood limitation of the conditioning data entails that statistical properties of the $(N + n)$ set of RV's will be reproduced only up to the maximum distance found in the neighborhood. For example, the search must extend at least as far as the distance to which the variogram is to be reproduced; this requires extensive conditioning as the sequence progresses from 1 to N. One solution is provided by the multiple grid concept which is to simulate the N nodal values in two or more steps [61]:

---

[3] "Closer" is not necessarily taken in terms of Euclidean distance, particularly if the original data set (n) and the N RV's include different attribute values. The data "closest" to each particular RV Z_i being simulated are those that have the most influence on its ccdf. This limitation of the conditioning data to the closest is also the basis of the Gibb's sampler as used in [56].

- First, a coarse grid, say each tenth node is simulated using a large data neighborhood. The large neighborhood allows the reproduction of large-scale variogram structures.

- Second, the remaining nodes are simulated with a smaller neighborhood.

• Theory does not specify the sequence in which the N nodes should be simulated. Practice has shown, however, that it is better to consider a random sequence [74]. Indeed, if the N nodes are visited row-wise, any departure from rigorous theory [4] may entail a corresponding spread of artifacts along rows.

V.1.4   Error Simulation

The smoothing effect of kriging, and more generally, of any low pass-type interpolation, is due to a missing error component. Consider the RF $Z(\mathbf{u})$ as the sum of the estimator and the corresponding error:

$$Z(\mathbf{u}) = Z^*(\mathbf{u}) + R(\mathbf{u}) \qquad (V.6)$$

Kriging, for example, would provide the smoothly varying estimator $Z^*(\mathbf{u})$. To restore the full variance of the RF model, one may think of simulating a realization of the random function error with zero mean and the correct variance and covariance. The simulated z-value would be the sum of the unique estimated value and a simulated error value:

$$z_c^{(l)}(\mathbf{u}) = z^*(\mathbf{u}) + r^{(l)}(\mathbf{u}) \qquad (V.7)$$

The idea is simple, but requires two rather stringent conditions:

1. The error component $R(\mathbf{u})$ must be independent or, at least, orthogonal to the estimator $Z^*(\mathbf{u})$. Indeed, the error value $r^{(l)}(\mathbf{u})$ is simulated independently, and is then added to the estimated value $z^*(\mathbf{u})$.

2. The RF $R(\mathbf{u})$ modeling the error must have the same spatial distribution or, at least, the same covariance as the actual error. Then, the simulated value in (V.7) will have the same covariance, hence the same variance, as the true values $z(\mathbf{u})$.

Condition (1) is met if the estimator $Z^*(\mathbf{u})$ is obtained by orthogonal projection of $Z(\mathbf{u})$ onto some (Hilbert) space of the data, in which case the error vector $Z^*(\mathbf{u}) - Z(\mathbf{u})$ is orthogonal to the estimator $Z^*(\mathbf{u})$ [94,105]. Condition (1) is met if $Z^*(\mathbf{u})$ is a *simple* kriging estimator of $Z(\mathbf{u})$.

Condition (2) is difficult to satisfy since the error covariance is unknown; worse, it is not stationary even if the RF model $Z(\mathbf{u})$ is stationary since the

---

[4] The limited data neighborhood concept coupled with ordinary kriging (as opposed to simple kriging) privileges the influence of close data.

data configuration used from one location $\mathbf{u}$ to another is usually different. The solution is to pick (simulate) the error values from a simulated estimation exercise performed with exactly the same data configuration ([94], p. 495; [106], p. 86).

More precisely, consider a non-conditional simulated realization $z^{(l)}(\mathbf{u})$ available at all nodal locations and actual data locations. This reference simulation shares the covariance of the RF model $Z(\mathbf{u})$. The simulated values $z^{(l)}(\mathbf{u}_\alpha), \alpha = 1, \ldots, n$, at the n data locations are retained and the estimation algorithm performed on the actual data is repeated. This provides a simulation of the estimated values $z^{*(l)}(\mathbf{u})$, hence also simulated errors $r^{(l)}\mathbf{u}) = z^{(l)}(\mathbf{u}) - z^{*(l)}(\mathbf{u})$. These simulated errors are then simply added to the actual estimated value $z^*(\mathbf{u})$ using the actual data values $z(\mathbf{u}_\alpha)$:

$$z_c^{(l)}(\mathbf{u}) = z^*(\mathbf{u}) + [z^{(l)}(\mathbf{u}) - z^{*(l)}(\mathbf{u})] \qquad (V.8)$$

The notation $z_c^{(l)}$ with subscript c differentiates the conditional simulation from the non-conditional simulation $z^{(l)}(\mathbf{u})$.

If the estimation algorithm used is exact, as in kriging, the data values are honored at the data locations $\mathbf{u}_\alpha$, thus; $z^{(l)}(\mathbf{u}_\alpha) = z^{*(l)}(\mathbf{u}_\alpha)$ entailing: $z_c^{(l)}(\mathbf{u}_\alpha) = z^*(\mathbf{u}_\alpha) = z(\mathbf{u}_\alpha), \forall \alpha = 1, \ldots, n$. Last, it can be shown that the variogram of $Z_c^{(l)}(\mathbf{u})$ is identical to that of the original RF model $Z(\mathbf{u})$; see [94], p.497.

The conditioning algorithm characterized by relation (V.8) requires a prior non-conditional simulation to generate the $z^{(l)}(\mathbf{u})$'s and two krigings to generate the $z^*(\mathbf{u})$ and $z^{*(l)}(\mathbf{u})$'s. Actually, only one set of kriging weights is needed since both the covariance function and the data configuration are common to both estimates $z^*(\mathbf{u})$ and $z^{*(l)}(\mathbf{u})$. A worthwhile addition to GSLIB would be a utility program to perform such kriging. Meanwhile the kriging programs okb2d and ktb3d can be adapted.

Relation (V.8) is used as a post-processor with non-conditional simulation algorithms such as the turning band algorithm (see section V.2.5 and program tb3d) and random fractals simulation [68].

V.1.5   Questions of Ergodicity

A simulation algorithm aims at drawing realizations that reflect the statistics modeled from the data. The question is how well these statistics should be reproduced. Given that the model statistics are inferred from sample statistics that are uncertain because of limited sample size, exact reproduction of the model statistics by each simulated realization may not be desirable, or possible.

Consider the example of drawing from a stationary Gaussian-related RF model $Z(\mathbf{u})$, fully characterized by its cdf $F(z)$ and the covariance $C_Y(\mathbf{h})$ of its normal score transform $Y(\mathbf{u}) = G^{-1}(F(Z(\mathbf{u})))$, where $G(y)$ is the standard normal cdf; see section V.2.1. Let $\{z^{(l)}(\mathbf{u}), \mathbf{u} \in A\}$ be a particular realization (number l) of the RF model $Z(\mathbf{u})$ over a finite field A, and $F^{(l)}(z)$, $C_Y^{(l)}(\mathbf{h})$ be the corresponding cdf and normal score covariance, the latter being the covariance of the Gaussian simulated values $y^{(l)}(\mathbf{u})$. "Ergodic fluctuations" refer to the discrepancy between the realization statistics $F^{(l)}(z)$, $C_Y^{(l)}(\mathbf{h})$ and the corresponding model parameters $F(z)$, $C_Y(\mathbf{h})$.

The stationary RF $Z(\mathbf{u})$ is said to be "ergodic" in the parameter μ if the corresponding realization statistics $\mu^{(l)}, \forall\, l$, tends toward μ as the size of the field A increases. The parameter μ is usually taken as the stationary mean of the RF, i.e., the mean of its cdf $F(z)$ [42]. The notion of ergodicity can be extended to any other model parameter, e.g., the cdf value $F(z)$ itself for a given cutoff z or the entire covariance model $C_Y(\mathbf{h})$.

Thus, provided that $Z(\mathbf{u})$ is stationary and ergodic and the simulation field is large enough,[5] one should expect the statistics of any realization l to reproduce the model parameters exactly.

The size of a simulation field A large enough to allow for ergodicity depends on the type of RF chosen and the range of its heterogeneities, as characterized, e.g., by the type and range[6] of the covariance model. The more continuous the covariance at $\mathbf{h} = 0$ and the larger the covariance range, the larger the field A has to be before ergodicity is achieved.

Figure V.2.a gives the distribution of the 20 variograms of 20 independent realizations of a standard Gaussian RF model with a zero mean, unit variance, and Gaussian variogram with a unit practical range ($a\sqrt{3} = 1$) and zero nugget effect. The simulated field is a square with each side 10 times the practical range. The discretization is 1/20th of the practical range, thus each simulation comprises 40,000 nodes. Although the field is 10 times the range, a very favorable case in practice, the fluctuations of the realization variograms are seen to be large.

Figure V.2.b gives a similar distribution but, now, the variogram of the standard Gaussian RF model is spherical with an actual range of 1 unit ($a = 1$) and a 20% nugget effect. This represents a more common model much less continuous than the previous one. The fluctuations of the realization variograms are seen to be reduced.

Figure V.3 considers a completely different, indicator-based, non-Gaussian RF model; however, the cdf and covariance are identical to those underlying Figure V.2.b. The RF model is that implicit to the median IK model, whereby all standardized indicator variograms and cross variograms are equal to each other and equal to the attribute variogram; see relation IV.30.

The fluctuations of the realization variograms of Figure V.3 for the median IK model are seen to be larger than those shown in Figure V.2.b for the Gaussian RF model, although both models share the same histogram and

---

[5] What matters is the relative dimensions of A with respect to the range(s) of the covariance of $Z(\mathbf{u})$, *not* the density of discretization of A

[6] Some random function models, for which the spatial variance keeps increasing with the field size, are modeled with power variogram models $\gamma(\mathbf{h}) = |\mathbf{h}|^\omega$, i.e., with an implicit infinite range.

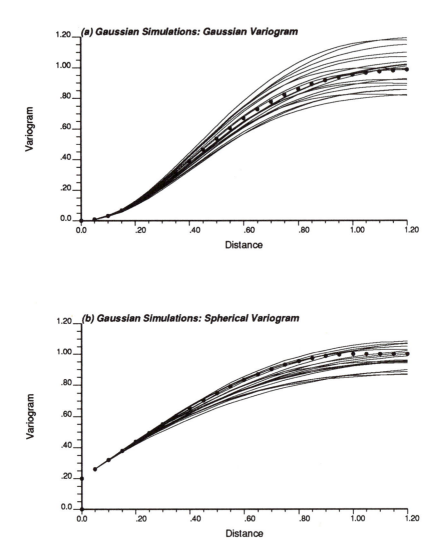

Figure V.2: Fluctuations of the variograms of 20 independent realizations of a Gaussian RF model. Figure V.2.a considers a Gaussian variogram with a unit practical range. Figure V.2.b considers a spherical variogram with a unit range and a non-zero nugget effect. In both cases the model variogram is shown by the large dots.

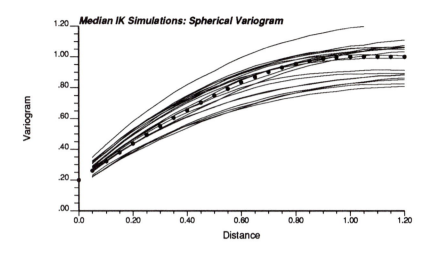

Figure V.3: Fluctuations of the variograms of 20 independent realizations of a median IK-type RF with a standard normal cdf and the same variogram as used for Figure V.2.b.

variogram. The median IK model, and in general all RF models implicit to the indicator simulation approach, have poorer ergodic properties than Gaussian-related RF models.[7] The larger fluctuations next to the origin seen in Figure V.3 are due to the discretization entailed by the indicator approach: within-class simulated values are drawn independently one from another.[8]

Observation of the fluctuations shown in Figure V.2 and Figure V.3 may lead to choosing naively the RF model with better ergodic properties, or even looking for the simulation algorithm that would reproduce exactly the imposed statistics. This inclination would be correct only if the original sample statistics are considered exact and unquestionable.

In most applications, the model statistics are inferred from sparse samples and cannot be deemed exactly representative of the population statistics. One must allow for departure from the model statistics. The statistical fluctuations shown by different realizations of the RF model are useful, and it would be an error to reduce[9] or remove such fluctuations. One could even

---

[7] Gaussian-related RF models are about the only practical models whose multivariate cdf is fully, and analytically, explicit. Most other RF models, including those implicit to indicator simulations, Boolean processes, and marked point processes, are not fully determined. Only some of their statistics are explicitly known (other statistics can be sampled from simulations). Be aware that an explicit analytical representation does not necessarily make a RF model more appropriate for a specific application.

[8] The nugget effect could be reduced by artificially lowering the nugget constant of the model.

[9] One example of artificial, and generally not recommended, reduction of fluctuations between simulated realizations consists of restandardizing each non-conditional Gaussian realization $\{y^{(l)}(\mathbf{u}), \mathbf{u} \in A\}$ to a zero mean and a unit variance before back transforming into the simulated values $\{z^{(l)}(\mathbf{u}), \mathbf{u} \in A\}$ of the original attribute. This sets all the z-

argue that, everything else being equal as between Figures V.2.b and V.3, one should prefer the RF model with poorer ergodic properties since it would provide the largest, hence most conservative, assessment of uncertainty [36].

It is worth noting that even a little data-conditioning significantly reduces ergodic fluctuations [106].

Selecting Realizations

Issues of ergodicity often arise because practitioners tend to retain only one realization, i.e., use stochastic simulation as an improved interpolation algorithm. In the worst case, the first realization drawn is retained with the user often complaining that the model statistics are not matched! The only match that simulation theory guarantees is in average (expected value) over a large number of realizations. The less ergodic the RF model chosen, the more realizations are needed to approach this expected value.

In other cases, several realizations are drawn and one (or very few) are retained for further processing, e.g., through a flow simulator. Various criteria are used to select the realization, from closeness of the realization statistics to the model parameters (see footnote 8), to subjective aesthetic appreciation of overall appearance or, better, selecting the realizations that match a datum or property that was not initially input to the model, e.g., travel times in a flow simulator using a simulated permeability field. Selecting realizations amounts to further conditioning by as yet unused information, be it subjective appreciation or hard production data for matching purposes. The realizations selected are better conditioned to actual data, whether hard or soft: in this sense they are better numerical models of the phenomenon under study. On the other hand, the selection process generally wipes out the measure of uncertainty[10] that the simulation approach was originally designed to provide.

How many realizations should one draw? The answer is definitely more than one to get some sense of the uncertainty. If two images, although both a priori acceptable, yield widely different results, then more images should be drawn. The number of realizations needed depends on how many are deemed sufficient to characterize (bracket) the uncertainty being addressed. Note that an evaluation of uncertainty need not require that each realization cover the entire field or site: simulation of a typical or critical sub-area or section may suffice.

In indicator-type simulations, a systematic poor match of the model statistics may be due to too many order relation problems, themselves possibly due to inconsistent indicator variograms from one cutoff to the next one, or to variogram models inconsistent with the (hard or soft) conditioning data used.

---

realization histograms equal: $F^{(l)}(z) = F^{(l')}(z), \forall l, l'$, thus giving a false sense of certainty about that unique cdf; see [106].

[10] An alternative to selection is ranking of the realizations drawn according to some property of each realization. From such ranking, a few quantile-type realizations can be selected which allows preserving a measure of uncertainty [11,165].

Interpolation or Simulation?

Although stochastic simulation was developed to provide measures of spatial uncertainty, simulation algorithms are increasingly used in practice to provide a single improved "estimated" map. Indeed, stochastic simulation algorithms have proven to be much more versatile than traditional interpolation algorithms in reproducing the full spectrum of data spatial variability *and* in accounting for data of different types and sources, whether hard or soft.

Inasmuch as a simulated realization honors the data deemed important, it can be used as an interpolated map for those applications where reproduction of spatial features is more important than local accuracy. The warning that there can be several such realizations applies equally to the interpolated map: there can be alternative estimated images depending on the interpolation/estimation algorithm used.

V.1.6 Going Beyond a Discrete CDF

Continuous cdf's and ccdf's are always informed at a discrete number of cutoff values. Regardless of whether a Gaussian-based method or an indicator-based method is used, the maximum level of discretization is provided by the following sample cumulative histogram with at most n step functions if all n data values $z(\mathbf{u}_\alpha)$ are different:

$$F^*(z) = \sum_{\alpha=1}^{n} a_\alpha i(\mathbf{u}_\alpha; z) \qquad (V.9)$$

$$\text{with: } a_\alpha \in [0,1], \ \sum_{\alpha=1}^{n} a_\alpha = 1$$

where $i(\mathbf{u}_\alpha; z)$ is the indicator datum, set to one if $z(\mathbf{u}_\alpha) < z$, to zero otherwise. a_α is the declustering weight attached to datum location $\mathbf{u}_\alpha$; see [37] and section VI.2.1. An equally weighted histogram would correspond to $a_\alpha = 1/n, \ \forall \ \alpha$.

If the sample size n is small, assumptions must be made for extrapolation beyond the smallest z-datum value (lower tail), the largest z-datum value (upper tail), and for interpolation between two consecutively ranked z-data. Being a data expansion technique, stochastic simulation will generate many more values than there are original data. Therefore, the maximum resolution provided by the sample cdf (V.9) is probably insufficient, particularly in the two tails.

This lack of resolution is particularly severe when using indicator-related algorithms with only a few cutoff values such as the nine deciles of the sample cdf (V.9). In this case, the procedures for interpolation between the IK-derived ccdf values and, *most importantly*, for extrapolating beyond the two extreme ccdf values are critical.

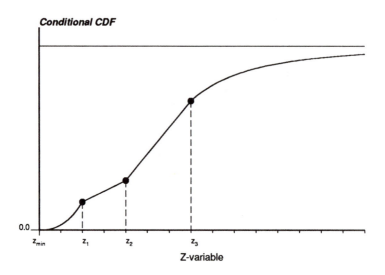

Figure V.4: Interpolation and extrapolation beyond the ccdf initially known at three cutoff values z_1, z_2, and z_3.

Figure V.4 shows an example of a ccdf $F(\mathbf{u}; z)$ known at 3 cutoff values (z_1, z_2, z_3), which could be the three quartiles of the marginal cdf. Linear cdf interpolation within the two central classes $(z_1, z_2], (z_2, z_3]$ amounts to assuming a uniform within-class distribution. The lower tail $(z_{min}, z_1]$ has been extrapolated towards a fixed minimum z_{min} using a power model (V.10) corresponding to a negatively skewed distribution. The upper tail $(z_3, +\infty]$ has been extrapolated up to a potentially infinite upper bound using a positively skewed hyperbolic model; see expression (V.11).

Correcting for higher nugget effect

Because the ccdf interpolation/extrapolation process does not account for any spatial correlation a noise is added to the resulting spatial structure of the z-simulated values. Aside from using more threshold values for the determination of the ccdf's, one can correct for the additional noise by artificially reducing the nugget constant of either the normal score variogram model (Gaussian-based methods) or the indicator variogram models.

Interpolation models

The within-class cdf interpolation models considered in GSLIB are:

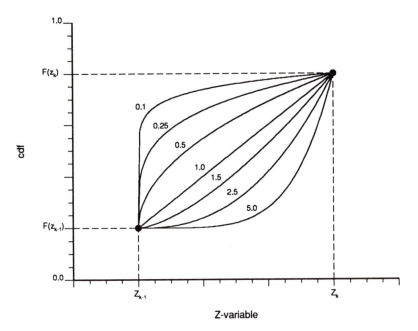

Figure V.5: Some power models for ω=0.1, 0.25, 0.5, 1.0, 1.5, 2.5 and 5.0. ω less than 1.0 leads to positive skewness, ω=1.0 gives the uniform distribution, and ω greater than 1.0 leads to negative skewness.

- **Power Model:** For a finite class interval $(z_{k-1}, z_k]$ and a parameter (the power) $\omega > 0$, this cdf model is written:

$$F^{\omega}_{z_{k-1}, z_k}(z) = \begin{cases} 0, & \forall z \leq z_{k-1} \\ \left[\frac{z - z_{k-1}}{z_k - z_{k-1}} \right]^{\omega}, & \forall z \in (z_{k-1}, z_k] \\ 1, & \forall z \geq z_k \end{cases} \quad (\text{V.10})$$

In practice, this cdf model is scaled between the calculated cdf values at z_{k-1} and z_k rather than between 0 and 1. Distributions with $\omega < 1$ are positively skewed, $\omega = 1$ corresponds to the linear cdf model (uniform distribution), and distributions with $\omega > 1$ are negatively skewed; see Figure V.5.

- **Linear interpolation between tabulated bound values:** This option considers a fixed number of sub-classes with given bound values within each class $(z_{k-1}, z_k]$. For example, the three bound values a_{k1}, a_{k2}, a_{k3} can be tabulated defining four sub-classes $(z_{k-1}, a_{k1}]$, $(a_{k1}, a_{k2}], (a_{k2}, a_{k3}], (a_{k3}, z_k]$ which share the probability p_k calculated for class $(z_{k-1}, z_k]$; p_k is shared equally unless specified otherwise. Then, linear cdf interpolation is performed separately within each sub-class.

This option allows the user to add detail to the distribution within the classes defined by the cutoffs z_k. That detail, i.e., the sub-classes bound values, can be attributed to some or all of the original data values falling within each class $(z_{k-1}, z_k]$ of the marginal (sample) distribution. Thus, some of the resolution lost through discretization by the z_k values can be recovered.[11]

More generally, the sub-class bound values a_k can be taken from any parametric model, e.g., beta or gamma distribution.

- **Hyperbolic model:** this last option is to be used only for the upper tail of a positively skewed distribution. Decisions regarding the upper tail of ccdf's are often the most consequential;[12] therefore, a great deal of flexibility is needed, including the possibility of a very long tail.

The hyperbolic cdf upper tail model for a strictly positive variable is a two parameter distribution:

$$F_{\omega,\lambda}(z) = 1 - \frac{\lambda}{z^\omega}, \ \ \omega \geq 1, \ \ z^\omega > \lambda > 0 \qquad (V.11)$$

The scaling parameter λ allows identification of any pre-calculated quantile value, for example, the p-quantile z_p such that $F_{\omega,\lambda}(z_p) = p$, then:

$$\lambda = z_p^\omega (1 - p).$$

The parameter $\omega > 1$ controls how fast the cdf reaches its upper limit value 1; the smaller ω, the longer the tail of the distribution.

The mean z-value above the p-quantile value z_p is: $m_p = \frac{\omega}{\omega-1} z_p > z_p$. Hence the smaller ω, the larger the mean above z_p. At its minimum value, $\omega = 1$ identifies the Pareto distribution which has an infinite mean $m_p, \forall p$, corresponding to a very long tail.

The parameter ω should be chosen to be realistic and conservative for the application being considered. When considering an environmental issue with z representing concentration of a pollutant, $\omega = 1$ is a safe choice because a long tail is usually a conservative (pessimistic) assessment. If a metal distribution is being considered, a value between $\omega = 2$ and $\omega = 3$ will cause the tail to drop off reasonably fast without being too optimistic about the occurrence of very large values. Figure V.6 shows hyperbolic distributions for ω between 1 and 5. Practice has shown that a distribution with $\omega = 1.5$ is a general purpose model that yields acceptable results in a wide variety of applications.

---

[11] However, original data values are marginal quantile values, whereas conditional quantile values, specific to each ccdf, would be needed.

[12] Even when using a multivariate Gaussian RF model (after normal score transform), a decision about the shape of the cdf in the upper tail is required for the back transform; see program sgsim. It is naive to believe that a fully parametric RF model allows escaping critical decisions about the distribution of extreme values; that decision is built into the RF model chosen. In the case of indicator-based methods that decision is made fully explicit.

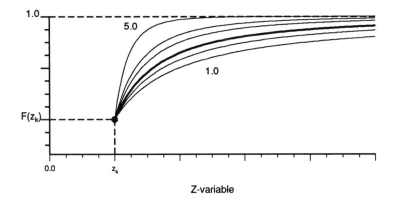

Figure V.6: The family of hyperbolic models; the distribution models for $\omega=1.0$, 1.25, 1.5, 2.0, 2.5, and 5.0 are shown. The models for $\omega=1.0$ and 5.0 are labeled and the model for $\omega=1.5$ is bolder than the rest. All models have a positive skewness.

The GSLIB programs allow different sets of options depending on whether interpolation is needed within the middle classes or extrapolation for the lower and upper tails. The available options are:

Lower Tail: below the first calculated cdf value:

1. Linear model or uniform distribution

2. Power model

3. Tabulated bound values

Middle: between any two calculated cdf values:

1. Linear model (uniform distribution)

2. Power model

3. Tabulated bound values

Upper Tail: above the last calculated cdf value:

1. Linear model (uniform distribution)

2. Power model

3. Tabulated bound values

4. Hyperbolic model

The user is asked for a specific model for each of these regions (the integer number identifying each model in the list above is used).

V.2   Gaussian-Related Algorithms

The Gaussian RF model is unique in statistics for its extreme analytical simplicity and for being the limit distribution of many analytical theorems globally known as "central limit theorems" [9,79].

In a nutshell, if the continuous[13] spatial phenomenon $\{z(\mathbf{u}), \mathbf{u} \in A\}$ is generated by the sum of a (not too large) number of *independent* sources $\{y_k(\mathbf{u}), \mathbf{u} \in A\}, k = 1, \ldots, K$, with similar spatial distributions then its spatial distribution can be modeled by a multivariate Gaussian RF model:

$$Z(\mathbf{u}) = \sum_{k=1}^{K} Y_k(\mathbf{u}) \approx \text{Gaussian}$$

The limiting constraint is not the number K, or the fact that the components $Y_k(\mathbf{u})$ are equally distributed, but the hypothesis of *independence* of the $Y_k(\mathbf{u})$'s.

If human and measurement errors can sometimes be considered as independent events or processes, in the earth sciences the diverse geological processes that have generated the observed phenomenon are rarely independent one from another, nor are they additive. Multivariate Gaussian models are extremely congenial, they are well understood, and they have an established record of successful applications. These heuristic considerations are enough to make the Gaussian model the privileged choice for modeling continuous variables, unless proven inappropriate.

The traditional definition of a multivariate Gaussian (or normal) distribution as expressed through the matrix expression of its multivariate pdf can be found in any multivariate statistics book; see, e.g., [9]. That expression is of little practical use. It is better to define the multivariate Gaussian RF model through its *characteristic* properties.

The RF $Y(\mathbf{u}) = \{Y(\mathbf{u}), \mathbf{u} \in A\}$ is multivariate normal if and only if:[14]

- all subsets of that RF, for example $\{Y(\mathbf{u}), \mathbf{u} \in B \subset A\}$ are also multivariate normal.

- all linear combinations of the RV components of $Y(\mathbf{u})$ are (univariate) normally distributed, e.g.,

$$X = \sum_{\alpha=1}^{n} \omega_{\alpha} Y(\mathbf{u}_{\alpha}) \text{ is normally distributed,} \qquad (\text{V.12})$$

$\forall$ the weights ω_{α}, as long as $\mathbf{u}_{\alpha} \in A$.

---

[13] "Continuous" as opposed to a phenomenon $z(\mathbf{u})$ characterized by discrete categorical values or by the juxtaposition of different populations.

[14] These characteristic properties are not all independent. For example, the second property of normality of all linear combinations suffices to characterize a multivariate normal distribution.

- zero covariance (or correlation) entails full independence:

    If $Cov\{Y(\mathbf{u}), Y(\mathbf{u}')\} = 0$, the two RV's $Y(\mathbf{u})$ and $Y(\mathbf{u}')$ (V.13)
    are not only uncorrelated, they are also independent.

- all conditional distributions of any subset of the RF $Y(\mathbf{u})$, given realizations of any other subset, are (multivariate) normal. For example, the conditional distribution of the K RV's $\{Y_k(\mathbf{u}'_k), k = 1, \ldots, K, \mathbf{u}'_k \in A\}$ given the realizations $y(\mathbf{u}_\alpha) = y_\alpha, \alpha = 1, \ldots, n$, is K-variate normal, $\forall k, \forall \mathbf{u}'_k, \forall \mathbf{u}_\alpha, \forall y_\alpha$.

The case of $K = 1, \mathbf{u}'_1 = \mathbf{u}_o$, where the RV $Y(\mathbf{u}_o)$ models the uncertainty about a specific unsampled value $y(\mathbf{u}_o)$ is of particular interest: the ccdf of $Y(\mathbf{u}_o)$, given the n data y_α, is normal and fully characterized by:

- its mean, or conditional expectation, identified to the SK (linear regression) estimate of $y(\mathbf{u}_o)$

$$E\{Y(\mathbf{u}_o)|y(\mathbf{u}_\alpha) = y_\alpha, \alpha = 1, \ldots, n\} \equiv [y(\mathbf{u}_o)]^*_{SK} \qquad (V.14)$$

$$= m(\mathbf{u}_o) + \sum_{\alpha=1}^{n} \lambda_\alpha [y_\alpha - m(\mathbf{u}_\alpha)]$$

where $m(\mathbf{u}) = E\{Y(\mathbf{u})\}$ is the expected value of the not necessarily stationary RV $Y(\mathbf{u})$. The n weights λ_α are given by the SK system (IV.2):

$$\sum_{\beta=1}^{n} \lambda_\beta C(\mathbf{u}_\beta, \mathbf{u}_\alpha) = C(\mathbf{u}_o, \mathbf{u}_\alpha), \alpha = 1, \ldots, n \qquad (V.15)$$

where $C(\mathbf{u}, \mathbf{u}') = Cov\{Y(\mathbf{u}), Y(\mathbf{u}'\}$ is the covariance, not necessarily stationary, of the RF $Y(\mathbf{u})$.

- its variance, or conditional variance, is the SK variance:

$$Var\{Y(\mathbf{u}_o)|y(\mathbf{u}_\alpha) = y_\alpha, \alpha = 1, \ldots, n\} = \qquad (V.16)$$

$$C(\mathbf{u}_o, \mathbf{u}_o) - \sum_{\alpha=1}^{n} \lambda_\alpha C(\mathbf{u}_o, \mathbf{u}_\alpha)$$

Note the homoscedasticity of the conditional variance: it does not depend on the data values y_α, but only on the data configuration and the covariance model.

For stochastic sequential simulation the normality of all ccdf's is a true blessing: indeed, the determination of the sequence of successive ccdf's reduces to solving a corresponding sequence of SK systems.

V.2.1   Normal Score Transform

The first necessary condition for the stationary RF $Y(\mathbf{u})$ to be multivariate normal is that its univariate cdf be normal, i.e.,

$$Prob\{Y(\mathbf{u}) \le y\} = G(y) \ \forall y \qquad (V.17)$$

where $G(\cdot)$ is the standard Gaussian cdf; $Y(\mathbf{u})$ is assumed to be standardized, i.e., with a zero mean and unit variance.

Unfortunately, most earth sciences data do not present symmetric Gaussian histograms. This is not a major problem since a non-linear transform can transform any continuous cdf into any other cdf, [41,78] and program **trans** in section VI.2.4.

Let Z and Y be two RV's with cdf's $F_Z(z)$ and $F_Y(y)$ respectively. The transform $Y = \varphi(Z)$ identifies the cumulative probabilities corresponding to the Z and Y p-quantiles:

$$F_Y(y_p) = F_Z(z_p) = p, \ \forall p \in [0, 1], \ \text{hence:}$$

$$y = F_Y^{-1}(F_Z(z)) \qquad (V.18)$$

with $F_Y^{-1}(\cdot)$ being the inverse cdf, or quantile function, of the RV Y: $y_p = F_Y^{-1}(p), \ \forall p \in [0, 1]$.

If Y is standard normal with cdf $F_Y(y) = G(y)$, the transform $G^{-1}(F_Z(\cdot))$ is the normal score transform.

In practice, the n z-sample data are ordered in increasing value: $z^{(1)} \le z^{(2)} \le \ldots \le z^{(n)}$. The cumulative frequency corresponding to the kth largest z datum is $F_Z(z^{(k)}) = k/n$, or $F_Z(z^{(k)}) = \sum_{j=1}^{k} \omega_j \in [0, 1]$ if a set of declustering weights ω_j have been applied to the n data.

Then, the normal score transform of $z^{(k)}$ is the k/n - quantile of the standard normal cdf, i.e.,

$$y^{(k)} = G^{-1}\left(\frac{k}{n}\right) \qquad (V.19)$$

Implementation details of the normal score transformation, including the treatment of the last value $G^{-1}\left(\frac{n}{n}\right) = +\infty$, and back transform, $Z \rightleftharpoons Y$, are given in sections VI.2.2 and VI.2.3; see programs **nscore** and **backtr**.

If the marginal sample cdf $F_Z(z)$ is too discontinuous due, for example, to too few data, one might consider smoothing it prior to normal score transform. Kernel functions could be used for this purpose; see Figure IV.2 and [141]. The same smoothed cdf should be considered for the back transform.

If weights are used to correct (decluster) the sample cdf $F_Z(z)$, it is the correspondingly weighted histogram of the normal score data which is standard normal, not the unweighted normal scores histogram.

V.2.2   Checking for Bivariate Normality

The normal score transform (V.18) defines a new variable Y which is, by construction, (univariate) normally distributed. This is a necessary but not a sufficient condition for the spatially distributed values $Y(\mathbf{u}), \mathbf{u} \in A$ to be multivariate normal. The next necessary condition is that the bivariate cdf of any pair of values $Y(\mathbf{u}), Y(\mathbf{u}+\mathbf{h})$, $\forall \mathbf{h}$, be normal.

There are various ways to check[15] for bivariate normality of a data set $\{y(\mathbf{u}_\alpha), \alpha = 1, \ldots, n\}$ whose histogram is already standard normal [9,16,79, 96,157].

The most discriminatory check directly verifies that the experimental bivariate cdf of any set of data pairs $\{y(\mathbf{u}_\alpha), y(\mathbf{u}_\alpha + \mathbf{h}), \alpha = 1, \ldots, N(\mathbf{h})\}$ is indeed standard bivariate normal with covariance function $C_Y(\mathbf{h})$. There exist analytical and tabulated relations linking the covariance $C_Y(\mathbf{h})$ with any standard normal bivariate cdf value [2,82,152,166]:

$$Prob\{Y(\mathbf{u}) \leq y_p, Y(\mathbf{u}+\mathbf{h}) \leq y_p\} = \qquad (V.20)$$

$$p^2 + \frac{1}{2\pi} \int_0^{arc\ sinC_Y(\mathbf{h})} exp\left(-\frac{y_p^2}{1+sin\theta}\right) d\theta$$

where $y_p = G^{-1}(p)$ is the standard normal p-quantile and $C_Y(\mathbf{h})$ is the correlogram of the standard normal RF $Y(\mathbf{u})$.

Now, the bivariate probability (V.20) is the non-centered indicator covariance for the threshold y_p:

$$Prob\{Y(\mathbf{u}) \leq y_p, Y(\mathbf{u}+\mathbf{h}) \leq y_p\} \qquad (V.21)$$

$$= E\{I(\mathbf{u}; p) \cdot I(\mathbf{u}+\mathbf{h}; p)\} = p - \gamma_I(\mathbf{h}; p)$$

where: $I(\mathbf{u}; p) = 1$, if $Y(\mathbf{u}) \leq y_p$; zero, if not; and $\gamma_I(\mathbf{h}; p)$ is the indicator semivariogram for the p-quantile threshold y_p.

The check consists of comparing the sample indicator semivariogram $\gamma_I(\mathbf{h}; p)$ to its theoretical bivariate normal expression (V.20). Program **bigaus** calculates the integral expression (V.20) for various p-quantile values. Simple parametric approximations to (V.20) are given in [96].

Remarks

- The indicator variable $I(\mathbf{u}; p)$ is the same whether defined on the original z-data or on the y-normal score data, as long as the cutoff values z_p and y_p are both p-quantiles of their respective cdf's. In other words, indicator variograms, i.e., bivariate cdf's parameterized in cdf values p are invariant by any linear or non-linear monotonic increasing transform.

---

[15] A check differs from a formal statistical test in that it does not provide any measure of accepting wrongly the hypothesis. Unfortunately, most formal tests typically require, in practice, independent data or data whose multivariate distribution is known a priori.

The normal score transform (V.18) and, more generally, any monotonic transform of the original variable z does *not* change the essence of its bivariate cdf:

$$Prob\{Y(\mathbf{u}) \leq y_p, Y(\mathbf{u} + \mathbf{h}) \leq y_p\} \qquad (V.22)$$

$$= Prob\{Z(\mathbf{u} \leq z_p, Z(\mathbf{u} + \mathbf{h}) \leq z_p\}, \forall p \in [0, 1],$$

for any monotonic increasing[16] transform $Y = \varphi(Z)$.

$y_p = \varphi(z_p)$ and z_p are the marginal p-quantiles of the Y and Z distributions.

Relation (V.22) is methodologically important since it indicates that it is naive to expect a univariate transform $\varphi(\cdot)$ to impart new bivariate or multivariate properties to the transformed y-data. Either those properties already belong to the original z-data, or they do not and the transform $\varphi(\cdot)$ will not help. Bivariate and multivariate normality are either properties of the z-data or they are not; the non-linear rescaling of the z-units provided by the normal score transform does not help.

- Symmetric normal quantiles are such that $y_p = -y_{p'}$, i.e., $y_p^2 = y_{p'}^2$, $\forall p' = 1 - p$. Simple arithmetic between relations (V.20) and (V.21) yields

$$\gamma_I(\mathbf{h}; p) = \gamma_I(\mathbf{h}; p'), \forall p' = 1 - p \qquad (V.23)$$

Moreover, as p tends towards its bound values, 0 or 1, $y_p^2 \rightarrow +\infty$, and expression (V.20) tends towards the product p^2 of the two marginal probabilities (independence). Consequently, the semivariogram $\gamma_I(\mathbf{h}; p)$ tends towards its sill value $p(1 - p)$ no matter how small the separation vector $\mathbf{h}$. Occurence of Gaussian extreme values, whether high or low,
. is purely random: extreme Gaussian values do not cluster in space.

A simple way to check for binormality is to check for the symmetric "destructuration" of the occurrence of extreme z-values: the practical ranges of the indicator variograms $\gamma_I(\mathbf{h}; p)$ should decrease symmetrically and continuously as p tends towards its bound values of 0 and 1; see [96].

Bivariate normality is necessary but not sufficient for multivariate normality. Beyond bivariate normality, one should check that all trivariate, quadrivariate, ..., K-variate experimental frequencies match theoretical Gaussian expressions of type (V.20) given the covariance model $C_Y(\mathbf{h})$. The problem does not reside in computing these theoretical expressions, but in the inference of the corresponding experimental multivariate frequencies: unless the

---

[16] If the transform φ is monotonic decreasing, then an equivalent relation holds:

$$Prob\{Y(\mathbf{u}) > y_p, Y(\mathbf{u} + \mathbf{h}) > y_p\}$$

$$= 1 - 2p + Prob\{Y(\mathbf{u} \leq y_p, Y(\mathbf{u} + \mathbf{h}) \leq y_p\}$$

$$= Prob\{Z(\mathbf{u} \leq z_p, Z(\mathbf{u} + \mathbf{h}) \leq z_p\}$$

data are numerous and gridded, there are rarely enough triplets, quadruplets, ..., K-tuples sharing the same geometric configuration to allow such inference. Therefore, in practice, if one cannot, from sample statistics, show that bivariate Gaussian properties are violated, the multivariate Gaussian RF model should be the prime choice for continuous variable simulation.

V.2.3 Sequential Gaussian Simulation (sGs)

The most straightforward algorithm for generating realizations of a multivariate Gaussian field is provided by the sequential principle described in section V.1.3. Each variable is simulated sequentially according to its normal ccdf fully characterized through an SK system of type (V.15). The conditioning data consist of all original data and all previously simulated values found within a neighborhood of the location being simulated.

The conditional simulation of a *continuous* variable $z(\mathbf{u})$ modeled by a Gaussian-related stationary RF $Z(\mathbf{u})$ proceeds as follows:

1. Determine the univariate cdf $F_Z(z)$ representative of the entire study area and not only of the z-sample data available. Declustering may be needed if the z-data are preferentially located; see problem set one and program **declus** in section VI.2.1.

2. Using the cdf $F_Z(z)$, perform the normal score transform of z-data into y-data with a standard normal cdf; see section V.2.1 and program **nscore** in section VI.2.2.

3. Check for bivariate normality of the normal score y-data; see section V.2.2. If the multivariate Gaussian model can not be retained, then consider alternative models such as a mixture of Gaussian populations [169] or an indicator-based algorithm for the stochastic simulation.

4. If a multivariate Gaussian RF model can be adopted for the y-variable, proceed with program **sgsim** and sequential simulation, i.e.,

   - Define a random path that visits each node of the grid (not necessarily regular) once. At each node $\mathbf{u}$, retain a specified number of neighboring conditioning data including both original y-data and previously simulated grid node y-values.

   - Use SK with the normal score variogram model to determine the parameters (mean and variance) of the ccdf of the RF $Y(\mathbf{u})$ at location $\mathbf{u}$.

   - Draw a simulated value $y^{(l)}(\mathbf{u})$ from that ccdf.

   - Add the simulated value $y^{(l)}(\mathbf{u})$ to the data set.

   - Proceed to the next node, and loop until all nodes are simulated.

5. Backtransform the simulated normal values $\{y^{(l)}(\mathbf{u}), \mathbf{u} \in A\}$ into simulated values for the original variable $\{z^{(l)}(\mathbf{u}) = \varphi^{-1}(y^{(l)}(\mathbf{u}), \mathbf{u} \in A\}$. Within-class interpolations and tail extrapolations are usually called for; see section V.1.6.

Multiple Realizations

If multiple realizations are desired $\{z^{(l)}(\mathbf{u}), \mathbf{u} \in A\}, l = 1, \ldots, L$, the previous algorithm is repeated L times with either of these options:

1. The same random path visiting the nodes. In this case the data configuration, hence the SK systems, are the same from one realization to another: they need be solved only once. CPU time is reduced considerably; see footnote 18, p 149.

2. A different random path for each realization. The sequence of data configurations is different, thus different SK systems must be set up and solved. CPU time is then proportional to the number of realizations. This last option has been implemented in program **sgsim**.

SK or OK?

The prior decision of stationarity requires that simple kriging (SK) with zero mean be used in step 4 of the sGs algorithm. If data are abundant enough to consider inference of a non-stationary RF model, one may

1. either consider an array of local SK means (different from zero) to be used for local SK estimates (V.14). This amounts to bypass the OK process by providing local non-stationary means obtained from some other information source,

2. or split the area into distinct sub-zones and consider for each sub-zone a different stationary RF model, which implies inference of a different histogram and a different normal score variogram for each sub-zone,

3. or consider a stationary normal score variogram, inferred from the entire pool of data, and a non-stationary mean for $Y(\mathbf{u})$. The non-stationary mean, $E\{Y(\mathbf{u})\}$, at each location $\mathbf{u}$, is implicitly re-estimated from the neighborhood data through ordinary kriging (OK); see section IV.1.2. The price of such local rescaling of the model mean is, usually, a poorer reproduction of the stationary Y-histogram and variogram model.

Similarly, sequential Gaussian simulation with an external drift can be implemented by replacing the simple kriging estimate and system (V.14, V.15) by a kriging estimate of type (IV.16) where the local normal score trend value is linearly calibrated to the collocated value of a smoothly varying secondary variable (different from z). Whether OK or KT with an external drift is used to replace the SK estimate, it is the SK variance (V.16) that should be used for the variance of the Gaussian ccdf; see [86]. An approximation would consist in multiplying each locally derived OK or KT (kriging) variance by a constant factor obtained from prior calibration.

Sequential Gaussian simulation with SK should be the preferred algorithm for the simulation of continuous variables unless proven inappropriate.

V.2.4 LU Decomposition Algorithm

When the total number of conditioning data plus the number of nodes to be simulated is small (fewer than a few hundred) and a large number of realizations is requested, simulation through LU decomposition of the covariance matrix provides the fastest solution [5,34,53].

Let $Y(\mathbf{u})$ be the stationary Gaussian RF model with covariance $C_Y(\mathbf{u})$. Let $\mathbf{u}_\alpha, \alpha = 1, \ldots, n$ be the locations of the conditioning data and $\mathbf{u}'_i, i = 1, \ldots, N$ be the N nodes to be simulated. The large covariance matrix $(n + N) \cdot (n + N)$ is partitioned into the data-to-data covariance matrix, the node-to-node covariance matrix, and the two node-to-data covariance matrices:

$$\mathbf{C}_{(n+N)(n+N)} = \begin{bmatrix} [C_Y(\mathbf{u}_\alpha - \mathbf{u}_\beta)]_{n \cdot n} & [C_Y(\mathbf{u}_\alpha - \mathbf{u}'_j)]_{n \cdot N} \\ [C_Y(\mathbf{u}'_i - \mathbf{u}_\beta)]_{N \cdot n} & [C_Y(\mathbf{u}'_i - \mathbf{u}'_j)]_{N \cdot N} \end{bmatrix} = \mathbf{L} \cdot \mathbf{U} \quad (V.24)$$

The large matrix $\mathbf{C}$ is decomposed into the product of a lower and an upper triangular matrix, $\mathbf{C} = \mathbf{L} \cdot \mathbf{U}$. A conditional realization $\{y^{(l)}(\mathbf{u}'_i), i = 1, \ldots, N\}$ is obtained by multiplication of $\mathbf{L}$ by a column matrix $\omega^{(l)}_{(N+n) \cdot 1}$ of normal deviates:

$$\mathbf{y}^{(l)} = \begin{bmatrix} [y(\mathbf{u}_\alpha)]_{n \cdot 1} \\ [y^{(l)}(\mathbf{u}'_i)]_{N \cdot 1} \end{bmatrix} = \mathbf{L} \cdot \omega^{(l)} = \begin{bmatrix} \mathbf{L}_{11} & \mathbf{0} \\ \mathbf{L}_{21} & \mathbf{L}_{22} \end{bmatrix} \cdot \begin{bmatrix} \omega_1 \\ \omega^{(l)}_2 \end{bmatrix} \quad (V.25)$$

where $[y(\mathbf{u}_\alpha)]_{n \cdot 1}$ is the column matrix of the n normal score conditioning data, $[y(\mathbf{u}'_i)]_{N \cdot 1}$ is the column matrix of the N conditionally simulated y-values.

Identification of the conditioning data is written as $\mathbf{L}_{11}\omega_1 = [y(\mathbf{u}_\alpha)]$, thus matrix ω_1 is set at:

$$\omega_1 = [\omega_1]_{n \cdot 1} = \mathbf{L}_{11}^{-1} \cdot [y(\mathbf{u}_\alpha)] \quad (V.26)$$

The column vector $\omega^{(l)}_2 = \left[\omega^{(l)}_2\right]_{N \cdot 1}$ is a vector of N *independent* standard normal deviates.

Additional realizations, $l = 1, \ldots, L$, are obtained at very little additional cost by drawing a new set of normal deviates $\omega^{(l)}_2$, then by applying the matrix multiplication (V.25). The major cost and memory requirement is in the upfront LU decomposition of the large matrix $\mathbf{C}$ and in the identification of the weight matrix ω_1.

The LU decomposition algorithm requires that *all* nodes and data locations be considered simultaneously in a single covariance matrix $\mathbf{C}$. The current practical limit of the number $(n + N)$ is no greater than a few hundred. The code lusim provided in GSLIB is a full 3-D implementation of the LU decomposition algorithm.

Implementation variants have been considered, relaxing the previous size limitation by considering overlapping neighborhoods of data locations [5]. Unfortunately, artifact discontinuities appear if the correlation between all simulated nodes is not fully accounted for.

The LU decomposition algorithm is particularly appropriate when a large number (L) of realizations is needed over a small area or block $(n + N$ is then small). A typical application is the evaluation of block ccdf's, a problem known in geostatistical jargon as "change of support"; see discussion in section IV.1.13. A block or any small subset of the study area, $B \subset C$, can be discretized into N points. The normal score values at these N points can be simulated repetitively $(l = 1, \ldots, L)$ through the LU decomposition algorithm and backtransformed into simulated point z-values: $\{z^{(l)}(\mathbf{u}'_i), i = 1, \ldots, N; \mathbf{u}'_i \in B \subset A\}, l = 1, \ldots, L$. Each set of N simulated point values can then be averaged to yield a simulated block value:

$$z_B^{(l)} = \frac{1}{N} \sum_{i=1}^{N} z^{(l)}(\mathbf{u}'_i)$$

The distribution of the L simulated block values $z_B^{(l)}, l = 1, \ldots, L$, provides a numerical approximation of the probability distribution (ccdf) of the block average, conditional to the data retained.

V.2.5   The Turning Band Algorithm

The turning band algorithm was the first large-scale 3-D Gaussian simulation algorithm actually implemented [83,106,109,115]. It provides only non-conditional realizations of a standard Gaussian field $Y(\mathbf{u})$ with a given covariance $C_Y(\mathbf{h})$. The conditioning to local normal score y-data calls for a post-processing of the non-conditional simulation by kriging (see relation (V.8) in section V.1.4).

The turning band algorithm is very fast, as fast as Fast-Fourier spectral domain techniques [18,64] or random fractal algorithms [68,159]. The additional kriging step needed to condition to local data make all these techniques less appealing. For comparison, the sequential Gaussian algorithm (sGs) also requires the solution of one kriging system per grid node, whether the simulation is conditional or not. In addition, the application of sGs is much more straightforward than that of the turning band algorithm.

The turning band algorithm is fast because it achieves a 3-D simulation through a series of 1-D simulations along lines which constitute a regular partitioning of the 3-D space; see [94], p. 499. Each node in the 3-D space is projected onto a particular point on each of the 15 lines, and the simulated value at that node is the sum of the simulated values at the 15 projected points. The covariance of the 1-D process to be simulated independently on each line is deduced by a deconvolution process from the imposed 3-D isotropic covariance. This deconvolution process is reasonably straightforward in 3-D but awkward in 2-D [22,109].

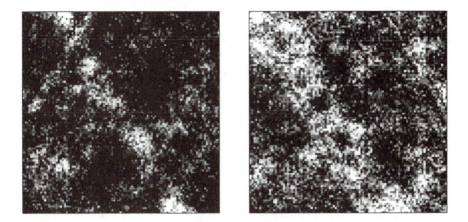

Figure V.7: Two 2-D slices through a 3-D realization generated by turning bands simulation.

Aside from being slow and cumbersome when used to generate *conditional* simulations, the turning band algorithm presents two important drawbacks:

1. The realizations show artifact banding due to the limitation to the maximum of 15 lines that provide a regular partitioning of the 3-D space [64]; see Figure V.7. There is no such maximum in 2-D.

2. Anisotropy directions other than the coordinate axes of the simulation grid cannot be handled easily.

The *non*-conditional simulation program **tb3d** given in section V.6.1. has been kept in GSLIB essentially for historical reasons.

V.2.6   Multiple Truncations of a Gaussian Field

If there are only two categories ($K{=}2$), one being the complement of the other, one can obtain non-conditional realizations of the unique indicator variable $i(\mathbf{u})$ by truncating the continuous realizations $\{y^{(l)}(\mathbf{u}), \mathbf{u} \in A\}$ of a Gaussian RF $Y(\mathbf{u})$:

$$
\begin{aligned}
i^{(l)}(\mathbf{u}) &= 1, \text{if } y^{(l)}(\mathbf{u}) \leq y_p \\
&= 0, \text{ if not}
\end{aligned}
\qquad (\text{V.27})
$$

with $y_p = G^{-1}(p)$ being the standard normal p-quantile and p being the desired proportion of indicators equal to one:

$$
E\{I(\mathbf{u})\} = p
$$

Since the Gaussian RF model is fully determined by its covariance $C_Y(\mathbf{h})$, there exists a one-to-one relation between that covariance $C_Y(\mathbf{h})$ and the indicator covariance after truncation at the p-quantile. By inverting relation (V.20) one can determine the covariance $C_Y(\mathbf{h})$ of the Gaussian RF that will yield the required indicator variogram [95].

Since the covariance $C_Y(\mathbf{h})$ is the unique parameter (degree of freedom) of the Gaussian RF model $Y(\mathbf{u})$, it cannot be used to reproduce more than one indicator covariance. Multiple truncations of the same realization $\{y^{(l)}(\mathbf{u}),$ $\mathbf{u} \in A\}$ at different threshold values would yield multiple categorical indicators with the correct marginal proportions [119], but with indicator covariances and cross correlations arbitrarily controlled by the Gaussian model. Uncontrolled indicator covariances imply realizations with uncontrolled spatial variability. Reliance on a Gaussian model limits the shapes of the boundaries that can be simulated; some complex boundaries require more parameters than the Gaussian model has available [106].

The series of threshold values which define the various categories can be made variable in space. This allows varying the proportions of each category, for example, within different depositional environments.

Another drawback of multiple truncations of the same continuous realization is that the spatial sequence of simulated categories is fixed. For example, along *any* direction the sequences of categories C_1, C_2, C_3, C_4 or C_4, C_3, C_2, C_1 will always be found. The algorithm does not allow generating discontinuous sequences, such as C_1, C_4, C_3, C_2. This constraint is useful in some cases, very limiting in other cases.

Last, since the Gaussian RF $Y(\mathbf{u})$ being simulated is continuous, conditioning to categorical data requires further approximations.

V.3   Indicator-Based Algorithms

Indicator random function models, being binary, are ideally suited for simulating categorical variables controlled by two-point statistics such as probabilities of the type:

$$Prob\{I(\mathbf{u}) = 1, I(\mathbf{u} + \mathbf{h}) = 1\} = E\{I(\mathbf{u}) \cdot I(\mathbf{u} + \mathbf{h})\} \qquad (V.28)$$

$I(\mathbf{u})$ is the RF modeling the binary indicator variable $i(\mathbf{u})$ set to 1 if a certain category prevails at location $\mathbf{u}$, zero if not.

Note that the probability of transition (V.28) is the non-centered indicator covariance. Thus, reproduction of an indicator covariance model through stochastic indicator simulation allows identifying a series of transition probabilities of type (V.28) for different distance vectors $\mathbf{h}$. The question is whether two-point statistics of the type (V.28) suffice to characterize the geometry of the category $\{\mathbf{u}$ such that $i(\mathbf{u}) = 1\}$.

If the answer to the previous question is no, then one needs to condition the indicator RF model $I(\mathbf{u})$ to higher order statistics such as three-point statistics:

$$Prob\{I(\mathbf{u}) = 1, I(\mathbf{u} + \mathbf{h}) = 1, I(\mathbf{u} + \mathbf{h}') = 1,\} = E\{I(\mathbf{u}) \cdot I(\mathbf{u} + \mathbf{h}) \cdot I(\mathbf{u} + \mathbf{h}')\}$$
$$(V.29)$$

or implicit multivariate statistics such as the shape and size of units s such that $i(\mathbf{u}) = 1, \forall \mathbf{u} \in s$.

Conditioning to explicit $3,4,\ldots,K$-variate statistics of the type (V.29) can be obtained through an annealing procedure (section V.5 and [36]), or generalized indicator kriging and normal equations the latter being beyond the scope of this manual [63,92]. Conditioning to implicit multivariate statistics, such as distributions of shape and size parameters, is better done by Boolean processes and other marked point processes; see below and section V.4.

The most important contribution of the indicator formalism is the direct evaluation of conditional probabilities as required by the sequential simulation principle. Indeed,

- If the variable to be simulated is already a binary indicator $i(\mathbf{u})$, set to 1 if the location $\mathbf{u}$ belongs to category $s \subset A$, to zero otherwise, then:

$$Prob\{I(\mathbf{u}) = 1|(n)\} = E\{I(\mathbf{u})|(n)\} \qquad (V.30)$$

- If the variable $z(\mathbf{u})$ to be simulated is continuous, its ccdf can also be written as an indicator conditional expectation; see also relation (IV.27):

$$Prob\{Z(\mathbf{u}) \le z|(n)\} = E\{I(\mathbf{u}; z)|(n)\} \qquad (V.31)$$

  with $I(\mathbf{u}; z) = 1$ if $Z(\mathbf{u}) \le z, = 0$ otherwise.

In both cases, the problem of evaluating the conditional probability is mapped onto that of evaluating the conditional expectation of a specific indicator RV. The evaluation of any conditional expectation calls for well-established regression theory, i.e., kriging. Kriging or cokriging applied to the proper indicator RV provides the "best[17]" estimates of the indicator conditional expectation, hence of the conditional probabilities (V.30) and (V.31).

Sequential indicator simulation for categorical variables is implemented in program **sisimpdf** presented in section V.7.1.

V.3.1 Simulation of Categorical Variables

Consider the spatial distribution of K mutually exclusive categories $s_k, k = 1, \ldots, K$. This list is also exhaustive, i.e., any location $\mathbf{u}$ belongs to one and only one of these K categories.

---

[17]Whenever the least-squared error criterion is used, i.e., kriging, the qualifier "best" is justified only when estimating an expected value or conditional expectation; then, the least-squares criterion *must* be used in preference to any other. Specific adaptations of this LS criterion for robustness and resistance of the estimator may be in order [72].

Let $i_k(\mathbf{u})$ be the indicator of class s_k, set to 1 if $\mathbf{u} \in s_k$, zero otherwise. Mutual exclusion and exhaustivity entails the following relations:

$$i_k(\mathbf{u}) i'_k(\mathbf{u}) = 0, \quad \forall\, k \neq k' \tag{V.32}$$

$$\sum_{k=1}^{K} i_k(\mathbf{u}) = 1$$

Direct kriging of the indicator variable $i_k(\mathbf{u})$ provides an estimate for the probability that s_k prevails at location $\mathbf{u}$. For example, using simple indicator kriging:

$$Prob^*\{I_k(\mathbf{u}) = 1 | (n)\} = p_k + \sum_{\alpha=1}^{n} \lambda_\alpha [I_k(\mathbf{u}_\alpha) - p_k] \tag{V.33}$$

where $p_k = E\{I_k(\mathbf{u})\} \in [0, 1]$ is the marginal frequency of category s_k inferred, e.g., from the declustered proportion of data of type s_k.

The weights λ_α are given by the SK system of type (IV.4) using the indicator covariance of category s_k.

In cases where the average proportions p_k vary locally, one can explicitly provide the simple indicator kriging systems with smoothly varying local proportions p_k's [8,119], or (implicitly) re-estimate these proportions from the indicator data available in the neighborhood of location $\mathbf{u}$. Such local re-estimation of p_k amounts to using ordinary kriging; see sections IV.1.2 and IV.1.9.

Sequential Simulation

See section V.1.3 and [7,61,82,93,136]. At each node $\mathbf{u}$ along the random path, indicator kriging followed by order relation correction provides K estimated probabilities $p_k^*(\mathbf{u}|(\cdot))$, $k = 1, \ldots, K$. The conditioning information $(\cdot)$ consists of both the original i_k-data and the previously simulated i_k-values for category s_k.

Next define *any* ordering of the K categories, say $1,2,\ldots,K$. This ordering defines a cdf-type scaling of the probability interval $[0,1]$ with K intervals, say:

$$[0, p_1^*(\cdot)], (p_1^*(\cdot), p_2^*(\cdot) + p_1^*(\cdot)], \ldots, (1 - \sum_{k=1}^{K-1} p_k^*(\cdot), 1]$$

Draw a random number p uniformly distributed in $[0,1]$. The interval in which p falls determines the simulated category at location $\mathbf{u}$. Update *all* K indicator data sets with this new simulated information, and proceed to the next location $\mathbf{u}'$ along the random path. The arbitrary ordering of the K probabilities $p_k^*(\cdot)$ does not affect which category is drawn nor the spatial distribution of categories [7].

Indicator Principal Component Kriging

In the IK approach the conditional probability of category s_k accounts only for the i_k-indicator data, i.e., the only information retained is that a datum location $\mathbf{u}_\alpha$ belongs to s_k or not. There are cases where the proximity of a datum $i_{k'}(\mathbf{u}_\alpha) = 1$ indicating presence of category $s_{k'}, k' \neq k$, leads to a higher probability of having s_k at $\mathbf{u}$: for example, the two categories s_k, $s_{k'}$ tend to be contiguous. Such cross-correlation between categories is ignored in IK.

The solution would be indicator cokriging (see section IV.1.10); but, as soon as $K > 3$, cokriging becomes impractical because of the inference of all indicator cross-covariances. A solution to shortcut this tedious inference is to perform indicator principal component kriging (IPCK) [154] as described in section IV.1.11 and implemented in program ipcsim.

In IPCK the principal components of the indicator data $i_k(\mathbf{u})$ are estimated by kriging. The linear back-transforms of the estimated principal components are the conditional probability estimates $p_k^*(\cdot)$ which are then corrected to match the order relations (IV.36).

The theoretically better cokriging approximation provided by IPCK is balanced by an increase in the number of order relation deviations [152].

V.3.2 Simulation of Continuous Variables

The spatial distribution of a continuous variable $z(\mathbf{u})$ discretized into K mutually exclusive classes $s_k : (z_{k-1}, z_k], k = 1, \ldots, K$ can be interpreted and simulated as the spatial distribution of K class indicators. The within-class resolution lost can be recovered in part by using some a priori within-class distribution, such as a power model; see section V.1.6.

One advantage of considering the continuous variable $z(\mathbf{u})$ as a paving (mosaic) of K classes is the flexibility to model the spatial distribution of each class by a different indicator variogram. For example, in the absence of any facies or mineralization-type information, the class of highest gold grades corresponding to a complex network of veinlets in fractured rocks, may be modeled by a zonal anisotropic indicator variogram with the maximum direction of continuity in the fracture direction; while the geometry of the classes of low-to-median gold grades would be modeled by more isotropic indicator variograms. The indicator formalism allows for modeling mixtures of populations loosely[18] defined as classes of values of a continuous attribute $z(\mathbf{u})$ [31,90].

---

[18] Recall the discussion of section V.1.1. Major heterogeneities characterized by actual categorical variables, such as lithofacies types, should be dealt with (simulated) first, e.g., through categorical indicator simulation or considering a Boolean process. There are cases and/or scales where the only property recorded is a continuous variable such as a mineral grade or acoustic log; yet experience tells us that this continuous variable is measured across heterogeneous populations. In this case the indicator formalism allows a "loose" separation of populations through discretization of the range of the continuous attribute measured [90].

The major advantage of the indicator formalism over most other approaches to estimation and simulation is the possibility of accounting for soft information; see section IV.1.12 and [85,99].

A single Gaussian RF model does not have the flexibility, i.e., enough free parameters, to handle a mixture of populations or account for soft information. Mixtures of Gaussian models with different covariances, e.g., one per lithotype, requires prior categorical data to separate the facies types: such categorical data may not be available at the scale required. Quadratic programming applied to include soft information, such as inequality constraints [47], is cumbersome, CPU intensive, and provides only a partial solution. For example, quadratic programming used to enforce the active constraint $z(\mathbf{u}_\alpha) \in (a_\alpha, b_\alpha]$ limits the solutions to the bounds of the interval, a_α or b_α, rather than allowing a solution within the interval itself [85,108].

Cumulative Class Indicators

As opposed to truly categorical populations which need not have any order, the classes $(z_{k-1}, z_k]$, $k = 1, \ldots, K$, of a continuous variable $z(\mathbf{u})$ are ordered sequentially. For this reason, it is better to characterize the classes with cumulative class indicators:

$$
\begin{aligned}
i(\mathbf{u}; z_k) &= 1, \text{ if } z(\mathbf{u}) \leq z_k \qquad\qquad (\text{V.34})\\
&= 0, \text{ otherwise}
\end{aligned}
$$

The class $(z_{k-1}, z_k]$ is defined by the product $i(\mathbf{u}; z_k)[1 - i(\mathbf{u}; z_{k-1})] = 1$.

Except for the first and last classes, inference of cumulative indicator variograms is easier than inference of class indicator variograms, particularly if the classes have small marginal probabilities. Also, considering cumulative (cdf-type) indicators instead of class (pdf-type) indicators allows using indicator data, that are cross-correlated across cutoffs; thus there is less need for indicator cokriging. Last, cumulative indicators are directly related to the ccdf of the continuous variable under study; see relation (V.31).

Variogram Reproduction

Simulation from an indicator-derived ccdf guarantees reproduction of the indicator variograms for the threshold z_k's considered, up to ergodic fluctuations; see section V.1.5 and [82], p. 35. It does not guarantee reproduction of the original z-variogram, unless a full indicator cokriging with a very fine discretization (large number of thresholds K) is implemented.

Rather, it is the madogram $2\gamma_M(\mathbf{h}) = E\{|Z(\mathbf{u}) - Z(\mathbf{u} + \mathbf{h})|\}$, as inferred from expression (III.9) which is reproduced since this structural function is the integral of all indicator variograms; see [6]. There is no a-priori reason to privilege reproduction of the variogram over that of the madogram.

If for some reason, the traditional z-variogram must be reproduced, one should use the median IK option (section IV.1.9 and relation (IV.30)) or revert to a Gaussian-based algorithm. Note that in this latter option it is the variogram of the normal score transforms of z which is reproduced. Another option is to post-process prior z-simulations by annealing to impose the required variogram; see later section V.5 and [36].

Recall also that the finite class discretization introduces an additional noise (nugget effect); see related discussion in section V.1.6.

Sequential Simulation

See section V.1.3 and [62,82,91]. At each node $\mathbf{u}$ to be simulated along the random path indicator kriging (SK or OK) provides a ccdf through K probability estimates:

$$F^*(\mathbf{u}; z_k|(n)) = Prob^*\{Z(\mathbf{u}) \le z|(n)\}, k = 1, \dots, K$$

Within-class interpolation, see section V.1.6, provides the continuum for all threshold values $z \in [z_{min}, z_{max}]$.

Monte-Carlo simulation of a realization $z^{(l)}(\mathbf{u})$ is obtained by drawing a uniform random number $p^{(l)} \in [0, 1]$ and retrieving the ccdf $p^{(l)}$-quantile:

$$z^{(l)}(\mathbf{u}) = F^{*-1}(\mathbf{u}; p^{(l)}|(n)) \tag{V.35}$$

$$\text{such that } : F^*(\mathbf{u}; z^{(l)}(\mathbf{u})|(n)) = p^{(l)}$$

The indicator data set (for all cutoffs z_k) is updated with the simulated value $z^{(l)}(\mathbf{u})$, and indicator kriging is performed at the next location $\mathbf{u}'$ along the random path.

Once all locations $\mathbf{u}$ have been simulated, a stochastic image $\{z^{(l)}(\mathbf{u}), \mathbf{u} \in A\}$ is obtained. The entire sequential simulation process with a new[19] random path can be repeated to obtain another independent realization $\{z^{(l')}(\mathbf{u}), \mathbf{u} \in A\}, l' \ne l$.

Sequential indicator simulation for continuous variables is implemented in program sisim; see description in section V.7.2. A combination of simple and ordinary indicator kriging is allowed. Soft information coded as missing indicator data (constraint intervals (IV.46)) or prior probability data (IV.47) can be considered, as long as the same indicator covariance $C_I(\mathbf{h}; z)$ is used for both hard and soft indicator data; see section IV.1.12. If a different indicator covariance is considered for the soft prior probability data, then the Markov-Bayes algorithm and program mbsim should be considered.

---

[19]The CPU time can be reduced considerably by keeping the same random path for each new realization. Then, the sequence of conditioning data configurations, hence the kriging systems, remain the same. The sequence of kriging weights can be stored and used identically for all realizations. This savings comes at the risk of drawing realizations that are too similar. Another major saving in CPU time is possible with median indicator kriging; see section IV.1.9 and system (IV.32). Provided there are only hard indicator data, median IK calls for solving only one kriging system (instead of K) per location $\mathbf{u}$. The price is the loss of modeling flexibility allowed by multiple indicator variograms.

Markov-Bayes Simulation

The Markov-Bayes implementation to account for soft indicator data amounts to an indicator cokriging, where the soft indicator data covariances and cross-covariances are calibrated from the hard indicator covariance models; see relations (IV.51) and (IV.52). In all other aspects program mbsim is similar to sisim. The soft indicator data, i.e., the prior probability cdf's of type (IV.47), are derived from calibration scattergrams using program mbcalib.
Warning: Artificially setting the accuracy measures $B(z)$ to high values close to one does not necessarily pull the simulated z-realizations closer to the secondary data map if, simultaneously, the variances of the pre-posterior distributions $y(\mathbf{u}_\alpha; z)$ derived from the secondary data are not lowered; see definition (IV.47). Indeed with $B(z)=1$, the cdf values $y(\mathbf{u}_\alpha; z)$ are not updated and the simulation at location $\mathbf{u}_\alpha$ amounts to draw from that cdf independently of neighboring hard or soft data: a *noisy* z-simulated realization may result.

V.4   Boolean Algorithms

Boolean algorithms cover a vast category of stochastic simulation algorithms and would require a whole book to discuss completely. Boolean algorithms and their extension, marked point processes [26,135,147], are beyond the scope of this guidebook. GSLIB is oriented more towards RF models characterized by two-point statistics. The practical importance of Boolean algorithms and more generally the study of spatial distributions of categorical variables deserves mention [13,65,128,151].

Boolean processes are generated by the distribution of geometric objects in space according to some probability laws. The spatial distribution of the object centroids constitutes a point process. A marked point process is a point process attached to (marked with) random processes defining the type, shape, and size of the random objects. For example, the point process generated by the impacts of German V2 bombs during WW II was studied by allied scientists for an indication of remote guidance. The distribution of impervious shales in a clastic reservoir can be modeled as a Boolean or marked point process of Poisson-distributed ellipsoids or parallelepipeds with a given distribution of volumes and elongation ratios [65].

Let U be a vector of coordinate RV's. Further, let $\mathbf{X}_k$ be a vector of parameter RV's characterizing the geometry (shape, size, orientation) of category k. The geometry could be defined by a parametric analytical expression or by a digitized template of points [151]. A wide variety of shapes could be generated by coordinate transformations of an original template.

The point process U is "marked" by the *joint* distribution of the shape random process $\mathbf{X}_k$ and the indicator random process for occurrence of category k. In geostatistical notation, one would consider the *joint* distribution of the 2 x K random functions (RF's) $\mathbf{X}_k(\mathbf{u}), I(\mathbf{u}; k), k = 1, \ldots, K, \mathbf{u} \in$ study area A, with $i(\mathbf{u}; k)$ set to 1 if $\mathbf{u}$ is the center of an object of category k, to zero if not.

In the earth sciences, the major problem with Boolean and other marked point processes is inference. Geological lithofacies or mineral bodies are rarely of a simple parametric shape, nor are they distributed at random within the reservoir or deposit. The available data, even as obtained from extensive 3-D outcrop sampling, rarely allows determination of the complex joint distribution of $\{\mathbf{X}_k(\mathbf{u}), I(\mathbf{u}; k), k = 1, \ldots, K\}$. Consequently the determination of a Boolean model is very much a trial-and-error process where various combinations of parameter distributions and interactions are tried until the final stochastic images are deemed satisfactory. Calibration of a Boolean model is more a matter of art (in the best sense of the term) than statistical inference. This allows for the extreme flexibility of Boolean models in reproducing very complex geometric shapes that are beyond the reach of better understood models based on only two-point statistics.

Other problems with Boolean processes are:

- They are difficult to condition to local data, such as the lithofacies series intersected by a well, particularly with closely spaced conditioning data. Specific ad-hoc solutions are available [151].

- The probability law (multivariate distribution) of most marked point processes is usually too complex to be analytically defined and understood. The same could be said about most RF models implicit to stochastic simulation algorithms, except the multivariate Gaussian model.

Boolean methods are typically custom-built[20] to reproduce a particular type of image. There cannot be a single general Boolean conditional simulation program. The program ellipsim provided in GSLIB is a very simple program for generating 2-D randomly distributed ellipses with specified distributions for the radius and the direction of ellipse elongation. This program could be easily modified to handle more difficult parametric shapes.

---

[20]Most marked point processes rely on congenial (rather than realistic) factorable exponential type probability distributions such as:

$$Prob\{\mathbf{X}_k = \mathbf{x}_k, k = 1, \ldots, K\} = C exp \sum_{k=1}^{K} F(\mathbf{x}_k)$$

which allows the conditional probability of $\mathbf{X}_0 = \mathbf{x}_0$ to be expressed as:

$$Prob\{\mathbf{X}_0 = \mathbf{x}_0 | \mathbf{X}_k = \mathbf{x}_k, \forall k \neq k_0\} = exp F(\mathbf{x}_{k_0})$$

C is a standardization factor and the function $F(\cdot)$ is some function chosen to match specific properties of the conditional probability [56,147].

V.5   Simulated Annealing

Generating alternate conditional stochastic images of either continuous or categorical variables with the aid of the numerical technique known as "simulated annealing" is a relatively new approach [1,36,52,56]. The technique has the potential of combining the reproduction of two-point statistics with complex multivariate spatial statistics implicit to, say, geometrical shapes. The technique can be extremely CPU-intensive unless used to modify prior rough stochastic images generated by very fast simulation algorithms, such as the sequential Gaussian algorithm (sGs) or median IK.

Two straightforward implementations are provided in GSLIB. In the first program (sasim) conditional simulations are achieved by imposing reproduction of the variogram model for specific distances. The second program (anneal) post processes or finishes prior simulated images by imposing reproduction of a series of transition probabilities as obtained from a training image.

V.5.1   Simulation by Simulated Annealing

An initial image is created by relocating the conditioning data to the nearest grid nodes and then drawing all remaining node values at random from the user-specified histogram. This initial image is sequentially modified by swapping the values in pairs of grid nodes not involving a conditioning datum. A swap is accepted if the objective function (average squared difference between the experimental and the model variogram) is lowered. Not all swaps which raise the objective function are rejected; the success of the method depends on a slow *cooling* of the image controlled by a temperature function which decreases with time [52,101,120]. The higher the temperature (or control parameter) the greater the probability that an unfavorable swap will be accepted. The simulation is complete when the image is frozen, i.e., when further swaps do not lower the objective function or when a specified minimum objective function is reached.

Data values are relocated at grid nodes and are never swapped, which allows their exact reproduction.

The following procedure is used to create an image which match the marginal distribution $F(z)$, the semivariogram $\gamma(h)$, and the conditioning data:

- For each node that does not represent a conditioning datum an outcome is randomly drawn from the univariate distribution $F(z)$. This generates an initial realization that matches the conditioning data and the univariate distribution, but does not likely match the semivariogram model $\gamma(h)$.

- The objective is for the semivariogram $\gamma^*(h)$ of the simulated realization to match the prespecified semivariogram model $\gamma(h)$. The following objective function should approach zero to achieve a conditional simulation:

$$O = \sum_h \frac{[\gamma^*(h) - \gamma(h)]^2}{\gamma(h)^2} \qquad (V.36)$$

The division by the square of the model semivariogram value at each lag standardizes the units and gives more weight to closely spaced (low variogram) values. The problem with this simple objective function is that the original data are overweighted by the sheer number of simulated values. Consequently, spatial discontinuities are often observed near the data locations. The user has the option to break the objective function into two parts: those pairs that involve an original conditioning datum and those that involve two simulated nodes; see [39] and hereafter Figure V.8.

- The initial image is modified by swapping pairs of nodal values z_i and z_j chosen at random, where neither node i nor node j are conditioning data. After each swap the objective value (V.36) is updated.

- One idea would be to reject all swaps that raise the objective O; see [44]. It turns out that this procedure might result in an image that is *frozen* before the objective is lowered close enough to zero. If the objective is approached more slowly, that is, if some swaps that increase the objective are accepted, the objective function can be brought extremely close to zero. As the metallurgical analogy suggests, the image must be cooled slowly (annealed), rather than cooled rapidly (quenched). In other words, accepting swaps that increase the objective function allows the possibility of getting out of a local minimum.

- The essential contribution of simulated annealing is a prescription for when to accept or reject a given perturbation. The acceptance probability distribution is given by the Boltzman distribution [1]:

$$P\{accept\} = \begin{cases} 1, & \text{if } O_{new} \leq O_{old} \\ e^{\frac{O_{old} - O_{new}}{t}}, & \text{otherwise} \end{cases} \qquad (V.37)$$

All favorable perturbations ($O_{new} \leq O_{old}$) are accepted and unfavorable perturbations are accepted with an exponential probability distribution. The parameter t of the exponential distribution is analogous to the "temperature" in annealing. The higher the temperature the more likely an unfavorable perturbation will be accepted.

The temperature t must not be lowered too fast or else the image may get trapped in a sub-optimal situation and never converge. However, if lowered too slowly then convergence may be unnecessarily slow. The specification of how to lower the temperature t is known as the "annealing schedule". There are mathematically based annealing schedules that guarantee convergence [1,56]; however, they are much too slow for a practical application. The following empirical annealing schedule is one practical alternative [52,132].

The idea is to start with an initially high temperature t_0 and lower it by some multiplicative factor λ whenever enough perturbations have been accepted (K_{accept}) or too many have been tried (K_{max}). The algorithm is stopped when efforts to lower the objective function become sufficiently discouraging. The following parameters describe this annealing schedule:

t_0: the initial temperature.

λ: the reduction factor $0 < \lambda < 1$.

K_{max}: the maximum number of attempted perturbations at any one temperature (on the order of 100 times the number of nodes). The temperature is multiplied by λ whenever K_{max} is reached.

K_{accept}: the acceptance target. After K_{accept} perturbations are accepted the temperature is multiplied by λ (on the order of 10 times the number of nodes).

S: the stopping number. If K_{max} is reached S times then the algorithm is stopped (usually set at 2 or 3).

ΔO : a low objective function indicating convergence.

An efficient coding of the objective function is essential to decrease the convergence time. For example, when a swap is considered the variogram lags are updated rather than recalculated, i.e., if a value z is a lag distance h from the location of z_i the previous contribution of z_i is subtracted from the variogram and the new contribution due to z_j is added:

$$\gamma^*_{new}(h) = \gamma^*(h) + \frac{1}{2N(h)} \left[(z - z_j)^2 - (z - z_i)^2 \right] \qquad \text{(V.38)}$$

As an example, consider the four conditional simulations shown on Figure V.8 generated by annealing simulation. Notice that with a one part objective function (top two realizations) there are artifact discontinuities next to the two vertical strings of conditioning data. The realizations generated with a two-part objective function (bottom two realizations) do not show the discontinuities. Figure V.9 illustrates the variogram reproduction at the beginning, half way through, and at the end of the simulation of the top left realization shown on Figure V.8.

The very simple objective function (V.36) considered in **sasim** is possibly better met by two-point statistics-based algorithms such as sGs or sequential indicator simulation. The interesting aspect of the simulated annealing algorithm is the ability to incorporate additional constraints into the objective

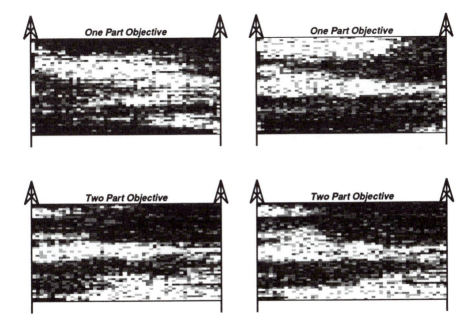

Figure V.8: Two annealing realizations generated with a one-part objective function (top) and two generated with a two-part objective function (bottom). The realizations generated with a one-part objective function show discontinuities next to the two strings of conditioning data.

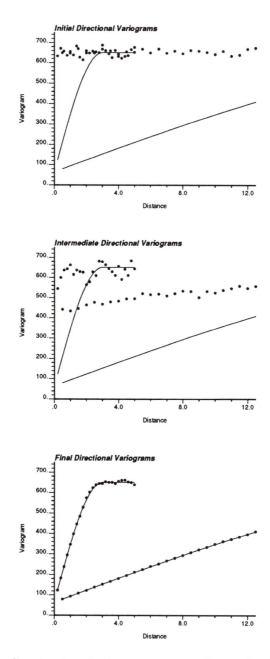

Figure V.9: The directional semivariograms corresponding to the simulation of the top left realization shown in Figure V.8. The solid lines are the model semivariograms in the two principal directions, and the black dots are the actual semivariogram values at the beginning, half way through, and at the end of the annealing conditional simulation.

function, e.g., specific n-variate statistics with $n > 2$. For example, a series of indicator connectivity functions [92] defined as:

$$\phi(n; z_c) = E\left\{\prod_{j=1}^{n} I(\mathbf{u} + (j-1)\mathbf{h}; z_c)\right\} \qquad (V.39)$$

$$= Prob\{Z(\mathbf{u} + (j-1)\mathbf{h}) \leq z_c, j = 1, \ldots, n\}$$

could be considered as part of the objective function

$$O = \sum_{k} [\phi^*(k; z_c) - \phi(k; z_c)]^2 \qquad (V.40)$$

An indicator connectivity function such as (V.39) describes the multiple-point connectivity in the direction of $\mathbf{h}$: it relates to the probability of having a string of n values z jointly lesser (greater) than a given cutoff value z_c.

V.5.2 Post Processing with Simulated Annealing

A useful application of simulated annealing is to post-process images that already have many desired spatial features, e.g., the result of sequential Gaussian or indicator simulation. The initial random image of program **sasim** is replaced by a realization that already possesses some of the desired features and an objective function is constructed that imposes additional spatial features.

The program **anneal** provided in GSLIB is meant to be a starter program for post processing with annealing. There are many possible control statistics that could be used in addition to the two-point histograms considered in **anneal**. For example, multiple-point covariances, multiple-point histograms, connectivity functions [92], fidelity to seismic data (through a calibration scatterplot), and fidelity to well test interpreted effective permeability [36]. Only two-point histograms of categorical variables are coded into **anneal**. Given a random variable Z that can take one of K outcomes $(k = 1, \ldots, K)$ the two-point histogram for a particular lag separation vector $\mathbf{h}$ is the set of all bivariate transition probabilities [51]:

$$p_{k,k'}(\mathbf{h}) = Prob\left\{\begin{array}{l} Z(\mathbf{u}) \in \text{ category } k, \\ Z(\mathbf{u}+\mathbf{h}) \in \text{ category } k' \end{array}\right\} \qquad (V.41)$$

independent of $\mathbf{u}$; for all $k, k' = 1, \ldots, K$. The objective function corresponding to this control statistic is as follows:

$$O = \sum_{h} \left(\sum_{k=1}^{K} \sum_{k'=1}^{K} \left[p_{k,k'}^{training}(\mathbf{h}) - p_{k,k'}^{realization}(\mathbf{h})\right]^2\right) \qquad (V.42)$$

where $p_{k,k'}^{training}(\mathbf{h})$ are the reference transition probabilities read directly from a training image and $p_{k,k'}^{realization}(\mathbf{h})$ are the corresponding frequencies of the realization image.

Figure V.10 illustrates an example application of the **anneal** program. The upper two figures are a training image and the corresponding normal score variogram used for sequential Gaussian simulation; the middle two figures are two Gaussian realizations (generated with **sgsim**) which are post processed into the two realizations shown on the bottom. The objective was to reproduce the two-point histogram of eight gray level classes, for 10 lags, in two orthogonal directions aligned with the sides of the image. Note how the general character, i.e., the positioning of highs and lows, has not been changed and yet more features from the training image have been imparted to the bottom realizations.

For post processing the annealing schedule can not be started at too high a temperature; otherwise, the initial image will be randomized before it starts converging. To avoid this problem and to speed convergence the MAP (maximum a posteriori) algorithm [15,39,44] is used to control the acceptance mechanism. That is, the temperature is fixed at zero and the program repeatedly cycles over all grid nodes along a random path. At each grid node location all of the possible K codes are checked and the one that reduces the objective function the most is kept. True simulated annealing could be implemented by accepting unfavorable code values with a certain probability. When starting from a fairly advanced image the problems are rarely difficult enough to warrant a full implementation of simulated annealing.

Program **anneal** does not allow for any local conditioning. The source code for conditioning to local data values may be taken from **sasim** if required.

V.6 Gaussian Simulation Subroutines

V.6.1 Turning Bands Simulation

The following files should be located on the diskette(s) before attempting to compile and execute this program/subroutine:

| tb3dm.f | an example driver program for **tb3d**

| tb3d.inc | an include file with maximum array dimensions

| tb3d.f | the subroutine **tb3d**

| tb3d.par | an example parameter file for **tb3dm**

Warning: **tb3d** yields *non*-conditional realizations of a Gaussian RF. These realizations should be conditioned to normal-score transforms of data through kriging prior to back transform; see section V.1.4 and [94], p. 509.

The stand alone normal scores transform program **nscore** and back transformation program **backtr** can be used since no transformation is performed within **tb3d**. These two utility programs are documented in sections VI.2.2 and VI.2.3.

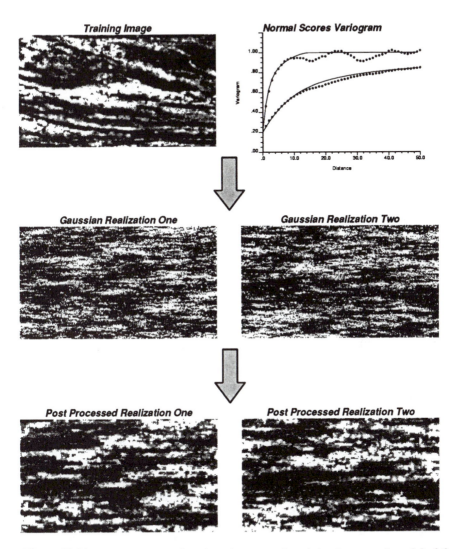

Figure V.10: An example application of anneal. A training image and model of the normal score variogram is shown at the top, two Gaussian realizations (generated with sgsim) are shown in the middle, and finally the outputs of anneal after post processing of the Gaussian realizations are shown at the bottom.

```
                    Parameters for TB3D
                    ********************

START OF PARAMETERS:
tb3d.out                              \output File for Realization(s)
1                                     \debugging level: 0,1,2,3
tb3d.dbg                              \output File for Debugging
112063                                \random number seed
2                                     \number of simulations
100  0.00   0.5                       \nx,xmn,xsiz
100  0.00   0.5                       \ny,ymn,ysiz
5    0.00   0.5                       \nz,zmn,zsiz
2    0.2                              \nst, nugget effect
1    5.0    0.4                       \it, aa, cc
0.0  0.0    0.0   1.0   1.0           \ang1,ang2,ang3,anis1,anis2:
1    15.0   0.4                       \it, aa, cc
0.0  0.0    0.0   1.0   1.0           \ang1,ang2,ang3,anis1,anis2:
```

Figure V.11: An example parameter file for **tb3d**.

The parameters required by the main program **tb3dm** are listed below and shown on Figure V.11:

- **outfl:** the output grid is written to this file. The output file will contain the unconditional simulated Gaussian values for the grid, cycling fastest on x, then y, then z, and then simulation by simulation.

- **idbg:** an integer debugging level between 0 and 3. The higher the debugging level the more output.

- **dbgfl:** a file for the debugging output.

- **seed:** random number seed (a large odd integer).

- **nsim:** the number of simulations to generate.

- **nx, xmn,** and **xsiz:** definition of the grid system (x axis). xmn is given to maintain consistency between the parameter listings of different programs; it is not used here.

- **ny, ymn,** and **ysiz:** definition of the grid system (y axis). ymn is given to maintain consistency between the parameter listings of different programs; it is not used here.

- **nz, zmn,** and **zsiz:** definition of the grid system (z axis). $nz > 1$, since **tb3d** should not be used to generate 2-D realizations.

- **nst** and **c0:** the number of variogram structures and the isotropic nugget constant.

- For each of the **nst** nested structures one must define **it** the type of structure, **aa** the a parameter, **cc** the c parameter, **ang1, ang2, ang3, anis1** and **anis2** the geometric anisotropy parameters. A detailed description of these parameters is given in section II.3. The variogram

```
                      Parameters for LUSIM
                      ********************

START OF PARAMETERS:
parta.dat                           \data file
1   2   0    3                      \column: x,y,z,vr - normal data
-1.0e21    1.0e21                    \data trimming limits
lusim.out                           \output file for realization(s)
3                                   \debugging level: 0,1,2,3
lusim.dbg                           \output file for debugging
112063                              \random number seed
100                                 \number of simulations
4    40.25   0.5                    \nx,xmn,xsiz
4    28.25   0.5                    \ny,ymn,ysiz
1     0.00   1.0                    \nz,zmn,zsiz
1     0.2                           \nst, nugget effect
1    10.0    0.8                    \it, aa, cc
0.0  0.0   0.0   1.0   1.0          \ang1,ang2,ang3,anis1,anis2:
```

Figure V.12: An example parameter file for **lusim**.

model refers to the normal scores. The kriging variance is directly interpreted as the variance of the conditional distribution; consequently, the nugget constant c_0 and c (sill) parameters should add to 1.0.

tb3d allows only isotropic spherical and exponential variogram models. The parameters *anis*1 and *anis*2 are given to maintain consistency, however, they must be set to 1.

V.6.2   LU Simulation

The following files should be located on the diskette(s) before attempting to compile and execute this program/subroutine:

| lusimm.f | an example driver program for **lusim**

| lusim.inc | an include file with maximum array dimensions

| lusim.f | the subroutine **lusim**

| lusim.par | an example parameter file for **lusimm**

The subroutine requires standard normal data and writes standard normal simulated values. The parameters for the main program **lusim** are documented below and shown on Figure V.12. Some dimensioning parameters are entered in an include file | lusim.inc |, and the input variables are passed in the argument list to **lusim**:

- **datafl:** the input data are in a simplified Geo-EAS formatted file.

- **icolx, icoly, icolz,** and **icolvr:** the column numbers for the x, y and z coordinates, and the variable to be simulated. Any or all of the coordinate column numbers can be set to zero which indicates that the simulation is 2-D or 1-D.

- **tmin** and **tmax:** all values strictly less than *tmin* and strictly greater than *tmax* are ignored.

- **outfl:** the output grid is written to this file. The output file will contain the grid, cycling fastest on x, then y, then z, and last per simulation.

- **idbg:** an integer debugging level between 0 and 3. The higher the debugging level the more output.

- **dbgfl:** a file for the debugging output.

- **seed:** random number seed (a large odd integer).

- **nsim:** the number of simulations to generate.

- **nx, xmn, xsiz:** definition of the grid system (x axis).

- **ny, ymn, ysiz:** definition of the grid system (y axis).

- **nz, zmn, zsiz:** definition of the grid system (z axis).

- **nst** and **c0:** the number of variogram structures and the isotropic nugget constant.

- For each of the **nst** nested structures one must define **it** the type of structure, **aa** the a parameter, **cc** the c parameter, **ang1, ang2, ang3, anis1** and **anis2** the geometric anisotropy parameters. A detailed description of these parameters is given in section II.3. The semivariogram model refers to the normal scores. The kriging variance is directly interpreted as the variance of the conditional distribution; consequently, the nugget constant $c0$ and c (sill) parameters should add to 1.0. Recall that the power model is not a legitimate model for a multiGaussian phenomenon.

V.6.3   Sequential Gaussian Simulation

The following files should be located on the diskette(s) before attempting to compile and execute this program/subroutine:

| sgsimm.f | an example driver program for **sgsim**

| sgsim.inc | an include file with maximum array dimensions

| sgsim.f | the subroutine **sgsim**

| sgsim.par | an example parameter file for **sgsim**

Parameters for SGSIM
****

```
START OF PARAMETERS:
cluster.dat                        \data file
1   2   0   3   0                  \column: x,y,z,vr,wt
-1.0e21    1.0e21                  \data trimming limits
0                                  \0=transform the data, 1=don't
sgsim.trn                          \   output transformation table
0.0    30.0                        \   zmin,zmax(tail extrapolation)
1      0.0                         \   lower tail option, parameter
4      2.0                         \   upper tail option, parameter
sgsim.out                          \output File for simulation
1                                  \debugging level: 0,1,2,3
sgsim.dbg                          \output File for Debugging
112063                             \random number seed
0                                  \kriging type (0=SK, 1=OK)
1                                  \number of simulations
50    0.5    1.0                   \nx,xmn,xsiz
50    0.5    1.0                   \ny,ymn,ysiz
1     0.0    1.0                   \nz,zmn,zsiz
0                                  \0=two part search, 1=data-nodes
1                                  \max per octant(0 -> not used)
20.0                               \maximum search radius
 0.0  0.0   0.0  1.0  1.0          \sang1,sang2,sang3,sanis1,2
0     8                            \min, max data for simulation
8                                  \number simulated nodes to use
1    0.2                           \nst, nugget effect
1   10.0    0.8                    \it, aa,   cc
0.0  0.0    0.0  1.0 1.0           \ang1,ang2,ang3,anis1,anis2:
```

Figure V.13: An example parameter file for **sgsim**.

The sequential Gaussian algorithm (sGs) is presented in section V.2.3. The dimensioning parameters in $\boxed{\text{sgsim.inc}}$ should be customized to each simulation study. The grid size should be set explicitly and enough storage should be allocated for the covariance look-up table. If the covariance look-up table is too small the program will still work; however, the reproduction of long range covariance structures may be poor.

The following parameters are required for **sgsim**; see also Figure V.13:

- **datafl:** the input data in a simplified Geo-EAS formatted file.

- **icolx, icoly, icolvr,** and **icolwt:** the column numbers for the x, y and z coordinates, the variable to be simulated and the declustering weight. Any or all of the coordinate column numbers can be set to zero which indicates that the simulation is 2-D or 1-D. If the declustering weight is unavailable then *icolwt* may be set to zero.

- **tmin** and **tmax:** all values strictly less than *tmin* and strictly greater than *tmax* are ignored.

- **igauss:** if set to 1 then the variable is already standard normal (the simulation results will be left unchanged); if *igauss=0*, transformation is required.

- **transfl:** output file for the transformation table if transformation is required (*igauss* = 0).

- **zmin** and **zmax** the minimum and maximum allowable data values. These are used in the back transformation procedure.

- **ltail** and **ltpar** specify the back transformation implementation in the lower tail of the distribution: *ltail* = 1 implements linear interpolation to the lower limit *zmin* and *ltail* = 2 implements power model interpolation, with $\omega = ltpar$, to the lower limit *zmin*.

   The middle class interpolation is linear.

- **utail** and **utpar** specify the back transformation implementation in the upper tail of the distribution: *utail* = 1 implements linear interpolation to the upper limit *zmax*, *utail* = 2 implements power model interpolation, with $\omega = utpar$, to the upper limit *zmax*, and *utail* = 4 implements hyperbolic model extrapolation with $\omega = utpar$. The hyperbolic tail extrapolation is limited by *zmax*.

- **outfl:** the output grid is written to this file. The output file will contain the results, cycling fastest on x, then y, then z, then simulation by simulation.

- **idbg:** an integer debugging level between 0 and 3. The larger the debugging level the more information written out.

- **dbgfl:** the file for the debugging output.

- **seed:** random number seed (a large odd integer).

- **ktype:** the kriging type (0 = simple kriging, 1 = ordinary kriging) used throughout the loop over all nodes. SK is required by theory, only in cases where the number of original data found in the neighborhood is large enough can OK be used without the risk of spreading data values beyond their range of influence [74].

- **nsim:** the number of simulations to generate.

- **nx, xmn, xsiz:** definition of the grid system (x axis).

- **ny, ymn, ysiz:** definition of the grid system (y axis).

- **nz, zmn, zsiz:** definition of the grid system (z axis).

- **sstrat:** if set to 0, the data and previously simulated grid nodes are searched separately: the data are searched with a super block search and the previously simulated nodes are searched with a spiral search (see section II.4). If set to 1, the data are relocated to grid nodes and a spiral search is used; the parameters *ndmin* and *ndmax* are not considered.

- **noct:** the number of original data to use per octant. If this parameter is set ≤ 0 then it will not be used; otherwise, it will override the *ndmax* parameter and the data will be partitioned into octants and the closest *noct* data in each octant will be retained for the simulation of a grid node.

- **radius:** the search radius. This radius can be made anisotropic with the following parameters:

- **sang1, sang2, sang3, sanis1,** and **sanis2:** parameters defining the 3-D anisotropy of the search ellipsoid. A detailed interpretation of these parameters is given in section II.3.

- **ndmin** and **ndmax:** the minimum and maximum number of original data that should be used to simulate a grid node. If there are fewer than *ndmin* data points the node is not simulated.

- **ncnode:** the maximum number of previously simulated nodes to use for the simulation of another node.

- **nst** and **c0:** the number of semivariogram structures and the isotropic nugget constant.

- For each of the **nst** nested structures one must define **it** the type of structure, **aa** the a parameter, **cc** the c parameter, **ang1, ang2, ang3, anis1** and **anis2** the geometric anisotropy parameters. A detailed description of these parameters is given in section II.3. The semivariogram model refers to the normal scores. The kriging variance is directly interpreted as the variance of the conditional distribution; consequently, the nugget constant *c0* and c (sill) parameters should add to 1.0. Recall that the power model is not a legitimate model for a multiGaussian phenomenon.

V.7 Sequential Indicator Simulation Subroutines

V.7.1 Categorical Variable (PDF) Simulation

The following files should be located on the diskette(s) before attempting to compile and execute this program/subroutine:

| sisimpdf.f | the main driver program and the subroutines for for `sisimpdf`

| sisimpdf.inc | an include file with maximum array dimensions

| sisimpdf.par | an example parameter file for `sisimpdf`

Parameters for SISIMPDF
****

```
START OF PARAMETERS:
nodata.dat                          \data file
1    2    0    3                    \column: x,y,z,vr
-1.0e21      1.0e21                 \data trimming limits
sisimpdf.out                        \output file for simulation
2                                   \debugging level: 0,1,2,3
sisimpdf.dbg                        \output File for Debugging
69069                               \random number seed
1                                   \number of simulations
50    0.5    1.0                    \nx,xmn,xsiz
50    0.5    1.0                    \ny,ymn,ysiz
1     1.0    10.0                   \nz,zmn,zsiz
1                                   \0=two part search, 1=data-nodes
0                                   \ max per octant(0 -> not used)
25.0                                \ maximum search radius
 0.0  0.0   0.0  1.0  1.0           \ sang1,sang2,sang3,sanis1,2
0    12                             \ min, max data for simulation
12                                  \number simulated nodes to use
0    2                              \0=full IK, 1=med approx(cat #)
0                                   \0=SK, 1=OK
3                                   \number of categories
1      0.75   1   0.00             \cat # , global pdf, nst, nugget
       1    20.0  1.00             \       it, aa, cc
       0.0  0.0   0.0  1.0  1.0    \       ang1,ang2,ang3,anis1,2
2      0.125  1   0.00             \cat # , global pdf, nst, nugget
       1    20.0  1.00             \       it, aa, cc
      -45.   0.0  0.0  0.1  1.0    \       ang1,ang2,ang3,anis1,2
3      0.125  1   0.00             \cat # , global pdf, nst, nugget
       1    20.0  1.00             \       it, aa, cc
       45.   0.0  0.0  0.1  1.0    \       ang1,ang2,ang3,anis1,2
```

Figure V.14: An example parameter file for sisimpdf.

The sequential indicator simulation algorithm (sis) for categorical variables is presented in section V.3.1. The **sisimpdf** program is for the simulation of integer-coded categorical variables. The following parameters are required; see also Figure V.14:

- **datafl:** the input data is a simplified Geo-EAS formatted file.

- **icolx, icoly, icolz,** and **icolvr:** the column numbers for the x, y, and z coordinates and the variable to be simulated. Any or all of the coordinate column numbers can be set to zero which indicates that the simulation is 2-D or 1-D.

- **tmin** and **tmax:** all values strictly less than *tmin* and strictly greater than *tmax* are ignored.

- **outfl:** the output grid is written to this file. The output file will contain the results, cycling fastest on x, then y, then z, then simulation by simulation.

- **idbg:** an integer debugging level between 0 and 3. The larger the debugging level the more information written out.

- **dbgfl:** the file for the debugging output.

- **seed:** random number seed (a large odd integer).

- **nsim:** the number of simulations to generate.

- **nx, xmn, xsiz:** definition of the grid system (x axis).

- **ny, ymn, ysiz:** definition of the grid system (y axis).

- **nz, zmn, zsiz:** definition of the grid system (z axis).

- **sstrat:** if set to 0, the data and previously simulated grid nodes are searched separately: the data are searched with a super block search and the previously simulated nodes are searched with a spiral search (see section II.4). If set to 1, the data are relocated to grid nodes and a spiral search is used; the parameters *ndmin* and *ndmax* are not considered.

- **noct:** the number of original data to use per octant. If this parameter is set ≤ 0 then it will not be used; otherwise, the closest *noct* data in each octant will be retained for the simulation of a grid node.

- **radius:** the search radius. This radius can be made anisotropic with the following parameters:

- **sang1, sang2, sang3, sanis1,** and **sanis2:** parameters defining the 3-D anisotropy of the search ellipsoid. A detailed interpretation of these parameters is given in section II.3.

- **ndmin** and **ndmax:** the minimum and maximum number of original data that will be used to simulate a grid node. If there are fewer than *ndmin* data points the node is not simulated.

- **ncnode:** the maximum number of previously simulated nodes to use for the simulation of another node.

- **mik** and **mikcat:** if *mik* is set to 0, then a full indicator kriging will be performed at each grid node location to establish the conditional distribution. If *mik* is set to 1, then the median approximation will be used, i.e., a single variogram is used for all categories; therefore, only one kriging system needs to be solved and the computer time is significantly reduced. The variogram at corresponding to category *mikcat* will be used.

- **ktype:** the kriging type (0 = simple kriging, 1 = ordinary kriging) used throughout the loop over all nodes. SK is required by theory, only in cases where the number of original data found in the neighborhood is large enough can OK be used without the risk of spreading data values beyond their range of influence [74]. The global pdf values (specified with each category) are used for simple kriging.

- **ncat:** the number of categories.

The following set of parameters are required for each category:

- **cat, cdf, nst,** and **c0:** the integer category, the global pdf value, the number of semivariogram structures, and the isotropic nugget constant.

- For each of the **nst** nested structures one must define **it** the type of structure, **aa** the a parameter, **cc** the c parameter, **ang1, ang2, ang3, anis1** and **anis2** the geometric anisotropy parameters. A detailed description of these parameters is given in section II.3. Each semivariogram model refers to the corresponding indicator transform. A Gaussian variogram with a small nugget constant is not a legitimate variogram model for a discontinuous indicator function. There is no need to standardize the parameters to a sill of one since only the relative shape affects the kriging weights.

V.7.2 Continuous Variable (CDF) Simulation

The following files should be located on the diskette(s) before attempting to compile and execute this program/subroutine:

sisimm.f an example driver program for `sisim`

sisim.inc an include file with maximum array dimensions

sisim.f the subroutine `sisim`

sisim.par an example parameter file for `sisim`

The sequential indicator simulation algorithm (sis) for continuous variables is presented in section V.3.2. The `sisim` program is for the simulation of continuous variables with indicator data defined from a cdf. The following parameters are required; see also Figure V.15:

- **datafl:** the input data in a simplified Geo-EAS formatted file.

- **icolx, icoly, icolz,** and **icolvr:** the column numbers for the x, y, and z coordinates and the variable to be simulated. Any or all of the coordinate column numbers can be set to zero which indicates that the simulation is 2-D or 1-D.

- **tmin** and **tmax:** all values strictly less than *tmin* and strictly greater than *tmax* are ignored.

- **zmin** and **zmax** are the minimum and maximum values that will be used for interpolation in the tails.

```
                        Parameters for SISIM
                        ********************

START OF PARAMETERS:
cluster.dat                             \data file
1    2    0    3                        \column: x,y,z,vr
-1.0e21     1.0e21                      \data trimming limits
0.0    30.0                             \minimum and maximum data value
1      2.0                              \lower tail option and parameter
1      1.0                              \middle    option and parameter
4      2.5                              \upper tail option and parameter
cluster.dat                             \tabulated values for classes
3    0                                  \column for variable, weight
direct.ik                               \direct input of indicators
sisim.out                              \output file for simulation
2                                       \debugging level: 0,1,2,3
sisim.dbg                               \output File for Debugging
0                                       \0=standard order relation corr.
69069                                   \random number seed
2                                       \number of simulations
50    0.5    1.0                        \nx,xmn,xsiz
50    0.5    1.0                        \ny,ymn,ysiz
1     1.0    10.0                       \nz,zmn,zsiz
1                                       \0=two part search, 1=data-nodes
0                                       \ max per octant(0 -> not used)
20.0                                    \ maximum search radius
 0.0  0.0   0.0  1.0  1.0               \ sang1,sang2,sang3,sanis1,2
0    12                                 \ min, max data for simulation
12                                      \number simulated nodes to use
0     2.5                               \0=full IK, 1=med approx(cutoff)
0                                       \0=SK, 1=OK
5                                       \number cutoffs
0.5   0.12   1    0.15                  \cutoff, global cdf, nst, nugget
      1     10.0   0.85                 \      it, aa, cc
      0.0  0.0   0.0  1.0  1.0          \      ang1,ang2,ang3,anis1,2
1.0   0.29   1    0.10                  \cutoff, global cdf, nst, nugget
      1     10.0   0.90                 \      it, aa, cc
      0.0  0.0   0.0  1.0  1.0          \      ang1,ang2,ang3,anis1,2
2.5   0.50   1    0.10                  \cutoff, global cdf, nst, nugget
      1     10.0   0.90                 \      it, aa, cc
      0.0  0.0   0.0  1.0  1.0          \      ang1,ang2,ang3,anis1,2
5.0   0.74   1    0.10                  \cutoff, global cdf, nst, nugget
      1     10.0   0.90                 \      it, aa, cc
      0.0  0.0   0.0  1.0  1.0          \      ang1,ang2,ang3,anis1,2
10.0  0.88   1    0.15                  \cutoff, global cdf, nst, nugget
      1     10.0   0.85                 \      it, aa, cc
      0.0  0.0   0.0  1.0  1.0          \      ang1,ang2,ang3,anis1,2
```

Figure V.15: An example parameter file for sisim.

- **ltail** and **ltpar** specify the extrapolation in the lower tail: $ltail = 1$ implements linear interpolation to the lower limit z_{min}; $ltail = 2$ implements power model interpolation, with $\omega = ltpar$, to the lower limit $zmin$; and $ltail = 3$ implements linear interpolation between tabulated quantiles.

- **middle** and **midpar** specify the interpolation within the middle of the distribution: $middle = 1$ implements linear interpolation; $middle = 2$ implements power model interpolation, with $\omega = midpar$; and $middle = 3$ allows for linear interpolation between tabulated quantile values.

- **utail** and **utpar** specify the extrapolation in the upper tail of the distribution: $utail = 1$ implements linear interpolation to the upper limit $zmax$, $utail = 2$ implements power model interpolation, with $\omega = utpar$, to the upper limit $zmax$, $utail = 3$ implements linear interpolation between tabulated quantiles, and $utail = 4$ implements hyperbolic model extrapolation with $\omega = utpar$. The hyperbolic tail extrapolation is limited by $zmax$.

- **tabfl:** If linear interpolation between tabulated values is the option selected for any of the three regions then this simplified Geo-EAS format file is opened to read in the values. One legitimate choice is exactly the same file as the conditioning data, i.e., $datafl$. Note that $tabfl$ specifies the tabulated values for all classes.

- **icolvrt** and **icolwtt:** the column numbers for the values and declustering weights in $tabfl$. Note that declustering weights can be used but are not required - just set the column number less than or equal to zero.

  If declustering weights are not used, then the class probability is split equally between the sub-classes defined by the tabulated values.

- **softfl:** if this data file exists then it contains already transformed indicator data. This allows the direct input of soft data and constraint intervals; see definitions (IV.46) and (IV.47). The input format of this file is fixed, i.e., it expects a standard simplified Geo-EAS header, then it expects all subsequent lines to contain three coordinates (x, y, z) and $ncut$ indicator values. Note that the indicator values should be real numbers between 0.0 and 1.0. Missing values should be strictly less than $tmin$.

- **outfl:** the output grid is written to this file. The output file will contain the results, cycling fastest on x, then y, then z, then simulation by simulation.

- **idbg:** an integer debugging level between 0 and 3. The larger the debugging level the more information written out.

- **dbgfl:** the file for the debugging output.

- **iorder:** if set to 0, then the standard order relations correction is performed, i.e., the corrected distribution is an average of the upward and downward corrected cdfs. If set to 1, then the number of data falling into each class is checked and a cutoff is discarded if there is an order relation violation *and* there are no data in the corresponding class; see Figure IV.3.

- **seed:** random number seed (a large odd integer).

- **nsim:** the number of simulations to generate.

- **nx, xmn, xsiz:** definition of the grid system (x axis).

- **ny, ymn, ysiz:** definition of the grid system (y axis).

- **nz, zmn, zsiz:** definition of the grid system (z axis).

- **sstrat:** if set to 0, the data and previously simulated grid nodes are searched separately: the data are searched with a super block search and the previously simulated nodes are searched with a spiral search (see section II.4). If set to 1, the data are relocated to grid nodes and a spiral search is used; the parameters *ndmin* and *ndmax* are not considered.

- **noct:** the number of original data to use per octant. If this parameter is set ≤ 0 then it will not be used; otherwise, the closest *noct* data in each octant will be retained for the simulation of a grid node.

- **radius:** the search radius. This radius can be made anisotropic with the following parameters:

- **sang1, sang2, sang3, sanis1,** and **sanis2:** parameters defining the 3-D anisotropy of the search ellipsoid. A detailed interpretation of these parameters is given in section II.3.

- **ndmin** and **ndmax:** the minimum and maximum number of original data that will be used to simulate a grid node. If there are fewer than *ndmin* data points the node is not simulated.

- **ncnode:** the maximum number of previously simulated nodes to use for the simulation of another node.

- **mik** and **mikcut:** if *mik* is set to 0, then a full indicator kriging will be performed at each grid node location to establish the conditional distribution. If *mik* is set to 1, then the median approximation will be used, i.e., a single variogram will be used for all cutoffs; therefore, only one kriging system needs to be solved and the computer time is significantly reduced. The variogram at the cutoff closest to *mikcut* will be used.

- **ktype:** the kriging type (0 = simple kriging, 1 = ordinary kriging) used throughout the loop over all nodes. SK is required by theory, only in cases where the number of original data found in the neighborhood is large enough can OK be used without the risk of spreading data values beyond their range of influence [74]. The global cdf values (specified with each cutoff) are used for simple kriging.

- **ncut:** the number of cutoffs discretizing the distribution.

The following specification of the indicator cutoffs and semivariograms must be ordered in increasing order of cutoff (and global cdf). The following set of parameters are required for each cutoff:

- **cut, cdf, nst,** and **c0:** the cutoff value, the global cdf value, the number of semivariogram structures, and the isotropic nugget constant.

- For each of the **nst** nested structures one must define **it** the type of structure, **aa** the a parameter, **cc** the c parameter, **ang1, ang2, ang3, anis1** and **anis2** the geometric anisotropy parameters. A detailed description of these parameters is given in section II.3. Each semivariogram model refers to the corresponding indicator transform. A Gaussian variogram with a small nugget constant is not a legitimate variogram model for a discontinuous indicator function. The variogram sill parameters have been standardized to sum to one on Figure V.15 so that it is easier to check for consistency between the variogram at different cutoffs. There is no need to standardize the parameters to a sill of one since only the relative shape affects the kriging weights.

V.7.3   Markov-Bayes Approach

The following files should be located on the diskette(s) before attempting to compile and execute this program/subroutine:

| mbsimm.f | an example driver program for **mbsim**

| mbsim.inc | an include file with maximum array dimensions

| mbsim.f | the subroutine **mbsim**

| mbsim.par | an example parameter file for **mbsim**

Prior to running **mbsim**, it is necessary to establish both the pre-posterior distributions corresponding to all classes of secondary data and the calibration values $B(z)$ which specify the y-auto and cross covariance based on the indicator auto covariance; see section IV.1.12. This calibration is straightforward with the program **mbcalib** given in section VI.2.8.

The parameters required by the **mbsim** program are almost identical to the **sisim** program with a number of additions. Only the additional required parameters are listed below. The format of the parameter file is illustrated on Figure V.16:

- **calibfl:** the calibration data file containing the pre-posterior distributions and the $B(z)$ values. The format of this file is specified with the **mbcalib** file in section VI.2.8.

- **secfl:** the data file containing the secondary data. This data file, in a simplified Geo-EAS format, must be specified with column numbers for the x, y, z, and value column numbers. Note that even with the accuracy measure set to 1.0 (perfectly accurate) these data are honored only within the class resolution; only the original hard data are exacty honored.

- **ncutv:** the number of cutoffs that apply to the secondary (soft) data. Note that this should correspond to the number used to create the calibration information in *calibfl*.

- **cutv():** the cutoffs that apply to the secondary data.

- **ndmaxv:** the maximum number of soft data to use for any one kriging system.

Program **mbsim** does not accept constraint intervals of type (IV.46), as opposed to programs **ik3d** and **sisim**. One possible approximation to a constraint interval $y(\mathbf{u}_\alpha) \in [a, b]$ would be to input a pre-posterior cdf $y(\mathbf{u}_\alpha; z)$ increasing linearly with z in that interval, i.e., a uniform pre-posterior pdf in the interval $[a, b]$. Another alternative would be to use the same implementation as coded in **sisim**.

V.7.4   Indicator Principal Components Approach

The following files should be located on the diskette(s) before attempting to compile and execute this program/subroutine:

| ipcsimm.f | an example driver program for **ipcsim**

| ipcsim.inc | an include file with maximum array dimensions

| ipcsim.f | the subroutine **ipcsim**

| ipcsim.par | an example parameter file for **ipcsim**

The **ipcsim** program is for simulation of **categorical** variables with indicator principal components kriging (IPCK) used to build the local conditional distributions; see sections IV.1.11 and V.3.1. Prior to running **ipcsim** it is necessary to perform the orthogonalization of the indicator covariance matrix and the modeling of the indicator principal components variograms. The

```
                        Parameters for MBSIM
                        ********************

START OF PARAMETERS:
cluster.dat                              \hard data file
1    2    0    3                         \  column: x,y,z,vr
-1.0e21    1.0e21                        \data trimming limits
mbsim.cal                                \soft data calibration File
cluster.dat                              \soft data file
1    2    0    4                         \  column: x,y,z,vr
-1.0e21    1.0e21                        \  trimming limits
5                                        \  number of soft cutoffs
0.5 1.0 2.5 5.0 10.0                     \  soft cutoffs
0.0   30.0                               \minimum and maximum data value
1    1.0                                 \lower tail option and parameter
1    1.0                                 \middle      option and parameter
4    2.0                                 \upper tail option and parameter
mbcalib.dat                              \tabulated values for classes
3    0                                   \  column for variable, weight
mbsim.out                                \output file for simulation
2                                        \debugging level: 0,1,2,3
mbsim.dbg                                \output file for debugging
0                                        \0=standard order relation corr.
69069                                    \random number seed
1                                        \number of simulations
25    1.0    2.0                         \nx,xmn,xsiz
25    1.0    2.0                         \ny,ymn,ysiz
1    1.0    1.0                          \nz,zmn,zsiz
1                                        \0=two part search, 1=data-nodes
0                                        \  max per octant (0-> not used)
20.0                                     \  maximum search radius
0.0  0.0  0.0  1.0  1.0                  \  sang1,sang2,sang3,sanis1,2
0    12    12                            \  min, max hard, max soft data
12                                       \number simulated nodes to use
0    2.5                                 \0=full IK, 1=med approx(cutoff)
0                                        \kriging type (0=SK, 1=OK)
4                                        \number of cutoffs
0.5   0.12   1    0.15                   \cutoff, global cdf, nst, nugget
      1    10.0   0.85                   \       it, aa, cc
      0.0  0.0  0.0  1.0  1.0            \       ang1,ang2,ang3,anis1,2
1.0   0.29   1    0.10                   \cutoff, global cdf, nst, nugget
      1    10.0   0.90                   \       it, aa, cc
      0.0  0.0  0.0  1.0  1.0            \       ang1,ang2,ang3,anis1,2
2.5   0.50   1    0.10                   \cutoff, global cdf, nst, nugget
      1    10.0   0.90                   \       it, aa, cc
      0.0  0.0  0.0  1.0  1.0            \       ang1,ang2,ang3,anis1,2
5.0   0.74   1    0.10                   \cutoff, global cdf, nst, nugget
      1    10.0   0.90                   \       it, aa, cc
      0.0  0.0  0.0  1.0  1.0            \       ang1,ang2,ang3,anis1,2
10.0  0.88   1    0.15                   \cutoff, global cdf, nst, nugget
      1    10.0   0.85                   \       it, aa, cc
      0.0  0.0  0.0  1.0  1.0            \       ang1,ang2,ang3,anis1,2
```

Figure V.16: An example parameter file for **mbsim**.

```
                        Parameters for IPCSIM
                        **********************

START OF PARAMETERS:
nodata.dat                              \data file
1    2    0    3                        \column: x,y,z,vr
-1.0e21    1.0e21                       \data trimming limits
ipcprep.svd                             \file with the orthogonal matrix
ipcsim.out                              \output file for simulation
1                                       \debugging level: 0,1,2,3
ipcsim.dbg                              \output file for debugging
69069                                   \random number seed
1                                       \number of simulations
5    1.0    1.0                         \nx,xmn,xsiz
5    1.0    1.0                         \ny,ymn,ysiz
1    1.0    10.0                        \nz,zmn,zsiz
1                                       \0=two part search, 1=data-nodes
0                                       \max per octant (0-> not used)
25.0                                    \maximum search radius
  45.0   0.0   0.0   0.5   1.0          \sang1,sang2,sang3,sanis1,2
0    12                                 \min, max data for simulation
12                                      \number simulated nodes to use
0                                       \kriging type (0=SK, 1=OK)
4    3                                  \# categories, # ipc's to krige
1    2    3    4                        \the categories
0.25 0.25 0.25 0.25                     \the ''p'' values
1 2 3                                   \ipc's to be kriged.
1   1   0.15                            \ipc #,   nst, nugget
    1   10.0   0.85                     \         it, aa, cc
    0.0   0.0   0.0   1.0   1.0         \         ang1,ang2,ang3,anis1,2
2   1   0.10                            \ipc #,   nst, nugget
    1   10.0   0.90                     \         it, aa, cc
    0.0   0.0   0.0   1.0   1.0         \         ang1,ang2,ang3,anis1,2
3   1   0.10                            \ipc #,   nst, nugget
    1   10.0   0.90                     \         it, aa, cc
    0.0   0.0   0.0   1.0   1.0         \         ang1,ang2,ang3,anis1,2
```

Figure V.17: An example parameter file for ipcsim.

construction of the transform matrix and the calculation of the principal component data is straightforward in the program **ipcprep** and is given in section VI.2.9.

ipcsim performs all of the necessary transformations and back transformations. Untransformed integer-coded data are input into both **ipcsim** and **ipcprep**; **ipcsim** writes out integer-coded realizations and **ipcprep** writes out the transformed indicator principal components. The variograms apply to the indicator principal components (computed from the output of **ipcprep**) and not to the original integer-coded data.

The following parameters are required by the **ipcsim** program; see also Figure V.17:

- **datafl:** the input data in a simplified Geo-EAS formatted file.

- **icolx, icoly, icolz,** and **icolvr:** the column numbers for the $x, y,$ and z coordinates and the variable to be simulated. Any or all of the coordinate column numbers can be set to zero which indicates that the simulation is 2-D or 1-D.

- **tmin** and **tmax:** all values strictly less than *tmin* and strictly greater than *tmax* are ignored.

- **svdfl:** a file containing the orthogonal matrix as computed in program `ipcprep`.

- **outfl:** the output grid is written to this file. The output file will contain the results, cycling fastest on x, then y, then z, then simulation by simulation.

- **idbg:** an integer debugging level between 0 and 3. The larger the debugging level the more information written out.

- **dbgfl:** the file for the debugging output.

- **seed:** random number seed (a large odd integer).

- **nsim:** the number of simulations to generate.

- **nx, xmn, xsiz:** definition of the grid system (x axis).

- **ny, ymn, ysiz:** definition of the grid system (y axis).

- **nz, zmn, zsiz:** definition of the grid system (z axis).

- **sstrat:** if set to 0, the data and previously simulated grid nodes are searched separately: the data are searched with a super block search and the previously simulated nodes are searched with a spiral search (see section II.4). If set to 1, the data are relocated to grid nodes and a spiral search is used; the parameters *ndmin* and *ndmax* are not considered.

- **noct:** the number of original data to use per octant. If this parameter is set ≤ 0 then it will not be used; otherwise, the closest *noct* data in each octant will be retained for the simulation of a grid node.

- **radius:** the search radius. This radius can be made anisotropic with the following parameters:

- **sang1, sang2, sang3, sanis1,** and **sanis2:** parameters defining the 3-D anisotropy of the search ellipsoid. A detailed interpretation of these parameters is given in section II.3.

- **ndmin** and **ndmax:** the minimum and maximum number of original data that will be used to simulate a grid node. If there are fewer than *ndmin* data points the node is not simulated.

- **ncnode:** the maximum number of previously simulated nodes to use for the simulation of another node.

- **ktype:** the kriging type (0 = simple kriging, 1 = ordinary kriging) used throughout the loop over all nodes. SK is required by theory, only in cases where the number of original data found in the neighborhood is large enough can OK be used without the risk of spreading data values beyond their range of influence [74]. The global pdf values (specified below) are used for simple kriging.

- **ncat** and **nipck:** the number of categories and the number of indicator principal components to krige, $nipck \leq ncat - 1$.

- **cat(i),i=1,. . .,ncat:** the integer categories.

- **pdf(i),i=1,. . .,ncat:** the global proportion of each category.

- **ipck(i),i=1,. . .,nipck:** the indicator principal components to krige.

The following set of parameters are required for each indicator principal component to be estimated by kriging:

- **ipc, nst,** and **c0:** the ipc number, the number of variogram structures, and the isotropic nugget constant.

- For each of the **nst** nested structures one must define **it** the type of structure, **aa** the a parameter, **cc** the c parameter, **ang1, ang2, ang3, anis1** and **anis2** the geometric anisotropy parameters. A detailed description of these parameters is given in section II.3. Each semivariogram model refers to the corresponding indicator principal component. There is no need to standardize the parameters to a sill of one since only the relative shape affects the kriging weights.

V.8 A Boolean Simulation Subroutine

The following files should be located on the diskette(s) before attempting to compile and execute this program/subroutine:

| ellipsim.f | the program `ellipsim`

| ellipsim.par | an example parameter file for `ellipsim`

This program provides a very simple example of Boolean simulation: ellipses of various sizes and anisotropies are dropped at random until a target proportion of points within-ellipse is met. The ellipses are allowed to overlap each other. The following parameters are required by the main program `ellipsim`; see also Figure V.18:

- **radiusfl:** a simplified Geo-EAS formatted file that contains the distribution of ellipse radius to be considered. Note that if only one value is

```
                  Parameters for ELLIPSIM
                  ************************

START OF PARAMETERS:
radius.dat                             \radius file
1   0                                  \columns: radius, weight
angles.dat                             \angles file
1   0                                  \columns: ang, weight
anis.dat                               \aniotropy ratios File
1   0                                  \columns: anis, weight
ellipsim.out                           \output file for simulation
69069                                  \random number seed
200      1.0                           \nx,xmn,xsiz
200      1.0                           \ny,ymn,ysiz
1                                      \number of simulations
0.25                                   \target proportion (in ellipses)
```

Figure V.18: An example parameter file for **ellipsim**.

given then all ellipses will be the same size. Further, note that the proportional weighting is used to establish the relative frequency of each radius.

- **icolvr** and **icolwt:** the column numbers for the radius and, optionally, for the relative proportion of each radius.

- **anglefl:** a simplified Geo-EAS formatted file that contains the distribution of ellipse angles to be considered. Note that the angles are in degrees measured clockwise from the y axis. Again, if only one value is in this file then all ellipses will be oriented in the same direction.

- **icolvr** and **icolwt:** the column numbers for the angle and for the relative proportion of each angle value.

- **anisfl:** a simplified Geo-EAS formatted file that contains the distribution of anisotropy ratios for the ellipses. Note that the anisotropy ratios are less than 1 if the ellipses are elongated in the principal direction defined by the angles; see above.

- **icolvr** and **icolwt:** the column numbers for the anisotropy ratio and for the relative proportion of each anisotropy ratio.

- **outfl:** the simulated realizations are written to this file. The output file will contain the unconditional simulated indicator values (1 if in an ellipse, zero otherwise) cycling fastest on x, then y, then z, and then simulation by simulation.

- **seed:** random number seed (a large odd integer).

- **nx, xsiz:** definition of the grid system (x axis).

- **ny, ysiz:** definition of the grid system (y axis).

- **nsim:** the number of simulations to generate.

- **targprop:** the target proportion of ellipse volume.

V.9 Simulated Annealing Subroutines

V.9.1 Straight Simulated Annealing

The following files should be located on the diskette(s) before attempting to compile and execute this program/subroutine:

| sasimm.f | an example driver program for `sasim`

| sasim.inc | an include file with maximum array dimensions

| sasim.f | the subroutine `sasim`

| sasim.par | an example parameter file for `sasim`

Program `sasim` presented in section V.5 allows conditional simulations of a continuous variable honoring an input histogram and variogram values for specified lags. The following parameters are required by the main program `sasimm`; see also Figure V.19:

- **condfl:** an input data file with the conditioning data (simplified Geo-EAS format). If this file does not exist then an unconditional simulation will be generated.

- **icolx, icoly, icolz,** and **icolvr:** the column numbers for the $x, y,$ and z coordinates and the variable to be simulated. Any or all of the coordinate column numbers can be set to zero which indicates that the simulation is 2-D or 1-D.

- **tmin** and **tmax:** all values strictly less than *tmin* and strictly greater than *tmax* are ignored.

- **igauss:** if set to 1 then a standard normal Gaussian deviate is simulated; otherwise, a non-parametric distribution defined by the next four input parameters will be used.

- **datafl:** contains the data that make up the non-parametric distribution (simplified Geo-EAS format). One common approach is to use the same file as *condfl*.

- **icolvr** and **icolwt:** the column numbers for variable and declustering weight to construct the univariate distribution.

- **zmin** and **zmax** the minimum and maximum data allowable data value. These may be used in the back transformation procedure.

```
                    Parameters for SASIM
                    ********************

START OF PARAMETERS:
cluster.dat                              \conditioning data (if any)
1    2    0    3                         \columns: x,y,z,vr
-1.0e21    1.0e21                        \data trimming limits
0                                        \0=non parametric; 1=Gaussian
cluster.dat                              \non parametric distribution
3    0                                   \columns: vr,wt
0.0    90.0                              \minimum and maximum data values
1      1.0                               \lower tail option and parameter
4      2.0                               \upper tail option and parameter
sasim.out                                \output File for simulation
sasim.var                                \output File for variogram
3      50                                \debug level, reporting interval
sasim.dbg                                \output file for debugging
1                                        \annealing schedule? (0=auto)
1.0 0.1 6250    2500  3 0.0001    \manual schedule: t0,lambda,ka,k,e,Omin
1                                        \1 or 2 part objective function
112063                                   \random number seed
1                                        \number of simulations
25    1.0    2.0                         \nx,xmn,xsiz
25    1.0    2.0                         \ny,ymn,ysiz
1     0.0    1.0                         \nz,zmn,zsiz
12                                       \max lags for conditioning
1     0.2  0                             \nst, nugget, (1=renormalize)
1    10.0  0.8                           \it,aa,cc:          STRUCTURE 1
0.0   0.0   0.0   1.0   1.0             \ang1,ang2,ang3,anis1,anis2:
```

Figure V.19: An example parameter file for **sasim**.

- **ltail** and **ltpar** specify extrapolation in the lower tail of the distribution: $ltail = 1$ implements linear interpolation to the lower limit $zmin$ and $ltail = 2$ implements power model interpolation, with $\omega = ltpar$, to the lower limit $zmin$.

  The middle class interpolation is linear.

- **utail** and **utpar** specify extrapolation in the upper tail of the distribution: $utail = 1$ implements linear interpolation to the upper limit $zmax$, $utail = 2$ implements power model interpolation, with $\omega = utpar$, to the upper limit $zmax$, and $utail = 4$ implements hyperbolic model extrapolation with $\omega = utpar$. The hyperbolic tail extrapolation is limited by $zmax$.

- **outfl:** the output grid is written to this file. The output file will contain the results, cycling fastest on x, then y, then z, then simulation by simulation.

- **varfl:** this output file will contain the model semivariogram, and the semivariogram of the final grid. If the two-part objective function is used the variogram of both parts will be in the file (see parameter *part* below).

- **idbg** and **report:** an integer debugging level between 0 and 3, and a reporting interval. After a fixed number of swaps (*report*) the objective function is written to the debugging file.

- **dbgfl:** the file for the debugging output.

- **isas:** the annealing schedule (next set of parameters) can be set explicitly or it can be set automatically (0=automatic,1=then use the following:).

- **sas(6):** the annealing schedule (read free format from one line of input): initial temperature, the reduction factor, the maximum number of swaps at any one given temperature, the target number of swaps, the stopping number, and a low objective function value indicating convergence.

- **part:** a one-part objective function can be used that considers the overall variogram (*part* = 1), or a two-part objective function that separates the pairs involving original conditioning data can be used (*part* = 2).

- **seed:** random number seed (a large odd integer).

- **nsim:** the number of simulations to generate.

- **nx, xmn, xsiz:** definition of the grid system (x axis).

- **ny, ymn, ysiz:** definition of the grid system (y axis).

- **nz, zmn, zsiz:** definition of the grid system (z axis).

- **maxlag:** the maximum number of lags to consider in the objective function. The closest *maxlag* lags, measured in terms of variogram distance, are retained.

- **nst, c0** and **istand:** the number of variogram structures, the isotropic nugget constant, and a flag specifying whether or not to standardize the sill of the semivariogram to the variance of the univariate distribution (*istand=1* will standardize).

  It is essential that the variance of the values of the initial random image matches the spatial (dispersion) variance implied by the variogram model. That dispersion variance should be equal to the total sill if it exists (i.e., a power model has not been used) and the size of the field is much larger than the largest range in the variogram model. Otherwise, the dispersion variance can be calculated from traditional formulae ([94], p. 61-67).

- For each of the **nst** nested structures one must define **it** the type of structure, **aa** the a parameter, **cc** the c parameter, **ang1, ang2, ang3, anis1** and **anis2** the geometric anisotropy parameters. A detailed description of these parameters is given in section II.3.

```
                         Parameters for ANNEAL
                         **********************

START OF PARAMETERS:
sisim.out                                \input image(s)
ellipsim.out                             \training Image
postsim.out                              \output file for simulation
postsim.hst                              \output file for 2-pt histogram
3    10                                  \debug level, reporting Interval
postsim.dbg                              \output file for debugging
69069                                    \random number seed
5    0.000001                            \maximum iterations, tolerance
1                                        \number of simulations
  25    25   1                           \nx, ny, nz
4                                        \ndir
1   0   0   10   1                       \ixl, iyl, izl, nlag, cross(1=y)
0   1   0   10   1                       \ixl, iyl, izl, nlag, cross(1=y)
1   1   0    5   0                       \ixl, iyl, izl, nlag, cross(1=y)
1  -1   0    5   0                       \ixl, iyl, izl, nlag, cross(1=y)
```

Figure V.20: An example parameter file for **anneal**.

V.9.2   An Annealing Post Processor

The following files should be located on the diskette before attempting to compile and execute this program:

anneal.f the program **anneal**

anneal.inc an include file for **anneal**.

anneal.par an example parameter file for **anneal**

This program requires an integer-coded training image, a series of initial integer-coded realizations (possibly only one) to post-process, and information about what components of a two-point histogram are to be reproduced. The integer coding could be derived naturally from a categorical variable or it could be derived from the discretization of a continuous variable. The following parameters are required by the main program **anneal**; see also Figure V.20:

- **datafl:** a file that contains the input realizations to post-process. The realizations must be integer-coded. The maximum number of different integer codes is specified in **anneal.inc**. The input file has one integer per line cycling fastest on x, then y, then z, then by realization.

- **trainfl:** a file with the training image. The training image *must* have the same integer code values as the realizations to post-process and cover the same spatial extent.

- **outfl:** the output file that will contain the post-processed realizations.

- **histfl:** the control and realization two-point histogram are written to this file before the post processing starts and after the post processing is finished.

- **idbg** and **report:** an integer debugging level between 0 and 3, and a reporting interval. After a fixed number of iterations (*report*) the realization is written to the output file.

- **dbgfl:** file for the debugging information.

- **seed:** random number seed (a large odd integer).

- **maxit** and **tol:** the maximum number of iterations over the entire grid network. Typically 99% of the improvement is reached in 5 iterations and the remaining 1% may be reached by 10 iterations. More than 10 iterations is not recommended. The initial objective function is normalized to one; when it is lowered to *tol* the simulation is stopped.

- **nsim:** the number of initial realizations to post process.

- **nx, ny,** and **nz:** definition of the grid system.

- **ndir:** the number of directions for the two-point histogram control. For each direction the following parameters are needed:

  ixl: the offset in the x direction for a unit lag

  iyl: the offset in the y direction for a unit lag

  izl: the offset in the z direction for a unit lag

  nlag: the number of lags

  icross: an integer flag specifying whether or not to consider classes of the two-point histogram involving two different classes ($k \neq k'$) (1=yes, 0=no).

V.10 Application Notes

- Marsaglia's random number generator [112] is implemented in all simulation programs except the annealing programs **sasim** and **anneal** which call for the ACORN random number generator [164].

- The sequential simulation programs in GSLIB use the random path specified by a linear congruential generator of the form [62]:

$$r_i = [5 \cdot r_{i-1} + 1] \, mod(2^n)$$

where r_i is the random index for node i and n is chosen large enough so that 2^n is greater than the number of grid nodes N. Note that each simulation follows the same random path, but starts from a different initial position r_1.

- All output files are currently written in simple ASCII format to facilitate machine independence and transfer of output to post-processing files (gray/color scale plotting, statistics, etc.). It is strongly advised that each installation consider some file compression scheme. More advanced approaches would consider a consistent packed format that would allow access by other programs without decompressing the entire file. A straightforward approach would be to apply a compression algorithm once the simulation is complete. With fast computers there is no need to store any realizations. The realizations are completely specified by the data, the algorithm, and the random number seed. Any time a realization is needed for display or input to a transfer function it could be re-created dynamically with the specified data and algorithm.

- Following the previous point, it may be an advantage to pipe certain realizations automatically through the **gam3** subroutine to check variogram reproduction. Similarly, one could automatically average the realizations (with, possibly, non-linear averaging), compute gray/color scale maps, etc.

- The procedure of drawing from a cdf constructed with indicator kriging (as implemented in **sisim**) involves simply generating a uniform random number between 0 and 1 and retaining the corresponding conditional quantile as the simulated value. In fact, the entire conditional distribution is not needed; one only needs to know the details near the drawn cdf value, i.e., the two bounding cutoffs or the tail to consider. Therefore, an efficient algorithm would draw the random number first and then focus on the z-interval of interest thereby reducing the number of kriging systems that must be solved to $log_2(ncut)$. This implementation has not been distributed with GSLIB since, theoretically, order relation problems must be corrected prior to drawing from the distribution; correcting order relations requires knowledge of the complete distribution.

- The statistics (cdf, variogram, etc.) of each realization will fluctuate around the model statistics; see section V.1.5. This fluctuation will be especially pronounced if the domain being simulated has either a small number of nodes or is small with respect to the range of correlation. Moreover, indicator methods tend to present more fluctuations than Gaussian methods. Depending on the goal of the study, this fluctuation may be reduced by resetting the univariate distribution to the reference cdf (see section V.2.1 and program **trans** in section VI.2.4) or by selection of realizations.

  Program **trans** is not suitable for categorical variables since there is no intrinsic ordering for integer-coded categories; applying **trans** would add *noise* to the realization by arbitrarily ranking the values in each category. In addition to selection of realizations, there are two alternatives to reduce fluctuations from the model or target statistics of categorical realizations:

1. In the sequential indicator simulation program **sisimpdf** the target proportion of each category is input to the program and used when simulating with the simple kriging option (*ktype=0*). These target proportions could be dynamically updated after the simulation of each grid node. For example, after assigning a location as code "1" the global proportion of code "1" used in subsequent simple kriging systems would be decremented. In this way, the final realizations will present less ergodic fluctuations from the target proportions.

2. The realizations which present unacceptable fluctuations could be post-processed with annealing so that both the proportions and the spatial structure are honored more closely. A starting point for such post-processing would be the **anneal** program.

- The question of which kriging method (OK or SK) to apply in the sequential Gaussian and sequential indicator methods is often asked. Theoretically and practically, simple kriging yields the most correct and the best realizations, in the sense of cdf and variogram reproduction. OK within moving data neighborhoods might be considered when there is a large number of *original* conditioning data and the local mean data value is seen to vary significantly with location [74,97].

- Virtually all of the simulation programs work with a limited data search neighborhood. The spatial structure beyond the limited neighborhood or search radius is uncontrolled and therefore random. In practice, there may be long range spatial structures that are important. The solution to this problem is to use the so-called *multiple grid concept* whereby a coarse grid is simulated first [61]. That coarse grid is then used as conditioning for a second, finer, grid simulation. This grid refinement can be performed multiple times leading to a better reproduction of the long range variogram structure. The current programs could perform this by multiple executions of the program. Routine application of the multiple grid concept, however, would require modifying the source code.

The multiple-grid concept is valid for both the sequential simulation programs and the simulated annealing program **sasim**. The limited number of lags entering the objective function of **sasim** implies that the variogram will be reproduced for a limited distance.

- The number of fixed cutoffs used by the indicator simulation algorithms must not be too great because of order relation problems. A common rule of thumb is that the number of data used in any indicator kriging should be three to five times the number of fixed cutoffs. This constraint makes it more likely that there are data within each class; when there are no data within a class order relation problems are more likely. On the other hand, when too many data are used negative kriging weights due to screening become significant, again causing order relations problems.

The median IK model (IV.30) allows a large number of cutoffs to be considered without incurring the related order relation problems.

- Indicator kriging does not provide resolution beyond the cutoff values retained for kriging. The additional class resolution is obtained through somewhat subjective decisions made through, e.g., parameters *ltail, middle,* and *utail* in sisim. An alternative, equally subjective, approach consists of replacing the hard indicator data seen as step functions of the cutoff z by continuous kernel functions of z; see [104,141].

- There are two sequential indicator simulation programs for categorical variables: sisimpdf and ipcsim. The sisimpdf program is the most straightforward since it calls for the indicator variogram of each category. An appreciation of the geometrical arrangement of each category may be used to help establish appropriate variogram models. The ipcsim program considers indicator principal components (IPCs) of the categorical variables; each IPC is a linear combination of the original indicator variables and, thus, simultaneously represents a number of categories. The advantage of the IPC approach is that it requires less CPU-time. A drawback of the IPC approach, however, is that the anisotropy of specific categories is smeared over the various IPCs resulting in a smearing of that anisotropy in the output realizations. Consider the reference image shown at the top of Figure V.21 which presents three categories (white background, circular gray regions, and black anisotropic regions). Direct indicator variograms and IPC variograms were computed from the exhaustive reference image, modeled, and used with sisimpdf and ipcsim to generate the four realizations. The sisimpdf realizations, shown at the left of Figure V.21, isolate the anisotropy of the black and gray regions much better than the ipcsim realizations shown to the right. It took 33% more CPU time to generate the sisimpdf realizations.

If the shapes are as well defined as circles or rectangles, one should consider a Boolean approach for their simulation; see section V.4 and program ellipsim.

- A useful diagnostic plot when working with the annealing simulation program sasim is the objective function versus the number of swaps; see Figure V.22. The vertical drops occur when the temperature is decreased. The magnitude of the drops is a function of the multiplicative factor λ; see Section V.5. The frequency of the drops is essentially a function of the parameter K; the parameter K_A does not play a role

Reference Image

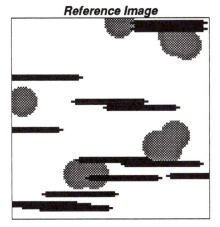

SISIMPDF Realization One

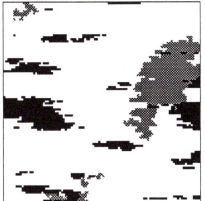

IPCSIM Realization One

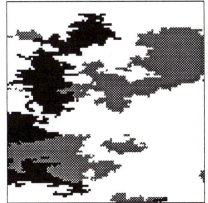

SISIMPDF Realization Two

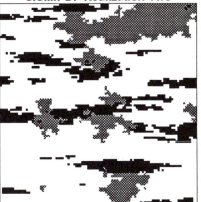

IPCSIM Realization Two

Figure V.21: A reference image, two realizations generated with `sisimpdf`, and two realizations generated with `ipcsim`. No conditioning data was retained.

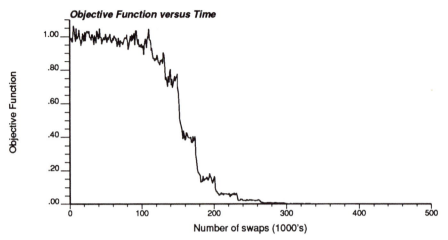

Figure V.22: An example plot of the objective function versus the number of swaps. The sudden vertical drops are when the temperature is lowered.

in the early part of the annealing procedure. The parameter K can be reduced if there is a long plateau before the objective function starts a fast decrease (as in Figure V.22). If the objective function does not decrease to near zero then K and λ should be increased.

V.11 Problem Set Six: Simulation

The goals of this problem set are to experiment with stochastic simulation techniques, to judge the important parameters, and to gain an appreciation of the applicability of the various algorithms. Most runs are based on the data sets presented in section II.5.

Part A: Gaussian Simulation

Consider the normal scores variogram computed in problem set 2 or the following isotropic semivariogram model:

$$\gamma(\mathbf{h}) = 0.30 + 0.70 Sph(\frac{|\mathbf{h}|}{10.0})$$

Use the 140 clustered data in cluster.dat and this variogram model to answer the following questions:

1. Describe briefly the sequential Gaussian simulation algorithm and the corresponding **sgsim** code with the aid of a simple flowchart.

2. Simulate point $x = 41, y = 29$ with a high debugging level and check that the solution is as expected. Use the data file parta.dat, defined

in problem set 3, for the conditioning data and set up a parameter file from the template provided in `sgsim.par`. Comment on the results.

3. Create two conditional simulations using the 140 data with `sgsim` (use declustering weights and **nx**=50, **xmn**=0.5, **xsiz**=1.0, **ny**=50, **ymn**=0.5, **ysiz**=1.0, **nz**=1, **zmn**=0.5, **zsiz**=1.0). Use your documented judgment to choose the other parameters. Generate and plot the following graphics to aid your interpretation:

   - Gray scale maps of the simulations. Use `gscale`.
   - A histogram of the simulated values. Use `histplt`.
   - A Q-Q plot of the declustered data distribution and the distribution of simulated points. Use `qpplt`.
   - Run the normal scores variogram of the simulated values in one direction. Use `gam2` and `vmodel` to compute the experimental variogram and the model respectively. The program `nscore` may be used to retrieve the normal scores from the already back transformed realization.

   Comment on the results.

4. Repeat question 3 without using declustering weights. Comment on the results.

5. Build your own semivariogram model, alter the number of conditioning data, alter the input parameters significantly and run `sgsim` to observe the consequences of your change. Comment on the sensitivity.

6. One common use of the LU approach is to build local block distributions. Discretize each block by a number of points (say 4 by 4 = 16 nodes), simulate these points simultaneously using the `lusim` program (with the surrounding data values as conditioning points), and obtain a simulated block value by averaging the simulated point values. This is repeated, say, 100 times to obtain 100 simulated block values. Considering the same block as in the block kriging example (see Figure IV.14 in problem set 3) and the `lusim` program, build the block probability distribution.

Part B: Indicator Simulation

The goal of this assignment is to gain an appreciation for sequential indicator simulation by experimenting with the same 140 data.

The three quartile cutoffs (z_1=0.34, z_2=1.20, z_3=2.75) of the declustered histogram of the 140 sample data are considered (see Figure A.5). The corresponding isotropic and standardized indicator semivariogram models:[21]

$$\gamma_I(\mathbf{h}; z_1) = 0.30 + 0.70Sph(|\mathbf{h}|/9.0)$$

---

[21] modeled from the 2500 exhaustive data values.

$$\gamma_I(\mathbf{h}; z_2) = 0.25 + 0.75 Sph(|\mathbf{h}|/10.0)$$
$$\gamma_I(\mathbf{h}; z_3) = 0.35 + 0.65 Sph(|\mathbf{h}|/8.0)$$

When performing a simulation study it is good practice to select one or two representative locations and have a close look at the simulation procedure, i.e., the kriging matrices, the kriging weights, the conditional distribution, and the drawing from the distribution. It is also a good idea to evaluate the sensitivity of the realizations to critical assumptions such as the models used for extrapolation in the tails. In addition, once enough realizations have been generated, one should also check for a reasonable (not exact) reproduction of the univariate distribution and indicator variograms; see the related discussion on ergodicity in section V.1.5.

1. Briefly describe the sequential indicator simulation algorithm and the corresponding **sisim** code with the aid of a simple flowchart.

2. The original 140 clustered data with an enlargement of the area around the point $x = 41$ and $y = 29$ are shown on Figure IV.11. Simulate point $x = 41, y = 29$ with a high debugging level and check that the solution is as expected. Use the data file $\boxed{\text{parta.dat}}$ for the conditioning data and set up a parameter file from the template provided in **simlib**. Comment on the results.

3. Create two conditional simulations using the 140 data with **sisim** (use **nx** = 50, **xmn** = 0.5, **xsiz** = 1.0, **ny** = 50, **ymn** = 0.5, **ysiz** = 1.0, **nz** = 1, **zmn** = 0.5, **zsiz** = 1.0). Use the data in $\boxed{\text{cluster.dat}}$ for the distribution within all four classes and the cutoffs and semivariogram models specified earlier; use your own documented judgement to choose all other parameters. Generate and plot the following graphics to aid your interpretation:

   - Gray scale maps of the simulations. Use **gscale**.
   - A histogram of the simulated values. Use **histplt**.
   - A Q-Q plot of the declustered 140 data distribution and the distribution of simulated points. Use **qpplt**.
   - The indicator variograms of simulated values in one direction. Use **gam2** and **vmodel** to compute the experimental variogram and the model respectively.

   Comment on the results.

4. Repeat question 3 using linear interpolation in the lower tail and a hyperbolic model in the upper tail. Comment on the results and compare to question 3.

5. The indicator simulation algorithm is extremely flexible in that it can handle different anisotropy at each cutoff. In this question simply expand the EW range to 90 (10:1 anisotropy) for the first cutoff and expand the NS range to 80 (10:1 anisotropy) for the last cutoff. Create two realizations, plot the gray scale maps, and comment.

6. Build your own variogram model(s), increase the number of cutoffs, alter the number of conditioning data, possibly including soft data such as constraint intervals, then run **sisim** and comment on the consequences of your changes.

   sisim allows a considerable amount of flexibility and the program will go ahead and attempt to reproduce whatever spatial structure is specified. Comment on what you consider practical limitations. For example, it may not be possible to impose a strong anisotropy and have the direction of continuity switch by 90 degrees at alternating cutoffs.

Part C: Markov-Bayes Indicator Simulation

The goal of this assignment is to gain an appreciation for situations when the Markov-Bayes extension to the sequential indicator simulation algorithm is warranted.

The secondary data in $\boxed{\text{ydata.dat}}$ are needed for the following questions:

1. Briefly describe the alterations to the **sisim** code that would be required to perform the Markov-Bayes variant of sequential indicator simulation?

2. Construct the pre-posterior distributions and the calibration parameters $B(z)$ values needed for **mbsim** using the 140 primary and secondary data in $\boxed{\text{cluster.dat}}$ and program **mbcalib**.

3. Run **mbsim** to create two alternate realizations. Compare them to realizations previously obtained in part B.

4. Everything else being equal, reduce drastically the number of hard primary data. Run **mbsim** again and compare the realizations to the ones obtained in question 3. Comment on the impact of the soft data.

Part D: Indicator Principal Components Simulation

The goal of this assignment is to gain an appreciation for situations when the indicator principal components (IPC) extension to the sequential indicator simulation algorithm is warranted.

The data in $\boxed{\text{cluster.dat}}$ will be used for the following problems. Note that in general, **ipcsim** is applied to truly categorical variables and not ones created arbitrarily from a continuous variable.

1. Briefly describe the alterations to the **sisim** code that would be required to perform the IPC variant of sequential indicator simulation.

2. Code the 140 primary data in ⎡cluster.dat⎤ as integer values of 1,2,3, or 4 depending on the quartile classes to which the data values belong.

3. Using the indicator covariance matrix at lag $\mathbf{h} = 0$ (default value assumed by `ipcprep`), construct the orthogonal transformation matrix and the 140 principal components data for the integer-coded values.

4. Compute and model the omnidirectional variograms of the indicator principal components. Comment on the results. How do they compare to the conventional indicator variograms for the quartile cutoffs whose models were given in the previous part B?

5. Apply the program `ipcsim` to the simulation of the integer-coded variable. Comment on the results.

Part E: Categorical Variable Simulation

The goal of this assignment is to experiment with the Boolean and simulated annealing simulation programs. The goal is to generate 2-D realizations of a binary variable that represents presence or absence of, say, shale. The shale bodies are approximated by ellipses elongated in the 90° direction with a 5:1 aspect ratio. All the shales are about 20 feet long in their major axis.

1. Use the Boolean simulation program `ellipsim` to simulate a 2-D field with 25% elliptical shales. Consider a 100 by 100 foot field discretized into unit square feet.

   The variogram for circles (or ellipses after a geometric transformation) positioned at random in a matrix can be calculated analytically [38, 113,142]. This semivariogram model, also called the *circular bombing model variogram*, is written:

   $$\gamma(\mathbf{h}) = p\left[1 - p^{Circ(\mathbf{h})}\right]$$

   where, $\mathbf{h}$ is the separation vector of modulus $|\mathbf{h}|$, a is the diameter of the circles, p is the fraction outside the circles, $(1 - p)$ is the fraction within the circles, $p(1 - p)$ is the variance of the indicator $I(\mathbf{u})$, and $Circ(\mathbf{h})$ is the circular semivariogram model ([106], p. 116):

   $$Circ(\mathbf{h}) = \begin{cases} \frac{2}{\pi a^2}\left[|h|\sqrt{a^2 - |h|^2} + a^2 sin^{-1}\left(\frac{|h|}{a}\right)\right], & |h| \leq a \\ 1, & |h| \geq a \end{cases}$$

   where a is the diameter of the circles. This indicator variogram model $\gamma(\mathbf{h})$ has the familiar $p(1 - p)$ variance form with the exponent corresponding to the hyperspherical variogram model for the dimension of the space being considered. In 1-D the exponent becomes the linear variogram up to the range a; in 2-D the exponent is the circular variogram model; and in 3-D the exponent becomes the spherical variogram model.

This variogram model can be written with the familiar geometric anisotropy if the circles become ellipses. Furthermore, if there are multiple ellipse sizes then the overall variogram can be written as a nested sum of elementary bombing model variograms.

2. Use that theoretical variogram (or consider an approximate model directly inferred from one of the Boolean realizations) and **sisimpdf** to generate two realizations.

3. Post-process these latter two realizations so that they share more of the features of the elliptical training image. Use the two-point histogram for 5 lags in the two orthogonal directions aligned with the sides of the training image.

Part F: Simulated Annealing Simulation

The goal of this part is to experiment with the simulated annealing simulation program **sasim**. Consider the following normal scores semivariogram model:

$$\gamma(\mathbf{h}) = 0.10 + 0.90 Sph(\frac{|\mathbf{h}|}{10.0})$$

1. Create two standard normal unconditional simulations. Consider 24 lags for conditioning and use your documented judgment for the other parameters.

   Generate and plot the following graphics to aid your interpretation:

   - Gray scale maps of the simulations. Use **gscale**.
   - A histogram of the simulated values. Use **histplt**.
   - Run the variogram of the simulated values in two directions. Use **gam2** and **vmodel** to compute the experimental semivariogram and the model respectively.

   Comment on the results.

2. Try conditioning to lesser (say 8) and greater (say 100) variogram lags and note the change in the realizations and the impact on CPU time.

3. Experiment with conditional simulation using the weighted non-parametric distribution corresponding to cluster.dat . Consider a one then a two-part objective function and comment on the results.

Chapter VI

Other Useful Programs

This chapter presents utility programs that depart from the mainstream geo-statistical algorithms presented in Chapters III through V, yet they are useful in producing the graphs and figures shown in the text. Many of the utilities documented in this chapter are available in statistical or numerical analysis software packages and textbooks.

Section VI.1 presents a set of programs that compute classical statistics and generate PostScript graphical displays. Facilities are available for histograms, normal probability plots, Q-Q/P-P plots, scatterplots, semivariogram plots, gray scale maps, and color scale maps.

Section VI.2 presents utility programs for spatial declustering, univariate data transformation, solving linear systems of equations, preparation for Markov-Bayes simulation, preparation for indicator principal components kriging (IPCK), and post processing of IK and simulation results.

VI.1 PostScript Display

The utility programs in this section generate graphics in the PostScript page description language [3,4,103]. The PostScript output from all of the programs are elementary plot units that are approximately 3 inches high by 4 inches wide. These elementary plot units require a header and footer before they can be printed on a PostScript laser printer or displayed on a PostScript previewer. The specific header and footer depend on how many elementary plot files are to be displayed on a single 8.5 by 11 inch sheet of paper.[1]

Figure VI.1 shows the possible arrangements of elementary plot units on a page. The plot units displayed on one page can be created from any of the utility programs.

The PostScript commands for the necessary header and footer are contained in template files named according to Figure VI.1, e.g., the template

---

[1] The plotting utilities work equally well with A4 paper.

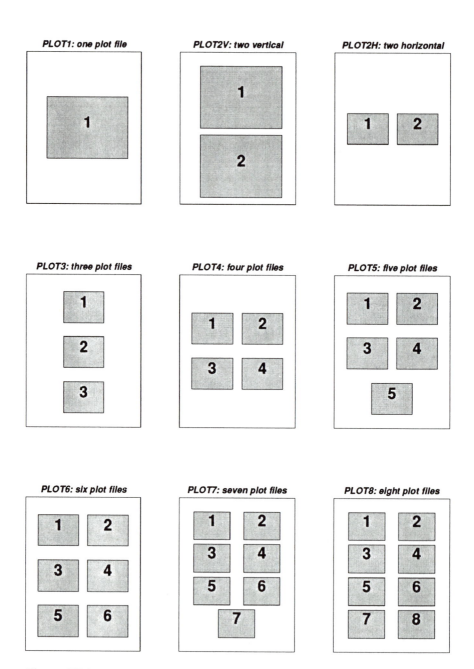

Figure VI.1: Arrangement of the elementary plot files (shaded areas) on 8.5 by 11 inch paper achieved by the scripts PLOT1 through PLOT8.

file for PLOT3 is named $\boxed{\text{plot3.tem}}$ in the **bin** directory on the diskette. An editor may be used to read the elementary plot files into the template file. The location for each file is noted by comments like "PLOT ONE STARTS HERE:", etc.

Special C-shell script commands have been provided for machines running under the UNIX operating system. The script commands are also named according to Figure VI.1, e.g., the command for PLOT3 is named $\boxed{\text{plot3}}$ in the **bin** directory on the diskette. The names of the elementary plot files are specified as command line arguments and finally the output file, e.g.,

plot3 histplt.out probplt.out scatplt.out plot3.ps

The three input files are $\boxed{\text{histplt.out}}$, $\boxed{\text{probplt.out}}$, and $\boxed{\text{scatplt.out}}$; the output file is $\boxed{\text{temp.ps}}$.

The scripts check that enough file names have been specified. e.g., PLOT3 must have four arguments - three input plot files and the output plot file name.

The motivation behind these PostScript display programs is to give initial users a way of getting quick graphical and statistical summaries of their data or results of other GSLIB subroutines. Inevitably, users will have to learn a little of the PostScript page description language if they want to customize the program output. We strongly recommend using familiar in-house or commercial statistics and graphics packages. These utilities have been included because they are useful to prepare report quality figures.

VI.1.1 Histograms and Statistics

The following files should be located on the diskette(s) before attempting to compile and execute this program/subroutine:

$\boxed{\text{histplt.f}}$ the program **histplt**

$\boxed{\text{histplt.par}}$ an example parameter file for **histplt**

histplt has been custom written to generate some univariate statistical summaries and a visual output that is compatible with a PostScript display device. The input data is a variable from a simplified Geo-EAS input file where the variable in another column can act as a weight (declustering weight, specific gravity, thickness for 2-D data,...). These weights do not need to add up to one. Minimum and maximum trimming limits can be set to remove missing values and outliers. The program will automatically scale the histogram. The user can choose to set the minimum and maximum histogram limits, number of classes, and whether or not to use a logarithmic scale.

The parameters required by the program **histplt** are listed below and shown in Figure VI.2:

- **datafl:** the data file in a simplified Geo-EAS format.

```
                  Parameters for HISTPLT
                  ***********************

START OF PARAMETERS:
cluster.dat                          \data file
3    0                               \column for variable and weight
histplt.out                          \output PostScript file
-0.99    999999.                     \trimming limits
0.0      20.0                        \histogram minimum and maximum
40                                   \number of classes
0                                    \1=log scale, 0=arithmetic
Clustered Data                       \title
```

Figure VI.2: An example parameter file for **histplt**.

- **ivr** and **iwt:** column number for the variable and the weight. If $iwt \le 0$ equal weighting is considered.

- **outfl:** the output elementary PostScript plot file. Note that this file must be inserted into a template ".tem" file or processed with a **plot** script before printing or displaying on a PostScript device.

- **tmin** and **tmax:** minimum and maximum trimming limits to remove missing values and/or erratic high values (all z-values within the interval $tmin \le z < tmax$ are accepted).

- **hmin** and **hmax:** minimum and maximum limits for the variable axis (abscissa). These do not have to be compatible with $tmin$ and $tmax$. Values that are less than $hmin$ or greater than $hmax$ will appear just beyond either end of the histogram plot. They contribute to the statistics calculated. Setting $hmin \ge hmax$ will cause the program to set the limits to the minimum and maximum data values encountered.

- **ncl:** the number of classes.

- **ilog:** if $ilog$ is set to 1 then a logarithmic scale is used (useful for looking at details of the low values); otherwise, an arithmetic scale is used.

- **title:** a 40-character title to appear at the upper left corner of the plot.

VI.1.2 Normal Probability Plots

The following files should be located on the diskette(s) before attempting to compile and execute this program/subroutine:

| probplt.f | the program **probplt**

| probplt.par | an example parameter file for **probplt**

```
                    Parameters for PROBPLT
                    ***********************
START OF PARAMETERS:
cluster.dat                       \data file
3    5                            \column for vr and wt
0.0    1.0e21                     \trimming limits
probplt.out                       \output PostScript file
1                                 \0=arith, 1=log
0.0   30.0    5.0                 \min,max,inc: arith
0.01    4                         \start,ncyc : log
Clustered Data                    \title
```

Figure VI.3: An example parameter file for probplt.

probplt generates either a normal or a lognormal probability plot. This plot displays all the data values on a chart that illustrates the general distribution shape and the behavior of the extreme values. Checking for normality or lognormality is a secondary issue. The input data is a variable from a simplified Geo-EAS input file where the variable in another column can act as a weight. The user must specify the plotting limits - only those data within the limits are shown.

The parameters required by the program probplt are listed below and shown in Figure VI.3:

- **datafl:** the data file in a simplified Geo-EAS format.

- **icolvr** and **icolwt:** column number for the variable and the weight. If $icolwt \leq 0$ equal weighting is considered.

- **tmin** and **tmax:** minimum and maximum trimming limits to remove missing values and/or erratic high values (all z-values within the interval $tmin \leq z < tmax$ are accepted).

- **outfl:** The output elementary PostScript plot file. Note that this file must be inserted into a template ".tem" file or processed with a **plot** script before printing or displaying on a PostScript device.

- **scale:** if set to 1 a logarithmic scale of base 10 (lognormal probability plot) is considered; otherwise, an arithmetic scale (normal probability plot) is considered.

- **pmin, pmax** and **pinc:** minimum, maximum, and increment for the variable axis. Only needed if *scale* not equal to 1.

- **start** and **ncyc:** starting value and number of base 10 logarithmic cycles if a lognormal probability plot is being constructed (*scale* = 1).

- **title:** a 40-character title to appear at the upper left corner of the plot.

```
                    Parameters for QPPLT
                    ********************

START OF PARAMETERS:
cluster.dat                          \first data file
3    0                               \column for vr and wt
true.dat                             \second data file
1    0                               \column for vr and wt
0.0        1.0e21                    \trimming limits
qpplt.out                            \output PostScript file
0                                    \0=Q-Q plot, 1=P-P plot
0.0      20.0                        \xmin  and  xmax
0.0      20.0                        \ymin  and  ymax
Clustered data                       \X label
Reference data                       \Y label
Q-Q Plot                             \title
```

Figure VI.4: An example parameter file for qpplt.

VI.1.3   Q-Q and P-P plots

The following files should be located on the diskette(s) before attempting to compile and execute this program/subroutine:

| qpplt.f | the program qpplt

| qpplt.par | an example parameter file for qpplt

qpplt takes univariate data from two data files and generates either a Q-Q or a P-P plot. Q-Q plots compare the quantiles of two data distributions, and P-P plots compare the cumulative probabilities of two data distributions. These graphical displays are used to compare two different data distributions, e.g., compare an original data distribution to a distribution of simulated points.

The parameters required by the program qpplt are listed below and shown in Figure VI.4:

- **datafl1:** The data file for the first variable (x axis) in a simplified Geo-EAS format.

- **icolvr** and **icolwt:** column number for the variable and the weight. If $icolwt \leq 0$ equal weighting is considered.

- **datafl2:** The data file for the second variable (y axis) in a simplified Geo-EAS format. Note: It is possible that $datafl2$ is the same as $datafl1$.

- **ivr** and **iwt:** column number for the variable and the weight. If $iwt \leq 0$ equal weighting is considered.

- **tmin** and **tmax:** minimum and maximum trimming limits to remove missing values and/or erratic high values (all z-values within the interval $tmin \leq z < tmax$ are accepted).

```
                    Parameters for SCATPLT
                    ***********************
START OF PARAMETERS:
cluster.dat                        \data file
3    4    5                        \column for X, Y, and weight
scatplt.out                        \output Postscript file
0.0      1.0e21                    \min and max variable X
0.0      1.0e21                    \min and max variable Y
0.0      20.0                      \xmin  and   xmax
0.0      20.0                      \ymin  and   ymax
Secondary versus Primary           \title
```

Figure VI.5: An example parameter file for **scatplt**.

- **outfl:** The output elementary PostScript plot file. Note that this file must be inserted into a template ".tem" file or processed with a **plot** script before printing or displaying on a PostScript display device.

- **qqorpp:** The flag to tell the program whether to construct a Q-Q plot ($qqorpp = 0$) or a P-P plot ($qqorpp = 1$).

- **xmin** and **xmax:** minimum and maximum limits for x axis.

- **ymin** and **ymax:** minimum and maximum limits for y axis.

- **xlab:** a 20-character label for the x axis (from $datafl1$).

- **ylab:** a 20-character label for the y axis (from $datafl2$).

- **title:** a 40-character title to appear at the upper left corner of the plot.

VI.1.4 Bivariate Scatterplots and Analysis

The following files should be located on the diskette(s) before attempting to compile and execute this program/subroutine:

| scatplt.f | the program **scatplt**

| scatplt.par | an example parameter file for **scatplt**

A scatterplot is the simplest yet often most informative way to compare pairs of values. **scatplt** displays bivariate scatterplots and some statistical summaries for PostScript display. The summary statistics are weighted, except for the rank order (Spearman) correlation coefficient. The parameters required by the program **scatplt** are listed below and are shown in Figure VI.5:

- **datafl:** The data file in a simplified Geo-EAS format.

- **icolx, icoly,** and **icolvr:** The columns for the x variable (horizontal axis), y variable (vertical axis), and weight (if less than or equal to zero then equal weighting is used).

- **outfl:** The output elementary PostScript plot file. Note that this file must be inserted into a template ".tem" file or processed with a **plot** script before printing or displaying on a PostScript

- **tminx** and **tmaxx:** minimum and maximum trimming limits on the x variable.

- **tminy** and **tmaxy:** minimum and maximum trimming limits on the y variable.

- **xmin** and **xmax:** minimum and maximum limits for the x axis. These do not have to be compatible with the trimming limits. Pairs within the trimming limits, but outside the plotting limits, are still used for computing the correlation coefficients. Setting $xmin \geq xmax$ will cause the program to use the minimum and maximum data values encountered.

- **ymin** and **ymax:** minimum and maximum limits for the y axis. Setting $ymin \geq ymax$ will cause the program to use the minimum and maximum data values encountered.

- **title:** a 40-character title to appear at the upper left corner of the plot.

VI.1.5　Variogram Plotting

The following file should be located on the diskette(s) before attempting to compile and execute this program/subroutine:

| vargplt.f | the program **vargplt**.

The program **vargplt** takes the special output format used by the variogram programs and creates graphical displays for PostScript display devices. This program is a straightforward display program and does not provide any facility for variogram calculation or model fitting.

This program is the only program in GSLIB that works interactively. A file name will be requested that contains the variogram results (from **gam2**, **gam3**, **gamv2**, **gamv3**, **sasim**, **bigaus**, or **vmodel**). The scaling of the distance and the variogram axes may be set automatically or manually. Multiple variograms may be displayed on a single plot or a new elementary PostScript plot file may be started at any time.

VI.1.6　Gray Scale Maps

The following files should be located on the diskette(s) before attempting to compile and execute this program/subroutine:

```
                    Parameters for GSCALE
                    *********************
START OF PARAMETERS:
true.dat                              \data file
gscale.out                            \output PostScript file
1                                     \column number
1                                     \format: 1=xyz, 2=200i1
50      1.0                           \nx, xsiz
50      1.0                           \ny, ysiz
1       1.0                           \nz, zsiz
1       1                             \igrid,iz
0.0     30.0                          \gmin,gmax
0                                     \legend bar: 0=yes, 1=no
2-D Data                              \title
```

Figure VI.6: An example parameter file for **gscale**.

| gscale.f | the program **gscale**

| gscale.par | an example parameter file for **gscale**

Gray scale maps are convenient to show 2-D spatial images. A horizontal 2-D slice available as output from one of the 3-D kriging or simulation programs can be plotted with **gscale**. The parameters required by the program **gscale** are listed below and are shown in Figure VI.6:

- **datafl:** The data file (in a simplified Geo-EAS format) containing the 2-D or 3-D grids with x cycling fastest, then y, then z, then per simulation.

- **outfl:** The output elementary PostScript plot file. Note that this file must be inserted into a template ".tem" file or processed with a **plot** script before printing or displaying on a PostScript display device.

- **icol:** column number in the input file to consider. This is typically one (1).

- **ifmt:** the format. Typically, this is set to 1 which means that the output has one value per line cycling fastest on x, then y, then z, then simulation by simulation. A more concise "(200i1)" format is allowed if *ifmt=2*.

- **nx** and **xsiz:** the number of nodes in the x direction and the grid node separation.

- **ny** and **ysiz:** the number of nodes in the y direction and the grid node separation.

- **nz** and **zsiz:** the number of nodes in the z direction and the grid node separation.

```
                    Parameters for CSCALE
                    *********************

START OF PARAMETERS:
true.dat                              \data file
cscale.out                            \output PostScript file
1                                     \column number
50      1.0                           \nx, xsiz
50      1.0                           \ny, ysiz
1       1.0                           \nz, zsiz
1       1                             \igrid,iz
0.0     12.0                          \cmin,cmax
2-D Reference Data                    \title
East                                  \X label
North                                 \Y label
```

Figure VI.7: An example parameter file for `cscale`.

- **igrid** and **iz:** the grid number and level number to be considered. The grid number corresponds to the realization number if a file contains multiple simulated realizations. The iz'th z level will be displayed.

- **gmin** and **gmax:** minimum and maximum limits for the gray scale. The minimum is associated to white, the maximum to black, with a linear scaling in between.

- **legend:** if set to 0 then a gray scale legend bar will appear to the right of the gray scale image, if $legend = 1$ then no legend will appear.

- **title:** a 40-character title that will appear at the upper center of the plot.

VI.1.7   Color Scale Maps

The following files should be located on the diskette(s) before attempting to compile and execute this program/subroutine:

cscale.f   the program `cscale`

cscale.par   an example parameter file for `cscale`

`cscale` plots a 2-D $x - y$ slice through gridded data. The parameters are listed below and shown in Figure VI.7:

- **datafl:** The data file containing the 2-D/3-D grids with x cycling fastest, then y, then z, then per simulation.

- **outfl:** The output elementary PostScript plot file. Note that this file must be inserted into a template ".tem" file or processed with a `plot` script before printing or displaying on a PostScript display device.

- **icol:** column number in the input file to consider. This is typically one (1).

- **nx** and **xsiz:** the number of nodes in the x direction and the grid node separation.

- **ny** and **ysiz:** the number of nodes in the y direction and the grid node separation.

- **nz** and **zsiz:** the number of nodes in the z direction and the grid node separation.

- **igrid** and **iz:** the grid number and level number to be considered. The grid number corresponds to the realization number if a file contains multiple simulated realizations. The iz'th z level will be displayed.

- **cmin** and **cmax:** minimum and maximum limits for the color scale. The color is scaled through the visual light spectrum (*cmin* corresponds to dark (violet) and *cmax* corresponds to light (red)).

- **title:** a 40-character title that will appear at the upper center of the plot.

- **xlab:** a 40-character label that will appear centered below the plot.

- **ylab:** a 40-character label that will appear vertically centered to the left of the plot.

VI.2 Utility Programs

This section presents some useful programs that complement the mainstream programs presented earlier. Once again, these are intended to be starting points for initial users.

VI.2.1 Cell Declustering

The following files should be located on the diskette(s) before attempting to compile and execute this program/subroutine:

| declus.f | the program `declus`

| declus.par | an example parameter file for `declus`

Data are often spatially clustered; yet, we may need to have a histogram that is representative of the entire area of interest. To obtain a representative distribution, one approach is to assign declustering weights whereby values in areas/cells with more data receive less weight than those in sparsely sampled areas. The program `declus` [37] provides an algorithm for determining 3-D

```
          Parameters for DECLUS
          *********************

START OF PARAMETERS:
cluster.dat                    \data file
1    2    0    3               \ix,iy,iz,ivr
0.0       1.0e21               \tmin,tmax
declus.sum                     \output with summary
declus.out                     \output with data & wts
1.0   1.0                      \anisotropy: y,z
0                              \0=look for min, 1 max
5  1.0  10.0                   \num, min and max size
5                              \num of origin offsets
```

Figure VI.8: An example parameter file for `declus`.

declustering weights in cases where the clusters are known to be clustered preferentially in either high or low valued-areas. In other cases, polygon-type declustering weights might be considered whereby the weight is made proportional to the sample area (polygon) of influence.

The declustering weights output by `declus` are standardized so that the sum to the number of data. A weight greater than 1.0 implies that the sample is being overweighted and a weight lesser than 1.0 implies that the sample is being downweighted (it is clustered with other samples).

For a more detailed discussion on declustering algorithms; see [37,76,88]. The parameters required by `declus` are listed below and are shown in Figure VI.8:

- **datafl:** a data file in simplified Geo-EAS format containing the variable to be declustered.

- **icolx, icoly, icolz,** and **icolvr:** the column numbers for the x, y, z coordinates and the variable.

- **tmin** and **tmax:** minimum and maximum trimming limits to remove missing values and/or erratic high values (all z-values within the interval $tmin \leq z < tmax$ are accepted).

- **sumfl:** the output file containing all the attempted cell sizes and the associated declustered mean values.

- **outfl:** The output elementary PostScript plot file. Note that this file must be inserted into a template ".tem" file or processed with a **plot** script before printing or displaying using PostScript display device.

- **anisy** and **anisz:** the anisotropy factors to consider rectangular cells. The cell size in the x direction is multiplied by these factors to get the cell size in the y and z directions, e.g., if a cell size of 10 is being considered and $anisy = 2$ and $anisz = 3$ then the cell size in the y direction is 20 and the cell size in the z direction is 30.

- **minmax:** an integer flag that specifies whether a minimum mean value ($minmax = 0$) or maximum mean value ($minmax = 1$) is being looked for.

- **ncell, cmin,** and **cmax:** the number of cell sizes to consider, the minimum size, and the maximum size. Note that these sizes apply directly to the x direction and the *anis* factors adjust the sizes in the y and z directions.

- **noff:** the number of origin offsets. Each of the *ncell* cell sizes are considered with *noff* different original starting points. This avoids erratic results caused by extreme values falling into specific cells.

VI.2.2 Normal Score Transformation

The following files should be located on the diskette(s) before attempting to compile and execute this program/subroutine:

| nscore.f | the main driver program for **nscore**

| nscore.par | an example parameter file for **nscore**

The Gaussian based simulation programs **tb3d, lusim2d,** and **sgsim** work with normal scores of the original data. The simulation is performed in the normal space, then the simulated values are back transformed. This section provides details of the normal scores transformation step. All simulation programs accept data that have already been transformed.

Consider the original data $z_i, i = 1, ..., n$, each with a specified probability $p_i, i = 1, ..., n$ (with $\sum_{i=1}^{n} p_i = 1.0$) to account for clustering. If clustering is not considered important then all the p_i's can be set equal to $\frac{1}{n}$. Tied z-data values are randomly ordered. When there is a large proportion of z-data in a tie, these tied values should be ranked prior to **nscore** with a procedure based on the physics of the data, or the ties may be broken according to the data neighborhood averages [157].

Because of the limited sample size available in most applications, one should consider a non-zero probability for values less than the data minimum or greater than the data maximum. Thus some assumptions must be made about the (unsampled) tails of the attribute distribution. Such assumptions should account for the specific features of the phenomenon under study.

One common solution ([94], p. 479) is to standardize all previous probabilities p_i to a sum slightly less than one, e.g., to $\frac{n}{n+1}$ if there are n data. This solution is sensitive to the number of data (sample size), and it does not offer any flexibility in modeling the distribution tails.

To avoid the problem of sensitivity to sample size, the cumulative probability associated to each data value is reset to the average between its cumulative probability and that of the next lowest datum. This allows finite probabilities to be lower than the data minimum and greater than the data

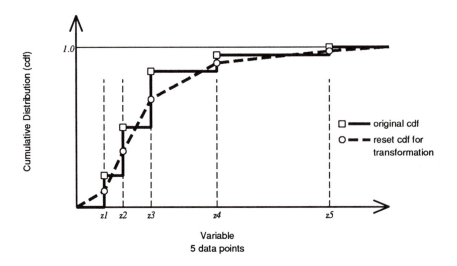

Figure VI.9: An illustration of how the sample cumulative probability values are adjusted for the transformation. The step increases are all the same unless some declustering has been used.

maximum. Figure VI.9 provides an illustration of this tail adjustment procedure.

For notation, let c_i be the cumulative probability associated with the ith largest data value z_i $(c_i = \sum_{j=1}^{i} p_j)$. The normal score transform y_i associated with z_i is then calculated as:

$$y_i = G^{-1}(\frac{c_i + c_{i-1}}{2})$$

with $G(y)$ being the standard normal cdf, $y_c = G^{-1}(c)$ being the corresponding standard normal c-quantile, and $c_0 = 0.0$. GSLIB utilizes the numerical approximation to $G^{-1}(\cdot)$ proposed by Kennedy and Gentle [100].

The limited sample size may yield an original sample cdf that is too discrete, as in Figure VI.9. This sample cdf may be smoothed by replacing the original indicator step functions $i(\mathbf{u}_\alpha; z)$ by continuous kernel functions $F(\mathbf{u}_\alpha; z)$ centered at the hard datum value $z(\mathbf{u}_\alpha)$; see Figure IV.2 and the related discussion. A worthwhile addition to GSLIB would be a utility program for such kernel functions.

The normal scores transformation is automatically performed in the programs that need normal data. There are times, however, when the transformed data are required (e.g., for variogram calculation) or when the procedure needs to be customized. The program **nscore** performs only the normal score transform. The parameters required by **nscore** are listed below and are shown in Figure VI.10:

Parameters for NSCORE
****

```
START OF PARAMETERS:
cluster.dat                        \data file
3    5                             \variable, weight
-1.0e21    1.0e21                  \trimming limits
nscore.dat                        \output file for data
nscore.trn                        \output transformation table
```

Figure VI.10: An example parameter file for **nscore**.

- **datafl:** a data file in simplified Geo-EAS format containing the variable to be transformed and optionally a declustering weight.

- **icolvr** and **icolwt:** the column numbers for the variable and the weight. If $icolwt \leq 0$, equal weighting is considered.

- **tmin** and **tmax:** minimum and maximum trimming limits to remove missing values and/or erratic high values (all z-values within the interval $tmin \leq z < tmax$ are accepted).

- **outfl:** the output file containing the normal scores.

- **transfl:** the output file containing the transformation look-up table, i.e., pairs of z and y values with z being the original data values and y being the corresponding normal scores values.

VI.2.3   Normal Score Back Transformation

The following files should be located on the diskette(s) before attempting to compile and execute this program/subroutine:

| backtr.f | the program **backtr**

| backtr.par | an example parameter file for **backtr**

Back transformation from the Gaussian or normal space to the original data space is discussed in section V.1.6. The back transformation z_i of the standard normal deviate y_i is given by

$$z_i = F^{-1}(G(y_i))$$

where $F(z)$ is the (declustered) cdf of the original data.

Almost always, the value $G(y_i)$ will not correspond exactly to an original sample cdf value F; therefore, some interpolation between the original sample z-values or extrapolation beyond the smallest and largest z-value will be required; see section V.1.6. Linear interpolation is always performed between two known values. A variety of options are available on how the tails of the distribution will be treated.

```
                    Parameters for BACKTR
                    *********************

START OF PARAMETERS:
sgsim.out                          \data file
1                                  \Gaussian variable
-1.0e21   1.0e21                    \trimming limits
sgsim.btr                          \output file for data
nscore.trn                         \input transformation table
0.0 45.0                           \minimum and maximum data value
1    0.0                           \lower tail option/parameter
4    1.5                           \upper tail option/parameter
```

Figure VI.11: An example parameter file for backtr.

This back transform aims at exactly reproducing the sample cdf $F(z)$, except for the within class interpolation and the two extreme class extrapolations. Hence details of the sample cdf that are deemed not representative of the population should be smoothed out prior to using programs **nscore** and **backtr**.

Back transformation is performed automatically in the programs that normal score transform the data (**sgsimm** and **lusim2d**). There are times when back transformation must be performed outside these programs. The program **backtrm** performs only the back transformation. The parameters required by **backtr** are listed below and are shown in Figure VI.11:

- **datafl:** a data file in simplified Geo-EAS format containing the normal scores.

- **ivr:** the column number for the normal scores.

- **tmin** and **tmax:** minimum and maximum trimming limits to remove missing values and/or erratic high values (all z-values within the interval $tmin \leq z < tmax$ are accepted).

- **outfl:** the output file containing the back transformed values.

- **transfl:** an input file containing the transformation look-up table, i.e., pairs of z and y values with z being the original data values and y being the corresponding normal scores values. This is an input file that corresponds to the table written out by **nscorem**, **sgsim**,...

- **zmin** and **zmax:** are the minimum and maximum values that will be used for extrapolation in the tails.

- **ltail** and **ltpar** specify the back transformation implementation in the lower tail of the distribution: $ltail = 1$ implements linear interpolation to the lower limit $zmin$ and $ltail = 2$ implements power model interpolation, with $\omega = ltpar$, to the lower limit $zmin$.

  Linear interpolation is used between known data values.

```
                        Parameters for TRANS
                        ********************

START OF PARAMETERS:
true.dat                              \file with ref distribution
1    0                                \ivr, iwt
-1.0e21  1.0e21                       \tmin, tmax
cluster.dat                           \file with uncorrected dists
3    0                                \ivr, iwt
-1.0e21  1.0e21                       \tmin, tmax
1000    1                             \nxyz, nsim
trans.out                            \file for revised bistributions
0.0    250.0                          \zmin, zmax
1    1.0                              \lower tail: option, parameter
4    1.5                              \upper tail: option, parameter
```

Figure VI.12: An example parameter file for **trans**.

- **utail** and **utpar** specify the back transformation implementation in the upper tail of the distribution: $utail = 1$ implements linear interpolation to the upper limit $zmax$, $utail = 2$ implements power model interpolation, with $\omega = utpar$, to the upper limit $zmax$, and $utail = 4$ implements hyperbolic model extrapolation with $\omega = utpar$.

VI.2.4 General Univariate Transformation

The following files should be located on the diskette before attempting to compile and execute this program:

| trans.f | the program **trans**

| trans.par | an example parameter file for **trans**

In certain cases it is necessary to have a close match between two univariate distributions. For example, a practitioner using only one simulated realization may want its univariate distribution to identify some original data distribution. This program will transform any set of values so that the resulting univariate distribution is the same as that of some reference distribution. Note that this transformation will undo the property of data honoring if applied to conditional simulations.

The transformation method is the conventional "graphical" transformation used for normal scores transformation, i.e., the p-quantiles of the data distribution are transformed to the p-quantiles of the reference distribution. Note that the reference distribution may be any arbitrary distribution.

The parameters required by **trans** are listed below and are shown in Figure VI.12:

- **refdist:** the input file with the reference data distribution and weights.

- **ivr** and **iwt:** the column for the values and the column for the (declustering) weight. If there are no declustering weights then set $iwt = 0$.

- **tmin** and **tmax**: minimum and maximum trimming limits to remove missing values and/or erratic high values (all z-values within the interval $tmin \leq z < tmax$ are accepted).

- **datafl**: the input file with the distribution(s) to be transformed.

- **ivrd** and **iwtd**: the column for the values and the declustering weights (0 if none).

- **tmin** and **tmax**: minimum and maximum trimming limits to remove missing values and/or erratic high values (all z-values within the interval $tmin \leq z < tmax$ are accepted).

- **nxyz** and **nsim**: the input data file may contain a number of simulated realizations and each may need to be transformed separately. Sets of $nxyz$ data will be read $nsim$ times and each set will be transformed separately. If there are fewer than $nxyz$ in the input data file then the available data will be transformed.

- **outfl**: output file for the transformed values.

- **zmin** and **zmax**: are the minimum and maximum values that will be used for extrapolation in the tails.

- **ltail** and **ltpar** specify the back transformation implementation in the lower tail of the distribution: $ltail = 1$ implements linear interpolation to the lower limit $zmin$ and $ltail = 2$ implements power model interpolation, with $\omega = ltpar$, to the lower limit $zmin$.

- **utail** and **utpar** specify the back transformation implementation in the upper tail of the distribution: $utail = 1$ implements linear interpolation to the upper limit $zmax$, $utail = 2$ implements power model interpolation, with $\omega = utpar$, to the upper limit $zmax$, and $utail = 4$ implements hyperbolic model extrapolation with $\omega = utpar$.

VI.2.5   Variogram Values from a Model

The following files should be located on the diskette(s) before attempting to compile and execute this program/subroutine:

| vmodel.f | the program **vmodel**

| vmodel.par | an example parameter file for **vmodel**

The details of entering a semivariogram model are given in section II.3. This program will take the semivariogram model and write out a file with the same format as the GAM programs so that it can be plotted with **vargplt**. The primary uses of **vmodel** are to overlay a model on experimental points and also to provide a utility to check the definition of the semivariogram

```
                    Parameters for VMODEL
                    *********************

START OF PARAMETERS:
vmodel.var                          \output file of variograms
2   22                              \ndir, nlag
1.0    0.0    0.0                   \xoff(id),yoff(id),zoff(id)
0.0    1.0    0.0                   \xoff(id),yoff(id),zoff(id)
1    0.00                           \nst, nugget effect
1    20.0   0.250                   \it,aa,cc:         STRUCTURE 1
90.  0.0    0.0  0.5 1.0            \ang1,ang2,ang3,anis1,anis2:
1    25.0   2.0                     \it,aa,cc:         STRUCTURE 2
90.0 0.0    0.0  99999999.  1.0     \ang1,ang2,ang3,anis1,anis2:
```

Figure VI.13: An example parameter file for **vmodel**.

model. The parameters required by **vmodel** are listed below and are shown in Figure VI.13:

- **outfl:** the output file that will contain the semivariogram values.

- **ndir** and **nlag:** the number of directions and the number of lags to be considered.

- **xoff, yoff** and **zoff:** for each of the *ndir* directions a unit lag offset in each coordinate direction is required. For example, north would be 0,1,0, east would be 1,0,0, vertical would be 0,0,1, and N60E would be 0.866,0.500,0.00.

- **nst** and **c0:** the number of structures and the nugget effect.

- **it, aa, cc, ang1, ang2, ang3, anis1** and **anis2:** required for each of the nested structures. See section II.3 for a detailed description of these parameters.

VI.2.6 Gaussian Indicator Variograms

The following files should be located on the diskette(s) before attempting to compile and execute this program/subroutine:

bigaus.f the program **bigaus**

bigaus.par an example parameter file for **bigaus**

This program allows checking for bivariate normality, see section V.2.2. A model of the normal scores variogram is input to **bigaus**. The corresponding indicator semivariograms for specified cutoffs, lags, and directions are output from the program. The idea is to see how far the experimental indicator variograms deviate from those implied by the bivariate Gaussian distribution.

A numerical integration algorithm is used to compute the Gaussian model-derived indicator variograms; see relation (V.20) and [82,166].

```
                    Parameters for BIGAUS
                    *********************

START OF PARAMETERS:
bigaus.out                              \output file of variograms
3                                       \number of cutoffs
0.25    0.50    0.75                    \the cutoffs (cdf values)
1   20                                  \ndir, nlag
1.0    0.0    0.0                       \xoff(id),yoff(id),zoff(id)
1 0.15                                  \nst, nugget effect
1    8.0    0.85                        \it,aa,cc:        STRUCTURE 1
0.0  0.0    0.0  1.0 1.0                \ang1,ang2,ang3,anis1,anis2:
1   30.0    0.5                         \it,aa,cc:        STRUCTURE 2
0.0  0.0    0.0  1.0 1.0                \ang1,ang2,ang3,anis1,anis2:
```

Figure VI.14: An example parameter file for **bigaus**.

The parameters required by **bigaus** are listed below and are shown in Figure VI.14:

- **outfl:** the output file for the theoretical Gaussian indicator semivariograms. The format is the same as that created by the GAM programs; therefore, the program **vargplt** could be used to plot these indicator variograms.

- **ncut:** the number of cutoffs.

- **zc(ncut):** *ncut* cutoff values are required. These cutoffs are expressed in units of cumulative probability, e.g., the lower quartile is 0.25, the median is 0.50, and so on. Note that the results are symmetric: the variogram for the 5th percentile (0.05) is the same as the 95th percentile (0.95).

- **ndir** and **nlag:** the number of directions and the number of lags to be considered.

- **xoff, yoff** and **zoff:** for each of the *ndir* directions a unit lag offset in each of the coordinate directions is required. For example, North would be 0,1,0, East would be 1,0,0, vertical would be 0,0,1, and N60E would be 0.866,0.500,0.00.

- **nst** and **c0:** the number of structures and the nugget effect.

- **it, aa, cc, ang1, ang2, ang3, anis1** and **anis2:** are required for each of the nested structures. See section II.3 for a detailed description of these parameters. Note that the sill of this variogram model should be one (the variance of the standard normal distribution). Also note that the power model is not a legitimate model for a multiGaussian phenomenon.

VI.2.7   Library of Linear System Solvers

Next to searching for data solving the kriging equations is the most time-consuming aspect of any kriging or simulation program. Depending on the application, a different solution algorithm is appropriate. Discussions are provided in references [58,60,71,132,139,140,161].

All of the linear system solvers documented below call for a small numerical tolerance parameter which is used to decide whether a matrix is singular. This parameter has been set between 10^{-10} and 10^{-4}. Ideally, this parameter should be linked to the units of the covariance function; the larger the magnitude of the covariance the larger the tolerance for a singular matrix. Code is provided for the following algorithms:

- **ksol:** A standard Gaussian elimination algorithm without any pivoting is adequate for SK and OK applications. When either the simple or ordinary kriging matrix is built with the covariance or pseudo-covariance, there is no need for pivoting. The computer savings can be significant.

- **smleq:** Another standard Gaussian elimination algorithm without any pivoting for SK and OK applications. The coding for SMLEQ uses double precision pivot/diagonal elements for greater accuracy.

- **ktsol:** A Gaussian elimination algorithm with partial pivoting is necessary for KT. The pivoting is necessary because of the zero elements on the diagonal arising from the unbiasedness conditions.

- **gaussj:** Gauss-Jordan elimination is a stable approach, but it is less efficient than Gaussian elimination with back substitution. The method is not recommended, but is included for completeness.

- **ludcmp:** An LU decomposition is used for certain simulation algorithms and can be used for any kriging system. This algorithm is general and provides one of the most stable solutions to any non-singular set of equations. Another advantage is that the right-hand side vectors are not required in advance.

- **cholfbs:** A Cholesky decomposition, with a forward and back substitution, is quite fast, but can only be used for symmetric positive definite matrices. One condition for a positive definite matrix is that it must have non-vanishing positive diagonal elements. The last diagonal element that comes from any unbiasedness constraint implies that the OK and KT kriging matrices do not fulfill that requirement. The Cholesky decomposition algorithm, or any of the methods requiring a positive definite symmetric matrix, will not work with the OK or KT matrices. A small ϵ cannot be used instead of zero because of the further condition that $a_{ik}^2 < a_{ii}a_{kk}$ for all i and k (positive definite condition). An alternative, which the authors did not investigate, consists of partitioning the OK or KT system and using a Cholesky decomposition only for the strict covariance matrix.

- **cholinv:** The SK covariance matrix can be inverted through a Cholesky decomposition, followed by an inverse of the lower or upper triangular decomposition and finally a matrix multiplication. The SK weights, the OK weights, and the weights for the OK estimate of the mean can be obtained simultaneously [34].

- **gsitrn:** A straightforward Gauss-Seidel iteration algorithm may be used to solve any of the kriging systems. A simple reordering of the equations (which destroys symmetry) is necessary to remove the zero from the diagonal of the OK system. This method seems particularly attractive when considering any of the indicator approaches because the solution at one cutoff provides the logical initial choice for the next cutoff. If the variogram model has not changed much, then the solution will be obtained very quickly. Lacking a previous solution, the Lagrange parameter can be estimated as zero and the weights as the normalized right hand side covariances.

To illustrate the relative speed of the algorithms, a system (which system only matters with an iterative approach) was solved 10,000 times. This took KSOL about a second for a 10 by 10 system on an Apollo 10000. The following table shows the times for the different subroutines with different SK matrix sizes. The time required for the various algorithms does not increase at the same rate as the size of the matrix.

Routine	10 samples	20 samples	40 samples	80 samples
ksol	1.0	4.8	28.4	184.2
smleq	1.0	5.9	39.5	292.1
ktsol	1.4	7.4	47.9	339.5
gaussj	6.2	44.8	337.6	2644.7
ludcmp	2.1	11.6	73.9	521.0
cholfbs	1.3	7.0	43.7	313.2
cholinv	8.7	60.5	447.4	3457.9
gsitrn	2.6	11.0	72.8	309.3

The times in the above table are only illustrative. Different computers, compilers, and coding will result in different numbers. The following comments can be made on the basis of the above results:

- The KSOL subroutine, which is commonly used in GSLIB, is an efficient direct method to solve either the OK or SK system of equations.

- The KTSOL subroutine seems the most efficient direct method to solve the KT equations (KSOL would not work).

- The Gauss-Jordan solution is a stable numerical method to solve a general system of equations. As illustrated above, the method is inefficient for kriging.

- The LUDCMP subroutine, as it stands in [132], is not an efficient way to solve kriging systems.

- The standard Cholesky LU decomposition with a forward and backward substitution is better than the general LU approach, but is not as fast as the KSOL subroutine. In principle the Cholesky decomposition should be quicker; after all, it is just a modification of the Gauss algorithm in which the elimination is carried out while maintaining symmetry. The reason for the slowness is the square root operation on all the diagonal elements.

- CHOLINV completes the inverse and solves the system explicitly to get simultaneously the SK estimate, the OK estimate, and the OK estimate of the mean. This method appears attractive, due to its elegance, until speed is considered. On the basis of the tabulated results it would be more efficient to solve two systems (the second one with two right-hand side vectors) with KSOL to obtain these three estimates.

- GSITRN is not an efficient method to solve a system without a good initial solution. Experimentation has shown that the final solution is obtained more quickly than it is obtained by KSOL when the initial solution is known to the first decimal place and the matrix is larger than 10 by 10. The time also depends on the kriging system and the ordering of the data points. Additional programming could result in an efficient iterative solver that optimally (with respect to speed) orders the equations and possibly accelerates convergence through some type of relaxation scheme [139].

VI.2.8 Markov-Bayes Calibration

The following files should be located on the diskette(s) before attempting to compile and execute this program/subroutine:

| mbcalib.f | the program `mbcalib`

| mbcalib.par | an example parameter file for `mbcalib`

The Markov-Bayes simulation program `mbsim`, documented in section V.7.3, the $B(z)$ calibration parameters and the prior distributions for various classes of secondary data; see section IV.1.12 for a more complete discussion. The program `mbcalib` computes the prior distributions and the $B(z)$ calibration parameters. These parameters are written to a file that must be passed unchanged to `mbsim`. Additional information is written to a report file which can be printed.

The parameters required by `mbcalib` are listed below and are shown in Figure VI.15:

```
                        Parameters for MBCALIB
                        **********************

START OF PARAMETERS:
cluster.dat                     \data file
3    4                          \column of primary and secondary
-1.0e21   1.0e21                \trimming limits
mbsim.cal                       \output file for input to MBSIM
mbcalib.rep                     \output file for report
5                               \number of cutoffs on primary
0.50 1.00 2.50 5.00 10.0        \cutoffs on primary
5                               \number of cutoffs on secondary
0.50 1.00 2.50 5.00 10.0        \cutoffs on secondary
```

Figure VI.15: An example parameter file for **mbcalib**.

- **datafl:** a data file in simplified Geo-EAS format containing pairs of primary and secondary data values (the calibration data).

- **ivru** and **ivrv:** the column numbers for the primary variable and the secondary variable.

- **tmin** and **tmax:** minimum and maximum trimming limits to remove missing values and/or erratic high values (all z-values within the interval $tmin \leq z < tmax$ are accepted).

- **calibfl:** the output file containing the calibration information for **mbsim**. The file is written in the following format:

    - A title line (default *The local prior distribution table:*)

    - The following $ncutv+1$ lines are taken to be the prior distributions of the primary variable for classes of the secondary variable, e.g., $tmin \leq v < cutv(1)$, $cutv(1) \leq v < cutv(2)$, ... $cutv(ncutv) \leq v < tmax$.

    - A second title line (default *The calibration parameters $B(i)$:*)

    - The following $ncutu$ lines are taken to be the $B(z)$ hardness parameters for each cutoff on the primary variable.

- **repfl:** an output file that contains a report of the pairs retained, some summary statistics, a bivariate distribution table, the prior distributions, and the $B(z)$ parameter values.

- **ncutu:** the number of cutoffs applied to the primary variable.

- **cutu(i),i=1,ncutu:** the cutoff values applied to the primary variable.

- **ncutv:** the number of cutoffs applied to the secondary variable.

- **cutv(i),i=1,ncutv:** the cutoff values applied to the secondary variable.

```
                    Parameters for IPCPREP
                    ***********************

START OF PARAMETERS:
temp.dat                          \data file
 1    2    0    4                 \column: x,y,z,vr
-1.0e21    1.0e21                 \missing value code
ipcprep.out                       \output file for trans. data
ipcprep.dbg                       \output file for debugging
ipcprep.svd                       \output file for orthog. matrix
4                                 \number of categories
1 2 3 4                           \categories
0.250 0.250 0.250 0.250           \''p'' values
```

Figure VI.16: An example parameter file for **ipcprep**.

When the calibration data pairs are few, the prior distributions calculated from **mbcalib** may be very coarse; a smoothing of these prior distributions by direct alteration of the values in the file *calibfl* or by using a kernel-type smoother (see Figure IV.2) may be in order.

VI.2.9 Orthogonal Matrix Decomposition

The following files should be located on the diskette(s) before attempting to compile and execute this program/subroutine:

| ipcprep.f | the program **ipcprep**

| ipcprep.par | an example parameter file for **ipcprep**

Before running the indicator principal components (IPCK) simulation program **ipcsim** documented in section V.7.4, the orthogonal matrix of principal components loadings are needed; see section IV.1.11 for a discussion on the IPCK algorithm. Moreover, the indicator principal components data are needed to compute and model the variograms needed by the **ipcsim** program. The program **ipcprep** computes the orthogonal decomposition of the indicator covariance matrix (at **h**=0) and computes optionally the indicator principal components of a set of indicator data. **ipcprep** is set up for categorical variables. The computation of the principal components loadings would be different for a continuous variable; see [152].

The parameters required by **ipcprep** are listed below and are shown in Figure VI.16:

- **datafl:** (optional) a data file, in simplified Geo-EAS format, containing categorical indicator data. When this file does not exist, then just the orthogonal matrix is computed; the principal components of the data are not computed.

- **icolx, icoly, icolz,** and **icolvr:** the column numbers for the x, y, z coordinates and the indicator variable.

- **tmin** and **tmax:** minimum and maximum trimming limits to remove missing values and/or erratic high values (all z-values within the interval $tmin \leq z < tmax$ are accepted).

- **outfl:** the output file for the principal components of the data in *datafl*. This file will not be created if the data file *datafl* does not exist.

- **dbgfl:** a debugging file that records the covariance matrix, the principal components loadings, etc.

- **svdfl:** the output file containing the orthogonal matrix for `ipcsim`.

- **ncat:** the number of categories.

- **cat(i),i=1,ncat:** the indicator values for each category. These integer-codes need not be consecutive.

- **pdf(i),i=1,ncat:** the global proportion within each category, declustered if necessary.

VI.2.10   Post Processing of IK Results

The following files should be located on the diskette(s) before attempting to compile and execute this program/subroutine:

| postik.f | the program `postik`

| postik.par | an example parameter file for `postik`

The output from `ik3d` requires additional processing before being used. The program `postik` performs order relations corrections, change of support calculations, and computes various statistics according to the following options:

1. Compute the "E-type" estimate, i.e., the mean value of the conditional distribution; see section IV.1.9.

2. Compute the probability of exceeding a fixed cutoff, the average value above that cutoff, and the average value below that cutoff.

3. Compute the value where a fixed conditional cumulative distribution function (cdf) value p is reached, i.e., the conditional p-quantile value.

Although order relations are corrected by `ik3d` they are checked again in `postik`.

A facility for volume support correction has been provided although we **strongly** suggest that volume support correction be approached through small-scale simulations; see section IV.1.13 and [74]. Either the affine correction or the indirect lognormal correction is available; see [76], p. 468-476.

```
                    Parameters for POSTIK
                    **********************
START OF PARAMETERS:
ik3d.out                        \input from IK3D
postik.out                      \output file
3    0.5                        \output option, output parameter
5                               \number of cutoffs
0.5  1.0   2.5  5.0  10.0       \the cutoffs
0    1        0.75              \volume support, type, varred
cluster.dat                     \global distribution
3    0    -1.0    1.0e21        \ivr,  iwt,  tmin,  tmax
0.0     30.0                    \minimum and maximum Z value
1    1.0                        \lower tail: option, parameter
1    1.0                        \middle    : option, parameter
1    2.0                        \upper tail: option, parameter
100                             \maximum discretization
```

Figure VI.17: An example parameter file for postik.

There are a number of consequential decisions that must be made about how to interpolate between and extrapolate beyond known values of the cumulative distribution function. These are discussed in section V.1.6. Both the conditional (E-type) mean and the mean above a specified cutoff are calculated by summing the z values at regularly spaced ccdf increments. A parameter *maxdis* specifies how many ccdf values to calculate. This can be increased depending on the precision required.

The parameters required by **postik** are listed below and are shown in Figure VI.17:

- **distfl:** the output IK file (from program **ik3d**) which is considered as input.

- **outfl:** the output file (output differs depending on output option, see below).

- **iout** and **outpar:** *iout* corresponds to the list given at the beginning of this section, i.e., *iout* = 1 computes the E-type estimate, *iout* = 2 computes the probability and mean above the value *outpar* (and mean below *outpar*), *iout* = 3 computes the z value (p-quantile) corresponding to the cdf value *outpar* = p.

- **ncut:** the number of cutoffs.

- **cut(i),i=1,ncut:** the cutoff values used in **ik3d**.

- **ivol, ivoltyp,** and **varred:** if *ivol* = 1 then volume support correction is attempted with the affine correction (*ivoltyp* = 1) or the indirect correction through permanence of a lognormal distribution (*ivoltyp* = 2). The variance reduction factor (between 0 and 1) is specified as *varred*.

- **datafl:** the data file containing z-data that provides the details between the IK cutoffs. This file is used only if table look-up values are called for in one of the interpolation/extrapolation options.

- **icolvr, icolwt, tmin,** and **tmax:** when either *ltail, middle* or *utail* is 3 then these variables identify the column for the values and the column for the (declustering) weight in *datafl* that will be used for the global distribution. *tmin* and *tmax* are the minimum and maximum trimming limits to remove missing values and/or erratic high values (all z-values within the interval $tmin \leq z < tmax$ are accepted).

- **zmin** and **zmax:** minimum and maximum data values allowed. Even when not using linear interpolation it is safe to control the minimum and maximum values entered. The lognormal or hyperbolic upper tail is constrained by $zmax$.

- **ltail** and **ltpar** specify the extrapolation in the lower tail: $ltail = 1$ implements linear interpolation to the lower limit $zmin$; $ltail = 2$ implements power model interpolation, with $\omega = ltpar$, to the lower limit $zmin$; and $ltail = 3$ implements linear interpolation between tabulated quantiles.

- **middle** and **midpar** specify the interpolation within the middle of the distribution: $middle = 1$ implements linear interpolation to the lower limit $zmin$; $middle = 2$ implements power model interpolation, with $\omega = midpar$, to the lower limit $zmin$; and $middle = 3$ allows for linear interpolation between tabulated quantile values;

- **utail** and **utpar** specify the extrapolation in the upper tail of the distribution: $utail = 1$ implements linear interpolation to the upper limit $zmax$, $utail = 2$ implements power model interpolation, with $\omega = utpar$, to the upper limit $zmax$; $utail = 3$ implements linear interpolation between tabulated quantiles; and, $utail = 4$ implements hyperbolic model extrapolation with $\omega = utpar$.

- **maxdis:** a maximum discretization parameter. The default value of 50 is typically fine; greater accuracy is obtained with a larger number and quicker execution time is obtained with a lower number.

VI.2.11   Post Processing of Simulation Results

The following files should be located on the diskette(s) before attempting to compile and execute this program/subroutine:

| postsim.f | the program `postsim`

| postsim.par | an example parameter file for `postsim`

```
                   Parameters for POSTSIM
                   ***********************
START OF PARAMETERS:
sgsim.out                          \simulated realizations
50                                 \number or realizations
-999.                              \minimum for missing values
20   20   1                        \nx, ny, nz
postsim.out                        \output File with E-type,...
3    0.0001                        \output option, output parameter

output option = 1   --->   E-type (average value)
              = 2   --->   prob > cutoff, mean > cutoff, mean < cutoff
              = 3   --->   Z quantile
```

Figure VI.18: An example parameter file for **postsim**.

postsim allows a number of summaries to be extracted from a set of simulated realizations:

1. the "E-type" estimates, i.e., the point-by-point average of the realizations; see relation (IV.37).

2. the probability of exceeding a fixed cutoff, the average value above that cutoff, and the average value below that cutoff.

3. the value where a fixed conditional cumulative distribution function (cdf) value p is reached, i.e., the conditional p-quantile value.

The parameters required by **postsim** are listed below and are shown in Figure VI.18:

- **simfl:** the output simulation file that contains all of the realizations. This file should contain the simulated values, one value per line, cycling fastest on x, then y, then z, and last per simulation.

- **nsim** specifies the number of realizations.

- **tmin:** low value to indicate a nodal location not simulated.

- **nx, ny,** and **nz** specify the number of nodes in the x, y, and z direction.

- **outfl:** an output file that contains the E-type estimate, the fraction and mean above cutoff, the fraction and mean below cutoff, or the conditional p-quantile values.

- **iout** and **outpar:** *iout* corresponds to the list given above, i.e., *iout* = 1 computes the E-type estimate, *iout* = 2 computes the probability and mean above the value *outpar* (and the mean below *outpar*), *iout* = 3 computes the z-value (p-quantile) corresponding to the cdf value *outpar* = p.

Appendix A

Partial Solutions to Problem Sets

The partial solutions given in this appendix illustrate the use of GSLIB programs, provide base cases for debugging, and serve as references to check program installations. As much as possible, bugs should be reported to the authors using the example data; this would make it simpler to document the problem.

The data, parameter, and output files are provided on the distribution diskettes.

The runs given in this appendix illustrate only some of the available options. It would be too cumbersome to exhaustively illustrate all program options and all solution alternatives. Whenever possible, the results of a program should be checked against analytical results or through an example small enough to allow calculation by hand.

A.1   Problem Set One: Data Analysis

Most summary statistics are affected by preferential clustering. In the case of cluster.dat the equal-weighted mean and median will be higher since clustering is in high concentration areas. The variance may also be increased because the clustered data are disproportionately valued high. Note that the effect of clustering is not always so predictable. For example, if the clustered samples are all close to the center of the distribution then the equal weighted mean and median may be unbiased but the variance could be too low.

An equal-weighted histogram and lognormal probability plot are shown in Figure A.1. The distribution is positively skewed with a coefficient of variation of 1.5. The parameter files required by **histplt** and **probplt** to generate the plots shown on Figure A.1 are presented on Figures A.2 and A.3. These parameter files are provided on the software distribution diskettes.

227

A quick-and-easy way to decluster these data would be to consider only the first 97 nearly regularly spaced samples. If cell declustering is considered a reasonable range of square cell sizes would be from 1 mile (no samples are spaced closer than 1 mile) to 25 miles (half of the size of the area of interest). A *natural* cell size would be the spacing of the underlying pseudo-regular grid, i.e., $\sqrt{\frac{2500}{97}} \simeq 5.0$. The resulting scatterplot of declustered mean versus cell size is shown in Figure A.4. The minimum declustered mean is found for a cell size of 5.0 miles; the corresponding declustered histogram and lognormal probability plot are shown in Figure A.5.

The parameters required by program `declus` are shown in Figure A.6. The 10 first lines and the 10 last lines of the `declus` output file are given in Figure A.7. The scatterplot parameter file is shown in Figure A.8.

A histogram and probability plot of the 2500 exhaustive measurements is shown in Figure A.9. Summary statistics for all 140 data equal weighted, the first 97 data equal weighted, 140 data with declustering weights, and the true exhaustive measurements are shown below:

	n	m	M	σ
Equal Weighting (140 data)	140	4.35	2.12	6.70
Equal Weighting (97 data)	97	2.21	1.02	3.17
Declustered (140 data)	140	2.52	1.20	4.57
Exhaustive	2500	2.58	0.96	5.15

The Q-Q and P-P plots of the sample data (equal weighted and declustered) versus the exhaustive data are shown in Figure A.10. Identical distributions should plot as a 45° degree line. When the points fall on any straight line, the shapes of the distributions are the same but with different means and variances.

The parameter file for the lower left Q-Q plot in Figure A.10 is shown in Figure A.11 and provided on the software distribution diskettes.

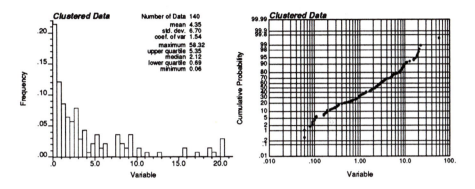

Figure A.1: An equal weighted histogram and lognormal probability plot of all 140 sample data.

```
                    Parameters for HISTPLT
                    **********************

START OF PARAMETERS:
cluster.dat                      \data file
3    0                           \column for variable and weight
histplt.out                      \output PostScript file
-0.99    999999.                 \trimming limits
0.0      20.0                    \histogram minimum and maximum
40                               \number of classes
0                                \1=log scale, 0=arithmetic
Clustered Data                   \title
```

Figure A.2: Parameter file for the histogram shown in Figure A.1.

```
                    Parameters for PROBPLT
                    **********************

START OF PARAMETERS:
cluster.dat                      \data file
3    0                           \column for vr and wt
0.0    1.0e21                    \trimming limits
probplt.out                      \output PostScript file
1                                \0=arith, 1=log
0.0   30.0    5.0                \min,max,inc: arith
0.01    4                        \start,ncyc : log
Clustered Data                   \title
```

Figure A.3: Parameter file for the probability plot shown in Figure A.1.

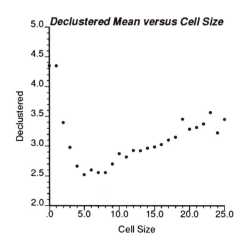

Figure A.4: Scatterplot of the declustered mean versus cell size.

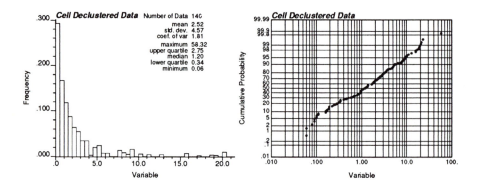

Figure A.5: Declustered histogram and lognormal probability plot of the 140 data.

```
                    Parameters for DECLUS
                    *********************

START OF PARAMETERS:
cluster.dat                          \data file
1    2    0    3                     \ix,iy,iz,ivr
0.0       1.0e21                     \tmin,tmax
declus.sum                           \output with summary
declus.out                           \output with data & wts
1.0    1.0                           \anisotropy: y,z
0                                    \0=look for min, 1 max
24   1.0   25.0                      \num, min and max size
5                                    \num of origin offsets
```

Figure A.6: Parameter file for the declustering results shown in Figure A.4.

```
Clustered 140 primary and secondary data
5
X
Y
Z
Value
Weight
        39.50        18.50        .00        .0600      1.2813
         5.50         1.50        .00        .0600      1.4001
        38.50         5.50        .00        .0800      1.6132
        20.50         1.50        .00        .0900      1.7974
        27.50        14.50        .00        .0900      1.4302
        40.50        21.50        .00        .1000      1.0869
        15.50         3.50        .00        .1000      1.2159
         6.50        25.50        .00        .1100      1.0834
        38.50        21.50        .00        .1100      1.0869
        23.50        18.50        .00        .1600      1.7974
• • •
         2.50        15.50        .00       2.7400       .6892
         3.50        14.50        .00       3.6100       .6889
        29.50        41.50        .00      58.3200       .4006
        30.50        40.50        .00      11.0800       .6786
        30.50        42.50        .00      21.0800       .3111
        31.50        41.50        .00      22.7500       .3675
        34.50        32.50        .00       9.4200       .4174
        35.50        31.50        .00       8.4800       .4327
        35.50        33.50        .00       2.8200       .3369
        36.50        32.50        .00       5.2600       .2878
```

Figure A.7: Sample output from **declus** created by the parameter file shown in Figure A.6.

```
                    Parameters for SCATPLT
                    ***********************

START OF PARAMETERS:
declus.sum                              \data file
1    2    0                             \column for X, Y, and weight
scatplt.out                             \output Postscript file
0.0     1.0e21                          \min and max variable X
0.0     1.0e21                          \min and max variable Y
0.0      25.0                           \xmin  and  xmax
2.0       5.0                           \ymin  and  ymax
Declustered Mean versus Cell Size       \title
```

Figure A.8: Parameter file for the scatterplot shown in Figure A.4.

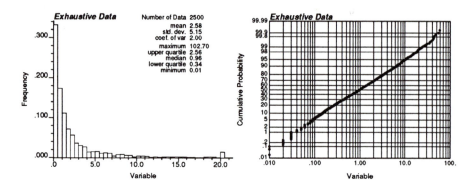

Figure A.9: Histogram and lognormal probability plot of the 2500 exhaustive measurements.

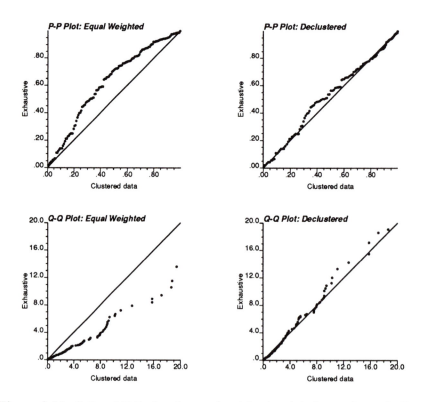

Figure A.10: Q-Q and P-P plots for equal weighted and declustered sample distributions (140 data) versus the exhaustive distribution.

```
                     Parameters for QPPLT
                     ********************

START OF PARAMETERS:
cluster.dat                              \first data file (X axis)
3    0                                   \column for vr and wt
true.dat                                 \second data file (Y axis)
1    0                                   \column for vr and wt
0.0         1.0e21                        \trimming limits
qpplt.03                                 \output PostScript file
0                                        \0=Q-Q plot, 1=P-P plot
0.0      20.0                            \xmin  and  xmax
0.0      20.0                            \ymin  and  ymax
Clustered data                           \X label
Exhaustive                               \Y label
Q-Q Plot: Equal Weighted                 \title
```

Figure A.11: Parameter file for the Q-Q plot shown in the lower left of Figure A.10.

A.2   Problem Set Two: Variograms

Preferential clustering of data in, say, high grade zones influences not only univariate statistics such as the sample histogram but also all two-point statistics such as the sample variogram.

Unfortunately there is no general declustering algorithm for bivariate and multivariate statistics, similar to the cell-declustering algorithm for univariate statistics. [1] Different measures of spatial variability, as presented in section III.1, will have different robustness properties with regard to data clustering and sparsity, presence of outliers, or any cause of sampling fluctuations and biases. Therefore, in the presence of sparse and preferentially clustered data, it is good practice to run a few other measures of spatial variability in addition to the traditional sample variogram (III.1). The modeling of the traditional sample variogram can take advantage of features, such as range and anisotropy, better seen on other more robust measures.

Figure A.12 gives eight experimental measures of spatial variability calculated on the 140 clustered sample values in cluster.dat . These measures are listed in section III.1. The omnidirectional $(0^\circ \pm 90^\circ)$, $NS(0^\circ \pm 22.5^\circ)$, and $EW(90^\circ \pm 22.5^\circ)$ directions are plotted.

Figures A.13 and A.14 give the corresponding input parameter file to program gamv2m and the 10 first and 10 last lines of the output file.

The traditional sample variogram, rodogram, and madogram [2] appear noisy while the other measures of continuity are much more stable, revealing an isotropic range between 10 and 20. The covariance/correlogram measures and relative variograms account for the data mean at each specific lag $\mathbf{h}$; see expressions (III.3) to (III.6). This allows for some correction of the preferential clustering; indeed, pairs with short separation $\mathbf{h}$ tend to be preferentially located in the high z-valued zones.

The sample traditional variogram can be modeled borrowing ranges and anisotropic features better revealed by other more robust measures of spatial variability.

One way around the problem of clustering is to ignore those data that are clustered; in the present case considering only the 97 first data. This is not always possible, particularly if drilling has not been done in recognizable

---

[1] Henning Omre [127] has proposed a variogram declustering algorithm that amounts to weighting each pair of data in the traditional variogram estimate:

$$2\gamma^*(\mathbf{h}) \ = \ \frac{1}{N(\mathbf{h})} \sum_{\alpha=1}^{N(\mathbf{h})} \omega_\alpha(\mathbf{h}) \cdot [\, z(\mathbf{u}_\alpha) \ - \ z(\mathbf{u}_\alpha \ + \ \mathbf{h})]^2$$

The weights $\omega_\alpha(\mathbf{h})$ are such that the marginal histogram built from the bivariate distribution of the $N(\mathbf{h})$ pairs of data $z(\mathbf{u}_\alpha), z(\mathbf{u}_\alpha + \mathbf{h})$ approximates best the prior declustered sample histogram built on all available z-data.

[2] The rodogram and madogram are robust with regard to outlier values, since the influence of such outlier is not squared as it is in the expression of the traditional variogram estimate. The problem in this data set, however, is not so much outlier data but clustered data.

sequences, with the first sequence being on a pseudo-regular grid. Also, ignoring data amounts to a loss of information; for example, ignoring clustered data may lead to ignoring most of the short scale information.

Figure A.15 gives the same eight experimental measures but now calculated on the first 97 data only. Fluctuations are seen to be somewhat reduced. Note the considerable reduction in variance due to the removal of the high-valued clustered data. The sample traditional semivariogram stabilizes around a sill value of about 11, a value consistent with the 97 data sample variance 10, but smaller than the 140 data declustered variance 20.9 and the actual variance 26.5. The naive, equal-weighted, 140 data sample variance is 44.9. These variance deviations stress the difficulty in estimating the correct (ordinate axis) scale for variograms.

Experimental variogram values for the first lag are often extremely unstable as seen in Figure A.15, because of the small number of pairs, preferential locations of close-spaced data, and the large averaging within that first lag.

Yet, with the help of the more robust experimental measures, an isotropic model can be inferred with range about 10 distance units and a relative nugget effect between 30 and 50%:

$$\gamma(\mathbf{h}) = K \left[.3 + .7 \, Sph \left(\frac{|\mathbf{h}|}{10}\right)\right] \,, \quad \forall \, |\mathbf{h}| > 0 \qquad (A.1)$$

The factor K defining the total sill of the semivariogram model can be identified to the 140 data declustered variance: $K = 20.9$. Fortunately, such proportional factor carries no influence on the kriging weights, hence on the kriging estimate. The proportional factor, however, does directly affect the kriging variance value: this is why non-linear kriging estimates, such as provided by lognormal kriging, which depend on the kriging variance value are particularly non-robust; see [130] and discussion in section IV.1.8.

Figure A.16 gives the same eight measures of spatial variability calculated from all 2500 reference data in the NS and EW directions. The additional thick continuous curve on the top left graph corresponds to the isotropic model:

$$\gamma(\mathbf{h}) = 10 + 16 \, Sph \left(\frac{|\mathbf{h}|}{8}\right) \qquad (A.2)$$

$$\approx 26 \left[.38 + .62 \, Sph \left(\frac{|\mathbf{h}|}{10}\right)\right] \,, \quad \forall \, |\mathbf{h}| > 0$$

It appears that, except for the variance factor K, the model (A.1) inferred from the various sample measures is reasonably close to the exhaustive model (A.2).

Figures A.17 and A.18 give the corresponding input parameter file to program **gam2m** and the 10 first and 10 last lines of the output file.

The first 97 data were normal score transformed using their own equal-weighted histogram. The parameter file and output summary for program **nscore** are shown in Figures A.19 and A.20. Then, the omnidirectional traditional semivariogram was calculated from these 97 normal score data. The results are shown in Figure A.21 together with the fit by the isotropic model:

$$\gamma_Y(\mathbf{h}) \;=\; Sph\left(\frac{|\mathbf{h}|}{10}\right) \tag{A.3}$$

As a measure of comparison, Figure A.22 gives the two directional (NS and EW) traditional semivariograms calculated from all 2500 reference normal score data. Also drawn in Figure A.22 is the previous model (A.3) inferred from the 97 sample data. The sample model appears to provide a reasonable fit of both exhaustive reference semivariograms.

Using the model (A.3) for $\gamma_Y(\mathbf{h})$ and assuming a bivariate Gaussian model for $RF\ Y(\mathbf{u})$, the theoretical indicator semivariograms for the three quartile y-threshold values were calculated. Program **bigaus** described in section VI.2.6 was used. The parameter file and output summary from **bigaus** are shown in Figures A.23 and A.24. These theoretical isotropic indicator semivariograms are given as the solid lines in Figure A.25. The corresponding 97 data sample indicator semivariograms are also plotted in Figure A.25. The reasonable fits indicate that the sample indicator semivariograms do not invalidate a multivariate Gaussian model for $Y(\mathbf{u})$; see related discussion in section V.2.2.

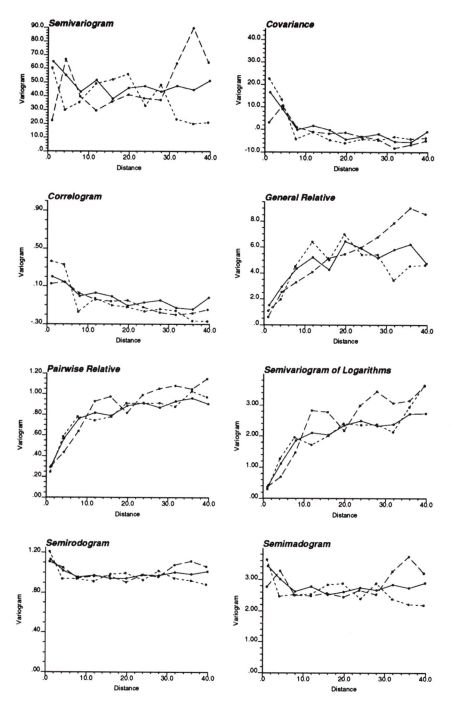

Figure A.12: Eight experimental measures of spatial variability calculated from the 140 sample values of cluster.dat .

```
                          Parameters for GAMV2M
                          **********************

START OF PARAMETERS:
cluster.dat                          \Data File in GEOEAS format
1   2                                \columns for x and y coordinates
1   3   0   0                        \nvar; column numbers...
-1.0e21     1.0e21                    \tmin, tmax (trimming limits)
gamv2.var                            \Output File for Variograms
10                                   \nlag - the number of lags
4.0                                  \xlag - unit separation distance
2.0                                  \xltol- lag tolerance
3                                    \ndir - number of directions
  0.0  90.0    50.0                  \azm(i),atol(i),bandw(i)i=1,ndir
  0.0  22.5    10.0                  \azm(i),atol(i),bandw(i)i=1,ndir
 90.0  22.5    10.0                  \azm(i),atol(i),bandw(i)i=1,ndir
8                                    \number of variograms
1   1   1                            \head, tail, variogram type
1   1   3                            \head, tail, variogram type
1   1   4                            \head, tail, variogram type
1   1   5                            \head, tail, variogram type
1   1   6                            \head, tail, variogram type
1   1   7                            \head, tail, variogram type
1   1   8                            \head, tail, variogram type
1   1   9                            \head, tail, variogram type
```

Figure A.13: Parameter file for **gamv2m** run.

Semivariogram		tail:Primary		head:Primary	direction 1
1	.000	.00000	280	4.35043	4.35043
2	1.216	65.17109	192	9.26318	9.26318
3	4.313	55.22798	838	6.12333	6.12333
4	7.951	43.06265	1344	4.44906	4.44906
5	11.877	51.64461	1776	4.44381	4.44381
6	15.899	38.07474	2020	4.22700	4.22700
7	19.831	45.92778	1828	3.77031	3.77031
8	23.938	47.25710	2190	4.00110	4.00110
9	27.918	43.29531	2138	4.08785	4.08785
• • •					
3	4.048	2.46444	98	6.11939	4.91316
4	7.585	2.52148	165	5.00309	2.90933
5	11.769	2.53368	205	4.52498	3.29615
6	15.902	2.84000	260	4.46558	4.63219
7	19.579	2.88253	245	3.92943	4.06902
8	24.013	2.38933	290	3.10497	3.87328
9	27.857	2.88331	298	3.33389	5.09527
10	31.707	2.38220	255	3.73208	3.64361
11	35.775	2.20961	179	2.66838	3.27084
12	39.441	2.18449	99	4.09172	1.94333

Figure A.14: Output of **gamv2m** corresponding to the parameter file shown in Figure A.13.

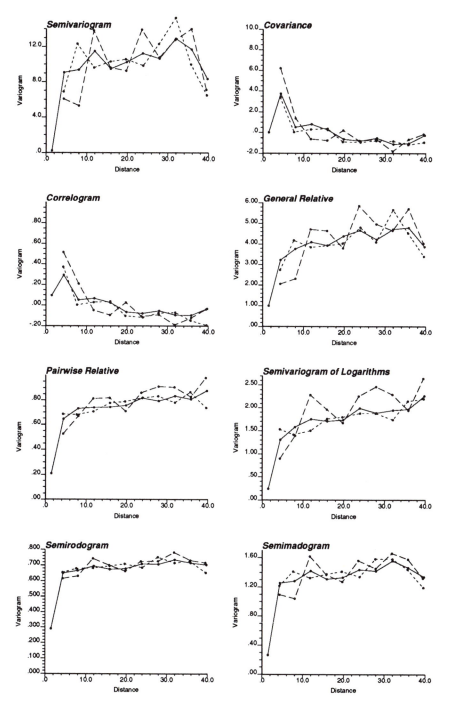

Figure A.15: Eight experimental measures of spatial variability calculated from the first 97 sample values of cluster.dat .

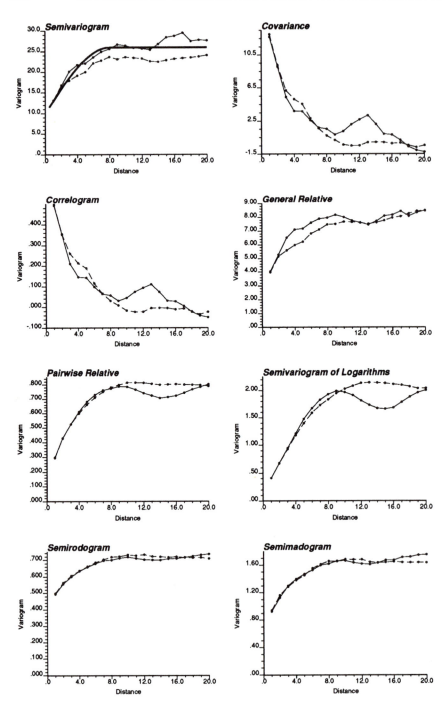

Figure A.16: Eight measures of spatial variability calculated on all 2500 reference data values of ⟨true.dat⟩.

```
                        Parameters for GAM2M
                        ********************

START OF PARAMETERS:
true.dat                                \data file
1   1   2   3 ·                         \nvar; column numbers...
-1.0e21     1.0e21                       \tmin, tmax (trimming limits)
gam2.var                                \output file for variograms
50    50    1                           \nx,    ny,     igrid
1.0  1.0                                \xsiz, ysiz
2  20                                   \ndir,nlag
 1  0                                   \ixd(i),iyd(i)  i=1,...,ndir
 0  1
8                                       \number of variograms
1   1   1                               \tail, head, variogram type
1   1   3                               \tail, head, variogram type
1   1   4                               \tail, head, variogram type
1   1   5                               \tail, head, variogram type
1   1   6                               \tail, head, variogram type
1   1   7                               \tail, head, variogram type
1   1   8                               \tail, head, variogram type
1   1   9                               \tail, head, variogram type
```

Figure A.17: Parameter file for **gam2m** run.

Semivariogram		tail:primary - Z		head:primary - Z	direction 1
1	1.000	13.03226	2450	2.53364	2.53909
2	2.000	16.64981	2400	2.49753	2.52300
3	3.000	20.16062	2350	2.46945	2.50771
4	4.000	21.73316	2300	2.45293	2.49312
5	5.000	22.07451	2250	2.46641	2.49146
6	6.000	23.60346	2200	2.48491	2.50075
7	7.000	24.93002	2150	2.50757	2.52079
8	8.000	25.62192	2100	2.53510	2.53821
9	9.000	26.70175	2050	2.55013	2.56459
10	10.000	26.42958	2000	2.57868	2.55968
• • •					
11	11.000	1.67927	1950	2.81330	2.13936
12	12.000	1.68288	1900	2.80745	2.14270
13	13.000	1.64556	1850	2.79272	2.11211
14	14.000	1.63687	1800	2.80217	2.05762
15	15.000	1.64110	1750	2.84072	2.01293
16	16.000	1.65112	1700	2.87931	1.94995
17	17.000	1.64072	1650	2.92879	1.89065
18	18.000	1.63764	1600	2.99539	1.77568
19	19.000	1.63833	1550	3.06566	1.67994
20	20.000	1.63538	1500	3.13565	1.63567

Figure A.18: Output of **gam2m** corresponding to the parameter file shown in Figure A.17.

```
                        Parameters for NSCORE
                        *********************

START OF PARAMETERS:
97data.dat                              \data file
3   0                                   \variable, weight
-1.0e21    1.0e21                        \trimming limits
97data.nsc                              \output file for data
97data.trn                              \output transformation table
```

Figure A.19: Parameter file for **nscore** run.

```
Normal Score Transform:First 97 data of cluster.dat
  6
Xlocation
Ylocation
Primary
Secondary
Declustering Weight
Normal Score value
   39.5000    18.5000       .0600      .2200    4.0000   -2.5653
    5.5000     1.5000       .0600      .2700    4.0000   -2.1580
   38.5000     5.5000       .0800      .4000    3.5000   -1.9469
   20.5000     1.5000       .0900      .3900    4.5000   -1.7981
   27.5000    14.5000       .0900      .2400    3.3330   -1.6809
   40.5000    21.5000       .1000      .4800    2.3330   -1.4984
   15.5000     3.5000       .1000      .2100    3.0000   -1.5831
    6.5000    25.5000       .1100      .3600    4.0000   -1.3555
   38.5000    21.5000       .1100      .2200    2.8330   -1.4233
   23.5000    18.5000       .1600      .3000    4.5000   -1.2359
• • •
    2.5000     9.5000      6.2600    17.0200     .7220    1.2934
   32.5000    36.5000      6.4100     2.4500     .8610    1.3555
     .5000     8.5000      6.4900    14.9500     .8110    1.4233
   31.5000    45.5000      7.5300    10.2100     .7560    1.4984
    9.5000    29.5000      8.0300     5.2100    1.1000    1.5831
   39.5000    31.5000      8.3400     8.0200     .8430    1.6809
   17.5000    15.5000      9.0800     3.3200    1.1000    1.7981
    2.5000    14.5000     10.2700     5.6700    1.0610    1.9469
   30.5000    41.5000     17.1900    10.1000     .8890    2.1580
   35.5000    32.5000     18.7600    10.7600     .6460    2.5653
```

Figure A.20: Output of **nscore** corresponding to the parameter file shown in Figure A.19.

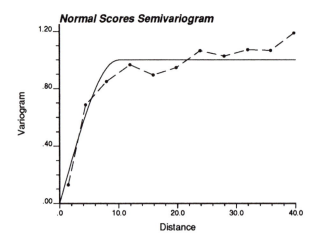

Figure A.21: Omnidirectional normal score semivariogram and its model fit (calculated from 97 data).

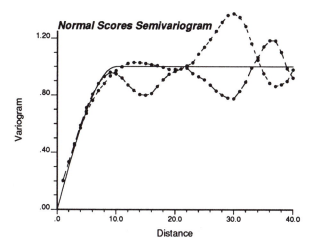

Figure A.22: Reference normal score semivariogram (calculated in the NS and EW directions) and their fit with the model calculated from 97 data.

```
                    Parameters for BIGAUS
                    *********************

START OF PARAMETERS:
bigaus.out                            \output file of variograms
3                                     \number of cutoffs
0.25    0.50     0.75                 \the cutoffs (cdf values)
1    80                               \ndir, nlag
0.5     0.0     0.0                   \xoff(id),yoff(id),zoff(id)
1    0.0                              \nst, nugget effect
1    10.0    1.00                     \it,aa,cc:        STRUCTURE 1
0.0  0.0    0.0  1.0 1.0              \ang1,ang2,ang3,anis1,anis2:
```

Figure A.23: Parameter file for **bigaus** run.

```
Model Indicator Variogram: cutoff =     -.674 Direction
  1    .000       .00000     1    1.17810      .00000
  2    .500       .04925     1     .86867      .26265
  3   1.000       .06980     1     .73955      .37225
  4   1.500       .08559     1     .64029      .45650
  5   2.000       .09888     1     .55684      .52734
  6   2.500       .11049     1     .48387      .58927
  7   3.000       .12085     1     .41875      .64455
  8   3.500       .13021     1     .35998      .69444
  9   4.000       .13870     1     .30664      .73972
 10   4.500       .14642     1     .25814      .78088
● ● ●
 71  35.000       .18750     1     .00000     1.00000
 72  35.500       .18750     1     .00000     1.00000
 73  36.000       .18750     1     .00000     1.00000
 74  36.500       .18750     1     .00000     1.00000
 75  37.000       .18750     1     .00000     1.00000
 76  37.500       .18750     1     .00000     1.00000
 77  38.000       .18750     1     .00000     1.00000
 78  38.500       .18750     1     .00000     1.00000
 79  39.000       .18750     1     .00000     1.00000
 80  39.500       .18750     1     .00000     1.00000
```

Figure A.24: Output of **bigaus** corresponding to the parameter file shown in Figure A.23.

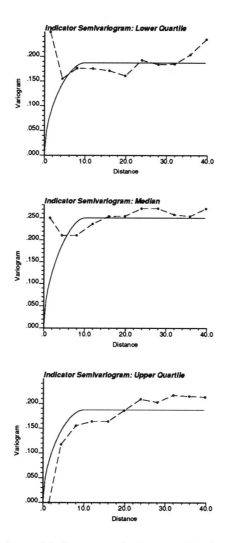

Figure A.25: Experimental indicator semivariograms (97 data) and their fits by Gaussian RF model-derived theoretical curves. The thick continuous line is the model.

A.3   Problem Set Three: Kriging

This solution set documents kriging results with some comments. The numerical results are less interesting than the interpretation.

Part A: Variogram Parameters

1. **Relative Nugget Effect:** When the relative nugget effect is small the data redundancy and the proximity to the point being estimated become important. This implies that close samples are weighted more and samples within clusters are treated as less informative. A low relative nugget effect may cause a wider variation in the weights (from negative to greater than one in some cases) which may cause outlier data values to have more of an effect. No general comment can be made about the magnitude of the estimate although a surface estimated with a large nugget effect will be smoother.

2. **Range:** When the range is smaller than the interdistance between any two data points then the situation is equivalent to a pure nugget effect model. As the range increases the continuity increases causing the proximity to the estimated location and data redundancy to become important. As the range becomes very large with a nugget effect then the estimation is equivalent to a pure nugget system. If there is no nugget effect then the kriging system behaves as if a linear variogram model were chosen.

3. **Scale:** The scaling of the variogram affects only the kriging (estimation) variance; only the relative shape of the variogram determines the kriging weights and estimate.

4. **Shape:** Both the exponential and spherical variograms have a linear shape near the origin. The exponential models growth from the origin is steeper, causing it to behave in a manner similar to a shorter range spherical model. The Gaussian model is parabolic near the origin, indicating a spatially very continuous variable. This may cause very large positive and negative weights (which are entirely reasonable given such continuity). If the actual data values are *not* extremely continuous and a Gaussian model is used, then erratic results can arise from the negative weights. Numerical precision problems also arise because of off-diagonal elements that have nearly zero variogram (maximum covariance) values. The problem is not so much the shape of the variogram as the lack of a nugget. The addition of a nugget does not change appreciably the shape of the variogram but improves numerical stability.

5. **Anisotropy:** The anisotropy factor may be applied to the range of the variogram, or equivalently to the coordinates (after rotation to identify

the main directions of continuity). It is instructive to think of geometric anisotropy as a geometric correction to the coordinates followed by application of an isotropic model.

Part B: Data Configuration

1. The proposed configuration is symmetric about the 45° degree line. As the second point approaches either endpoint the redundancy with the far point decreases and the redundancy with the near point increases. The rate of increase and decrease of the kriging weight differs depending on the variogram model. In the first case (low nugget effect $C_0 = 0.1$) the increase in redundancy with the near point is less than the decrease in redundancy with the far point; hence, the weight attributed to the second point increases as the point nears an endpoint.

2. With a high nugget effect ($C_0 = 0.9$) the increase in redundancy with the near point is greater than the decrease in redundancy with the far point; hence, the weight decreases as the point nears an endpoint. All the kriging weights in this case are almost equal.

Part C: Screen Effect

1. The first point gets more weight in all cases (unless a hole effect variogram model is considered). The weight applied to the second point decreases as it falls behind the first point. It is important to note that this weight may become negative, in which case the second point actually carries a significant information content. For example, a large negative weight (with an even larger positive weight applied to the closer data) would effectively impose the *trend* of the two data values (e.g., if the close datum is smaller than the far datum then the estimate will be less than either datum).

2. The screen effect is reduced when the nugget is large. Both samples have an almost equal contribution to the estimate.

Part D: Support Effect

1. Discretization with a 1 by 1 point is *equivalent* to point kriging; a greater level of discretization is required to perform block kriging. When the block size increases with respect to the average data spacing, the estimation variance drops considerably because of short scale averaging of errors. The discretization level is not as important as the relative size of the block.

2. The block is effectively larger than the scale of variability and its average smooths out most of the spatial variability. The estimation variance becomes small and the resulting surface (spatial distribution of block estimates) smooth.

Part E: Cross Validation

- **Cross Validation with Ordinary Kriging:** The scatterplots of the true values and kriging estimates exhibit the typical spatial smoothing of kriging. Conditional unbiasedness, i.e., overestimation of low values and underestimation of high values, can be observed from the scatterplot.

- **Histogram of Errors:** The histogram of the kriging errors shows a fairly symmetric distribution that is typically more "peaked" than a normal distribution. Note that any reasonable interpolation method can achieve global unbiasedness.

- **Map of Errors:** The errors are most significant near the high values but do not show any correlation (a variogram of the errors would confirm this).

- **Kriging Variance:** If the kriging variance was a good measure of local accuracy there should be a positive correlation between the actual absolute errors and the kriging variance. The significant negative correlation found with the sample of 140 is due to the fact that there is clustering in the high values. Near the high values the kriging variance is lowest because extra samples were preferentially located near these highs. Conversely, the actual error happens to be greatest near the high values because the variable is heteroscedastic (its local variance depends on its magnitude). The kriging variance is generally not a good measure of local accuracy because the data values are not taken into account. Further, the SK variance is always less than the OK variance which in-turn is always less than the KT variance. This is due to the additional uncertainty in estimating the mean (surface), which is assumed known in SK.

- **Kriging Parameters:** Often the choice of kriging parameters (search radius, minimum and maximum number of data) is based more on data availability than on the physical process underlying the data. Enough data must be found to provide a stable estimate; too many data will unnecessarily smooth the kriging estimates and, possibly, create matrix instability.

- **Simple Kriging:** SK requires knowledge of the stationary mean. In this case the mean is given a considerable weight (due to the high nugget effect), causing significant smoothing and conditional bias.

- **Kriging with a Trend:** KT allows the locally estimated trend (mean) surface to be fit with monomials of the coordinates. In actuality it behaves much like OK in interpolation conditions and may yield erratic results in extrapolation (unless the trend model is appropriate); see also [97].

Part F: Kriging on a Grid

- **Simple Kriging:** the mean and variance of the kriging estimates are considerably less than the true values. The general trends are reproduced quite well.

- **Minimum Number of Data:** This effectively prevents the kriging algorithm from estimating in areas where there are too few data. The fewer data retained the more (artificially) variable the estimate becomes.

- **Maximum Number of Data:** The more data that are used the smoother the estimates are appear. In general, enough data should be used so that the estimate is stable and does not present any artificial discontinuities. Ordinary kriging allows accounting for all data in the search neighborhood, even if they are not correlated with the point being estimated, because the mean is implicitly re-estimated locally. For that reason it may not be appropriate to use a large maximum limit *ndmax* even if the computer can handle it. Note that the computation cost of matrix inversion increases as the number of samples cubed.

- **Search Radius:** The search radius has to be large enough to allow a stable estimate. If the search radius is set too large, however, the resulting surface may be too smooth. Note that the search radius and the maximum number of data interact with one another. If *ndmax* is set small enough it does not matter how large the search radius is set and vice versa. It is generally an error to set a priori the search radius smaller than the correlation range, because in OK remote data contribute to the local re-estimation of the mean.

- **Ordinary Kriging:** OK results appear similar to SK except for less smoothing and conditional bias. By accumulating the kriging weights applied to each datum and then normalizing these weights to sum to one, reasonable declustering weights are obtained.

- **Kriging with a Trend:** KT results appear similar to OK except that they are more erratic near the borders. In fact there are some negative estimates due to the trend surface being estimated unreasonably low (the kriging systems do not know that contaminant concentration cannot be negative).

- **Block Kriging:** The block kriging results are similar to the point kriging results because the block size is small. The kriging variance is less due to the averaging of errors. The smoothing effect of block kriging increases with larger blocks.

- **Trend Surface:** The shape of the trend surface in interpolation areas is similar to that indicated by a contour map of block kriging estimates with a large block size.

A.4   Problem Set Four: Cokriging

Cokriging is a form of kriging where the data can relate to attribute(s) different from that being estimated; see section IV.1.7.

The problem proposed here consists of estimating the 2500 reference z-values, see Figure A.26, from the sparse sample of 29 z-data in file $\boxed{\text{data.dat}}$ and 2500 related y-data in file $\boxed{\text{ydata.dat}}$. The gray scale map of the y-data is given in Figure A.27.

Figure A.28 gives the histograms of the 2500 reference z-values to be estimated, the 29 z-data, the 2500 secondary y-data and the 29 y-data collocated with the z-data.

Figure A.29 gives the scattergram of the 29 pairs of collocated z and y-data. The rank correlation (0.88) is seen to be larger than the linear correlation (0.70). Also heteroscedasticity is observed: the variance of z-values varies depending on the class of y-values. These features indicate that there is potentially more information to gather from the secondary y-information than the mere linear correlation and regression; see [99] and section IV.1.12.

The variograms were calculated from the 140 collocated $(z$-$y)$ data. Figure A.30 gives the two experimental direct semivariograms and the experimental cross semivariogram and their model fits by the linear model of coregionalization (A.4). For each variogram, the two NS and EW directions with tolerance $\pm 22.5^o$ and the omnidirectional curve (tolerance $\pm 90^o$) were considered. The traditional expression (III.1) for the direct semivariograms and (III.2) for the cross semivariogram were used. The range (10) was better picked from the pairwise relative semivariograms (III.6) for both z and y-variable, not shown here. The isotropic linear model of coregionalization consists of a nugget effect and a spherical structure of range 10:

$$\begin{cases} \gamma_Z(\mathbf{h}) & = & 10.00 & + & 32.0 \; Sph \; (|\mathbf{h}|/10) \\ \gamma_Y(\mathbf{h}) & = & .10 & + & 19.0 \; Sph \; (|\mathbf{h}|/10) \\ \gamma_{ZY}(\mathbf{h}) & = & .01 & + & 16.0 \; Sph \; (|\mathbf{h}|/10) \; , \; \text{for} \; |\mathbf{h}| > 0. \end{cases} \qquad (A.4)$$

The determinants of the nugget coefficients and the spherical structure factors are positive, thus this model is legitimate ([76], p. 391):

$$\begin{vmatrix} 10.0 & .01 \\ .01 & .10 \end{vmatrix} = 1 - .0001 > 0 \; ; \; \begin{vmatrix} 32.0 & 16.0 \\ 16.0 & 19.0 \end{vmatrix} = 608 - 256 > 0$$

Ordinary kriging with program okb2d was performed to estimate the 2500 z-reference values using only the 29 primary z-data and the corresponding γ_Z-semivariogram model given in (A.4). There were no negative kriging estimates. The resulting gray scale estimated map is given in Figure A.31 and the histogram of estimation errors $(z^* - z)$ is given in Figure A.32. The OK estimated map can be compared with the reference gray scale map of Figure A.26.

Figures A.33 and A.34 give the corresponding input parameter file to program **okb2d** and the 10 first and 10 last lines of the output file.

Cokriging using both the 29 primary z-data and the 2500 secondary data was performed with program **cokb3d**. The corresponding input parameter file and 10 first and 10 last lines of the output file are given in Figures A.35 and A.36. The traditional ordinary cokriging option was considered with the two unbiasedness conditions (IV.22). Because the y-data are much more numerous than the z-data, a different search strategy was considered for z and y-data. Figure A.37 gives the gray scale map of the cokriging estimates and Figure A.38 gives the histogram of the cokriging errors $(z^* - z)$. The cokriging map is to be compared with the reference map of Figure A.26 and the OK map of Figure A.31. The improvement brought by the large amount of secondary y-data is seen to be marginal, and possibly not worth the additional modeling and computation effort demanded by cokriging.

Because the y-data were generated as moving averages of z-values taken at a different level, the external drift concept is ideally suited to integrate such information. Program **ktb3d** with the external drift option was used; see section IV.1.5.

Figures A.39 and A.40 provide the input parameter file and the 10 first and 10 last lines of the output file. The resulting set of 2500 z-estimates include 65 negative values which were reset to zero.[3] This large number of negative estimates is explained by the fact that kriging with an external drift amounts to extrapolating the y-trend to the z-values, thus incurring a risk of extrapolating beyond the limit zero z-value. Figure A.41 gives the gray scale map of the z-estimates and Figure A.42 gives the histogram of errors (z^*-z). The results appear significantly better than those provided by kriging using only the 29 primary z-data; see Figure A.31 and the reference map of Figure A.26.

---

[3] This resetting to zero of negative estimates corresponds to the solution that would be provided by quadratic programming with the condition $z^*(x) \geq 0$; see [108].

Reference z-values

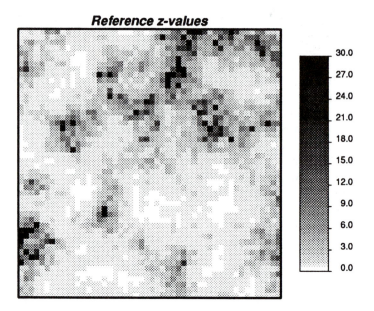

Figure A.26: Reference z-values to be estimated.

Secondary data

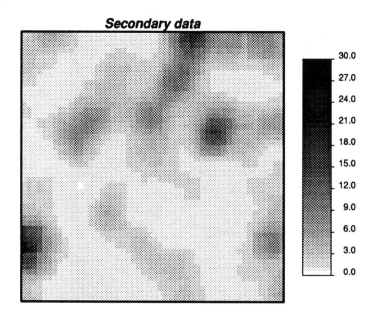

Figure A.27: Secondary information (2500 y-values).

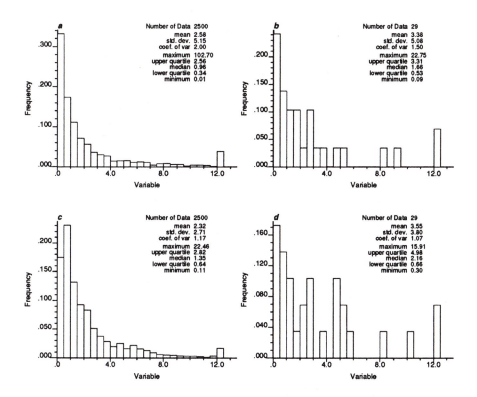

Figure A.28: Histograms and statistics. **a:** 2500 reference z-values, **b:** 29 z-data, **c:** 2500 secondary y-data, **d:** 29 y-data collocated with the z-data.

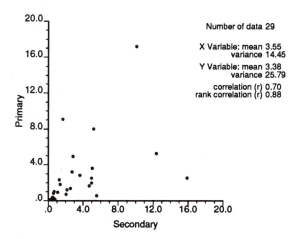

Figure A.29: Scattergram of collocated z-primary and y-secondary data.

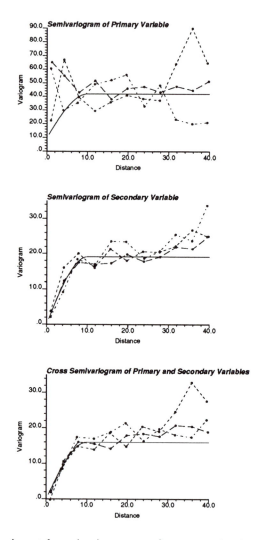

Figure A.30: Experimental semivariograms and cross semivariogram and their model fits (140 collocated z-y data). The thick continuous line is the model.

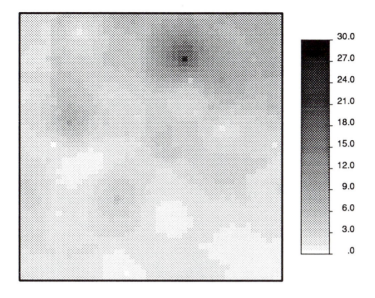

Figure A.31: Gray scale map of OK z-estimated values (29 data).

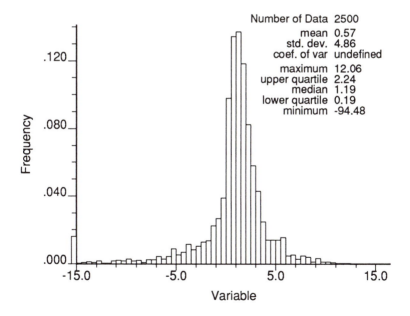

Figure A.32: Histogram of OK kriging errors $(z^* - z)$.

```
                        Parameters for OKB2D
                        ********************

START OF PARAMETERS:
data.dat                                    \data file
1    2    3                                 \columns for x,y, and variable
-1.0e21    1.0e21                           \data trimming limits
okb2d.out                                   \output File of Kriged Results
1                                           \debugging level: 0,1,2,3
okb2d.dbg                                   \output file for debugging
50    0.5    1.0                            \nx,xmn,xsiz
50    0.5    1.0                            \ny,ymn,ysiz
1    1                                      \x and y block discretization
4    16                                     \min and max data for kriging
25.0                                        \maximum search radius
1    10.0                                   \nst, nugget effect
1    10.0    32.0    0.00    1.00           \it,aa,cc,ang1,anis: structure 1
```

Figure A.33: Parameter file for **okb2d**.

```
OKB2D ESTIMATES WITH: Irregular Spaced 29 primary and secondary data
2
Estimate
Estimation Variance
    2.048    42.183
    1.874    42.147
    1.874    42.147
    1.874    42.147
    1.874    42.147
    1.874    42.147
    1.524    42.118
    1.663    42.123
    1.663    42.123
    1.663    42.123
  • • •
    6.721    42.191
    5.794    42.160
    6.358    42.191
    6.358    42.191
    6.358    42.191
    6.358    42.191
    6.358    42.191
    6.358    42.191
    6.358    42.191
    6.358    42.191
```

Figure A.34: Output of **okb2d** corresponding to the parameter file shown in Figure A.33.

```
                    Parameters for COKB3D
                    *********************

START OF PARAMETERS:
ydata.dat                            \data file
2                                    \number of variables primary+oth
1    2    0    3    4                 \columns for x,y,z and variables
-0.5        1.0e21                    \data trimming limits
2                                    \ktype (0=SK,1=OK,2=OK-trad)
3.38  2.32  0.00  0.00               \mean(i),i=1,nvar
cokb3d.out                           \output file for kriging results
1                                    \debugging level: 0,1,2,3
cokb3d.dbg                           \output file for debugging
50    0.5    1.0                     \nx,xmn,xsiz
50    0.5    1.0                     \ny,ymn,ysiz
1     0.0    1.0                     \nz,zmn,zsiz
1     1      1                       \x, y, and z block discret
1     12     8                       \min, max primary, max secondary
25.0  10.0                           \maximum search radius: prim,sec
1     1                              \semivariogram for "i" and "j"
1     10.0                           \  number of structures, nugget
1     10.0   32.0                    \  it,aa,cc:        structure 1
0.0   0.0    0.0   1.0 1.0           \  ang1,ang2,ang3,anis1,anis2:
1     2                              \semivariogram for "i" and "j"
1     0.01                           \  number of structures, nugget
1     10.0   16.0                    \  it,aa,cc:        structure 1
0.0   0.0    0.0   1.0 1.0           \  ang1,ang2,ang3,anis1,anis2:
2     2                              \semivariogram for "i" and "j"
1     0.10                           \  number of structures, nugget
1     10.0   19.0                    \  it,aa,cc:        structure 1
0.0   0.0    0.0   1.0 1.0           \  ang1,ang2,ang3,anis1,anis2:
```

Figure A.35: Parameter file for cokb3d.

```
COKB3D with:All of the secondary data plus 29 primar
2
estimate
estimation variance
        2.6481       46.3685
        2.1062       44.5336
        1.7395       44.5291
        2.0076       44.5336
        1.3728       44.5429
        1.3908       44.5546
        1.4648       43.8448
        1.6393       43.5830
        1.7466       43.5866
        1.7040       43.5866
• • •
        7.7541       46.4266
        8.3422       48.3173
        4.8839       46.4266
        7.7715       46.4266
        7.5002       46.4266
        7.0268       46.4266
        6.9845       46.4266
        7.9966       46.4266
        6.3431       46.4266
        5.5095       46.7878
```

Figure A.36: Output of cokb3d corresponding to the parameter file shown in Figure A.35.

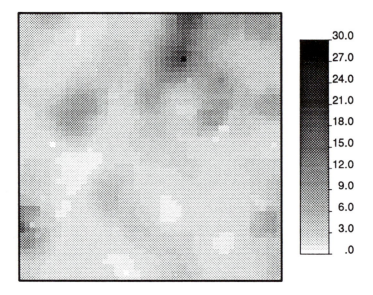

Figure A.37: Gray scale map of cokriging z-estimated values.

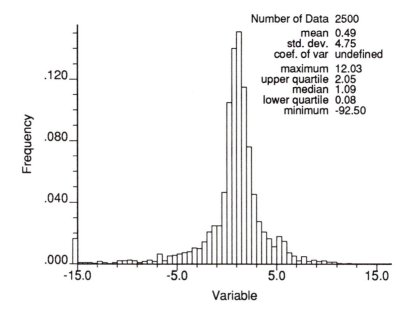

Figure A.38: Histogram of cokriging errors $(z^* - z)$.

```
                          Parameters for KTB3D
                          ********************

START OF PARAMETERS:
data.dat                              \data file
1    2    0    3                      \column for x,y,z and variable
-1.0e21   1.0e21                      \data trimming limits
ktb3d.out                            \output file of kriged results
1                                     \debugging level: 0,1,2,3
ktb3d.dbg                            \output file for debugging
50   0.5    1.0                       \nx,xmn,xsiz
50   0.5    1.0                       \ny,ymn,ysiz
1    0.5    1.0                       \nz,zmn,zsiz
1    1     1                          \x,y and z block discretization
1    16                               \min, max data for kriging
0                                     \max per octant (0-> not used)
25.0                                  \maximum search radius
0.0  0.0  0.0  1.0  1.0               \search: ang1,2,3,anis1,2
0    2.302                            \1=use sk with mean, 0=ok+drift
0 0 0 0 0 0 0 0 0                      \drift: x,y,z,xx,yy,zz,xy,xz,zy
0                                     \0, variable; 1, estimate trend
1                                     \1, then consider external drift
4                                     \column number in original data
ydata.dat                            \Gridded file with drift variabl
4                                     \column number in gridded file
1    10.0                             \nst, nugget effect
1    10.0   32.0                      \it,aa,cc:          structure 1
0.0  0.0    0.0  1.0  1.0             \ang1,ang2,ang3,anis1,anis2:
```

Figure A.39: Parameter file for **ktb3d**.


```
KTB3D ESTIMATES WITH: Irregular Spaced 29 primary and secondary data
2
Estimate
EstimationVariance
      2.017    49.702
      1.827    48.212
      1.776    48.353
      1.728    48.576
      1.604    49.552
      1.579    49.821
      1.205    48.106
      1.380    48.321
      1.390    48.245
      1.391    48.236
• • •
      8.206    51.238
      8.563    52.018
      7.147    50.294
      9.203    53.641
      9.574    54.649
      9.559    54.604
      8.902    52.913
      9.404    54.172
      7.835    50.992
      7.247    50.370
```

Figure A.40: Output of **ktb3d** corresponding to the parameter file shown in Figure A.39.

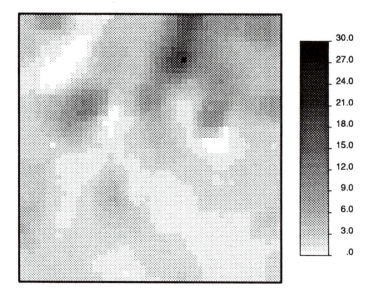

Figure A.41: Gray scale map of kriging with external drift z-estimated values.

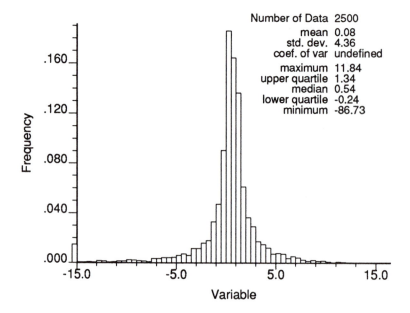

Figure A.42: Histogram of errors of kriging with external drift $(z^* - z)$.

A.5 Problem Set Five: Indicator Kriging

The main objective of indicator kriging is not to provide an indicator value estimate, but to provide a model of uncertainty about the unsampled original value. That model of uncertainty takes the form of a conditional cumulative distribution function (*ccdf*), recall relation (IV.26):

$$\text{Prob}^* \left\{ Z(\mathbf{u}) \le z \mid (n) \right\} = E^* \left\{ I(\mathbf{u}; z) \mid (n) \right\}$$

The first column of Table A.1 gives the 9 decile cutoffs of the declustered marginal *cdf* $F^*(z_k)$ of the 140 z-data. The third column gives the corresponding *non*-declustered *cdf* values $\hat{F}(z_k)$ related to the raw indicator data variances:

$$\widehat{Var} \left\{ I(\mathbf{u}; z_k) \right\} = \hat{F}(z_k) \left[1 - \hat{F}(z_k) \right]$$

Figures A.43 and A.44 give the corresponding 9 omnidirectional sample indicator semivariograms and their model fits. All semivariograms have been standardized by the non-declustered indicator data variance $\hat{F}(z_k)[1 - \hat{F}(z_k)]$. All nine models are of the same type:

$$\gamma_I(\mathbf{h}; z_k) = C_0(z_k) + C_1(z_k)\, Sph(|\mathbf{h}|/11) + C_2(z_k)\, Sph(|\mathbf{h}|/30)$$

$$\text{with} : \quad C_0(z_k) + C_1(z_k) + C_2(z_k) = 1 \ , \ \forall\, k = 1, \ldots, 9$$

$$\text{(A.5)}$$

Table A.1 gives the list of the sill parameter values $C_0(z_k), C_1(z_k), C_2(z_k)$. The corresponding model fits should be evaluated globally over all nine cutoffs z_k, because the parameter values were adjusted so that they vary smoothly with cutoff.

Note that, except for three extreme cutoffs, the model is the same. The lowest sample data (indicator at cutoff z_1) appear more spatially correlated, whereas the highest sample data (indicators at cutoffs z_8 and z_9) appear less correlated. In the absence of ancillary information confirming this different behavior of extreme values and given the uncertainty due to small sample size and preferential data clustering, one could adopt the same model for all relative indicator variograms, i.e. a median IK model; see relation (IV.30) and hereafter.

Indicator kriging using program **ik3d** was performed with the 9 cutoffs z_k and semivariogram models specified by model (A.5) and Table A.1. The resulting *ccdf*'s are corrected for order relations problems (within **ik3d**). Figure A.45 gives the input parameter file for the **ik3d** run and Figure A.46 gives the first and last 10 lines of output.

A correct handling of within-class interpolation and extreme classes extrapolation is critical when using IK. Interpolation using the marginal declustered sample *cdf* was adopted for all classes.

Program **postik** provides the E-type estimate for the unsampled z-value, i.e., the mean of the calculated *ccdf*. Figure A.47 gives the input parameter file for the **postik** run; the 10 first and 10 last lines of the **postik** output

	z_k	$F^*(z_k)$	$\bar{F}(z_k)$	$C_0(z_k)$	$C_1(z_k)$	$C_2(z_k)$
$k=1$	0.159	0.1	0.079	0	0.5	0.5
	0.278	0.2	0.150	0	1.0	0
	0.515	0.3	0.221	0	1.0	0
	0.918	0.4	0.300	0	1.0	0
	1.211	0.5	0.379	0	1.0	0
	1.708	0.6	0.443	0	1.0	0
	2.325	0.7	0.521	0	1.0	0
	3.329	0.8	0.643	0.4	0.6	0
$k=9$	5.384	0.9	0.757	0.9	0.1	0

Table A.1: Parameters for the indicator variogram models.

file are shown in Figure A.48. Figure A.49 gives the gray scale map of the 2500 E-type estimates and the corresponding histogram of the error (z^*-z). This E-type map compares favorably to the reference map of Figure A.26. The location of the conditioning data, see Figure II.10, are apparent on the E-type map; the configuration of the clustered data appears as dark "plus" signs. Figure A.49a may not be compared directly to the ordinary kriging, cokriging, or kriging with an external drift of problem set four because more data have been used in this case and a different gray scale has been adopted.

The discontinuities apparent at the data locations are due to the preferential clustering of the 140 data combined with the large nugget effect adopted for the high cutoff values; see Table A.1. One way to avoid such discontinuities is to shift the estimation grid slightly so that the data do not fall on grid node locations. Another solution is to arbitrarily set all nugget constants to zero. Discontinuities are also present but much less apparent on the figures of problem set four because only 29 unclustered data were used.

Program **postik** also provides any p-quantile value derived from the *ccdf*, for example the .5 quantile or conditional median, also called M-type estimate: this is the estimated value that has equal chance to be exceeded or not exceeded by the actual unknown value. Figure A.50 provides gray scale maps for the M-type estimate and the .1 quantile. Locations with high .1 quantile values are almost certainly high in the sense that there is a 90% chance that the actual true values would be even higher.

Last, program **postik** also provides the probability of exceeding any given threshold value z_0, which need not be one of the original cutoffs used in **ik3d**. Figure A.51 provides the isopleth maps of the probability of *not* exceeding the low threshold value $z_1 = .16$ and the probability of exceeding the high threshold value $z_9 = 5.39$. Dark areas indicate zones that are most likely low z-valued on the first map, and zones most likely high z-valued in the second map. These dark areas relate well to the corresponding low and high z-valued zones of the reference map of the Figure A.26. At the data locations there is no uncertainty and the gray scale is either white (the data is below 5.39) or black (the data is above 5.39).

Sensitivity to Indicator Variogram Models

The last probability map, that of exceeding the high cutoff $z_9 = 5.39$, was produced again starting from

- a median IK model where all indicator semivariograms are made equal to the median indicator model:

$$\gamma_I(\mathbf{h}; z_k) \,/\, F(z_k)[1 - F(z_k)] \;=\; Sph(|\mathbf{h}|/11) \qquad (A.6)$$

- a model stating a stronger continuity of high z-values with a NW-SE direction of maximum continuity. The model is of type (A.6), except for the last three cutoffs z_7, z_8, z_9. For these cutoffs z_7, z_8, z_9, the minor range in direction SW-NE is kept at 11. The major range is direction NW-SE is increased to 22 for z_7, 44 for z_8, and 88 for z_9. The largest range 88 is greater than the dimension of the field and, for all practical purposes, the corresponding indicator variogram is linear without reaching its sill.

Figure A.52-a gives the probability map (of exceeding z_9) using the median IK model. When compared to Figure A.51-b this map shows essentially the same features although with much more contrast. Median IK is much faster CPU-wise.

Figure A.52-b gives the probability map corresponding to a model with artificial greater continuity of high z-values in the diagonal direction NW-SE. The difference with Figure A.51-b is dramatic: the dark areas are now all tightly connected diagonally.

Figure A.53 gives the probability map corresponding to a model in all points similar to that used for Figure A.52-b except that the major direction of continuity is now NE-SW. Figure A.52-b and A.53 are conditional to the same 140 sample data; their differences are strictly due to the "soft" structural information introduced through the indicator variogram models.

As opposed to Gaussian-related models and the median IK model where only one variogram is available for the whole range of z-values, multiple IK with as many different models as cutoff values allows a greater flexibility in modeling different behaviors of different classes of z-values. Recall though from relation (IV.38) that a measure of continuum in the variability of indicator variogram model parameters should be preserved from one cutoff to the next.

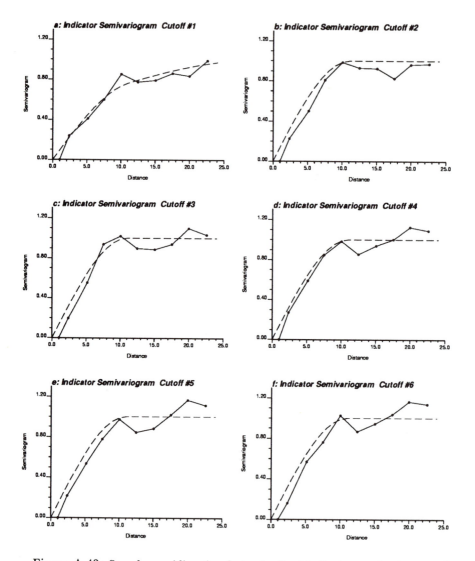

Figure A.43: Sample omnidirectional standardized indicator semivariograms (140 data) and their model fits.

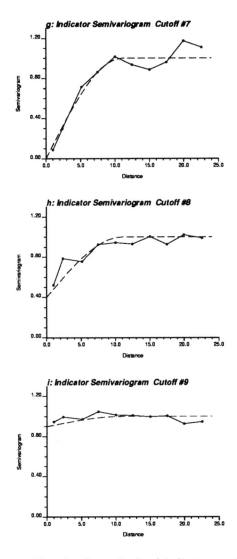

Figure A.44: Sample omnidirectional standardized indicator semivariograms (140 data) and their model fits.

Parameters for IK3D
****

```
START OF PARAMETERS:
cluster.dat                          \data file
1    2    0    3                     \column for x,y,z and variable
direct.ik                            \direct indicator input (soft)
-1.0e21    1.0e21                     \data trimming limits
ik3d.out                             \output file of kriged results
1                                    \debugging level: 0,1,2,3
ik3d.dbg                             \output file for debugging
50    0.5    1.0                     \nx,xmn,xsiz
50    0.5    1.0                     \ny,ymn,ysiz
1     0.0    1.0                     \nz,zmn,zsiz
1    16                              \min, max data for kriging
25.0                                 \maximum search radius
0.0  0.0  0.0  1.0  1.0              \search: ang1,2,3,anis1,2
0                                    \max per octant (0-> not used)
0    1.211                           \0=full IK, 1=Med IK (cutoff)
1                                    \0=SK, 1=OK
9                                    \number cutoffs
0.159  0.10 2  0.00                  \cutoff, global cdf, nst, nugget
       1   11.0  0.50                \       it, aa, cc
       0.0  0.0   0.0  1.0  1.0      \       ang1,ang2,ang3,anis1,2
       1   30.0  0.50                \       it, aa, cc
       0.0  0.0   0.0  1.0  1.0      \       ang1,ang2,ang3,anis1,2
0.278  0.20 1  0.00                  \cutoff, global cdf, nst, nugget
       1   11.0  1.00                \       it, aa, cc
       0.0  0.0   0.0  1.0  1.0      \       ang1,ang2,ang3,anis1,2
0.515  0.30 1  0.00                  \cutoff, global cdf, nst, nugget
       1   11.0  1.00                \       it, aa, cc
       0.0  0.0   0.0  1.0  1.0      \       ang1,ang2,ang3,anis1,2
0.918  0.40 1  0.00                  \cutoff, global cdf, nst, nugget
       1   11.0  1.00                \       it, aa, cc
       0.0  0.0   0.0  1.0  1.0      \       ang1,ang2,ang3,anis1,2
1.211  0.50 1  0.00                  \cutoff, global cdf, nst, nugget
       1   11.0  1.00                \       it, aa, cc
       0.0  0.0   0.0  1.0  1.0      \       ang1,ang2,ang3,anis1,2
1.708  0.60 1  0.00                  \cutoff, global cdf, nst, nugget
       1   11.0  1.00                \       it, aa, cc
       0.0  0.0   0.0  1.0  1.0      \       ang1,ang2,ang3,anis1,2
2.325  0.70 1  0.00                  \cutoff, global cdf, nst, nugget
       1   11.0  1.00                \       it, aa, cc
       0.0  0.0   0.0  1.0  1.0      \       ang1,ang2,ang3,anis1,2
3.329  0.80 1  0.40                  \cutoff, global cdf, nst, nugget
       1   11.0  0.60                \       it, aa, cc
       0.0  0.0   0.0  1.0  1.0      \       ang1,ang2,ang3,anis1,2
5.384  0.90 1  0.90                  \cutoff, global cdf, nst, nugget
       1   11.0  0.10                \       it, aa, cc
       0.0  0.0   0.0  1.0  1.0      \       ang1,ang2,ang3,anis1,2
```

Figure A.45: Input parameter file for ik3d.

```
IK3D Estimates with:Clustered 140 primary and secondary data
   9
Cutoff:  1 at        .15900
Cutoff:  2 at        .27800
Cutoff:  3 at        .51500
Cutoff:  4 at        .91800
Cutoff:  5 at       1.21100
Cutoff:  6 at       1.70800
Cutoff:  7 at       2.32500
Cutoff:  8 at       3.32900
Cutoff:  9 at       5.38400
     .0000   .0653   .0653   .1058   .1058   .1058   .1058   .3398   .5451
     .0176   .0768   .0768   .1029   .1029   .1029   .1029   .3389   .5553
     .1309   .1520   .1520   .1714   .1714   .1714   .1714   .3705   .5656
     .3823   .3823   .3823   .4021   .4021   .4021   .4021   .4948   .6270
     .6435   .6435   .6435   .6435   .6435   .6435   .6435   .6435   .6558
     .7463   .7580   .7580   .7662   .7662   .7662   .7662   .7662   .7662
     .7458   .7951   .7951   .8187   .8187   .8187   .8187   .8187   .8187
     .7030   .7979   .8118   .8219   .8219   .8219   .8219   .8219   .8219
     .6547   .7992   .8214   .8214   .8214   .8214   .8214   .8214   .8214
     .6182   .8369   .8369   .8415   .8415   .8415   .8415   .8415   .8415
     .5302   .8807   .8807   .8807   .8807   .8807   .8807   .8807   .8807
     .4982   .8807   .8807   .8807   .8807   .8807   .8807   .8807   .8807
     .4994   .8710   .8710   .8950   .8950   .8950   .8950   .8950   .8950
 • • •
     .0000   .0000   .0000   .0000   .0000   .0000   .0000   .4823   .5805
     .0000   .0000   .0000   .0000   .0000   .0000   .0006   .5438   .6324
     .0000   .0000   .0000   .0000   .0000   .0000   .0000   .4964   .6796
     .0000   .0000   .0000   .0000   .0000   .0000   .0000   .4555   .7303
     .0000   .0000   .0000   .0000   .0000   .0000   .0000   .4085   .7276
     .0000   .0000   .0000   .0000   .0000   .0000   .0000   .3119   .7788
     .0000   .0000   .0000   .0000   .0000   .0000   .0000   .2692   .7773
     .0000   .0000   .0000   .0000   .0000   .0000   .0000   .2303   .7660
     .0000   .0000   .0000   .0000   .0000   .0000   .0000   .2085   .7640
     .0000   .0000   .0000   .0000   .0000   .0000   .0000   .1995   .7621
```

Figure A.46: Output of ik3d corresponding to the parameter file shown in Figure A.45.

```
                 Parameters for POSTIK
                 ********************

START OF PARAMETERS:
ik3d.out                            \input from IK3D
postik.out                          \output file
1    0.5                            \output option, output parameter
9                                   \number of cutoffs
0.159 0.278 0.515 0.918 1.211 1.708 2.325 3.329 5.384 \the cutoffs
0    1       0.75                   \volume support, type, varred
cluster.dat                         \global distribution
3    5    -1.0    1.0e21            \ivr, iwt, tmin, tmax
0.0     30.0                        \minimum and maximum Z value
3    1.0                            \lower tail: option, parameter
3    1.0                            \middle    : option, parameter
3    1.0                            \upper tail: option, parameter
100                                 \maximum discretization
```

Figure A.47: Input parameter file for postik.

```
E-type mean values
1
mean
        7.2586
        7.1817
        6.8632
        5.5479
        4.4362
        3.0028
        2.3608
        2.3328
        2.3382
        2.0746
 • • •
        7.0011
        6.4576
        6.0982
        5.7669
        5.8546
        5.5762
        5.6518
        5.8007
        5.8482
        5.8775
```

Figure A.48: Output of **postik** corresponding to the parameter file shown in Figure A.47.

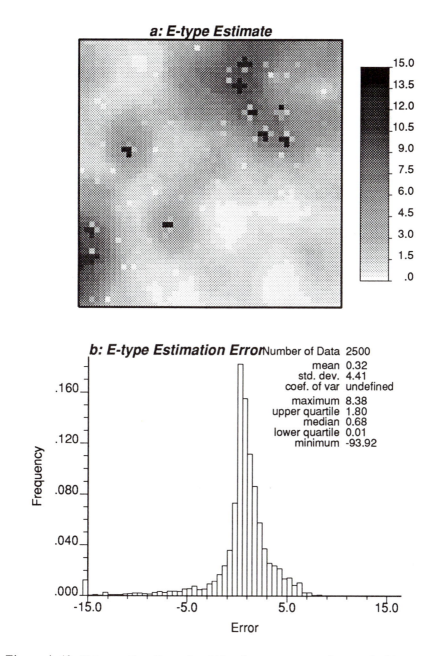

Figure A.49: E-type estimation using 140 z-data. **a:** gray scale map, **b:** histogram of errors (z^*-z).

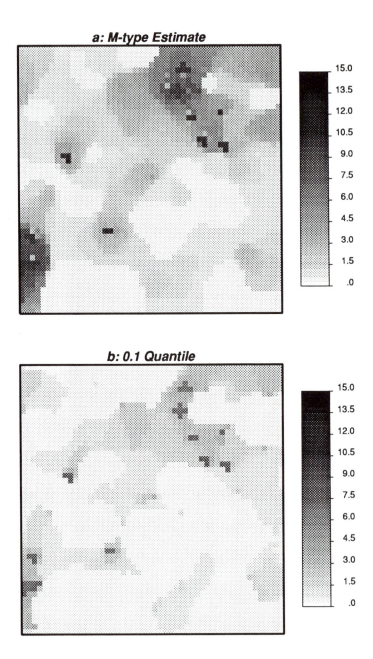

Figure A.50: Conditional quantile maps using 140 z-data. **a:** .5 quantile or M-type estimate map, **b:** .1 quantile map.

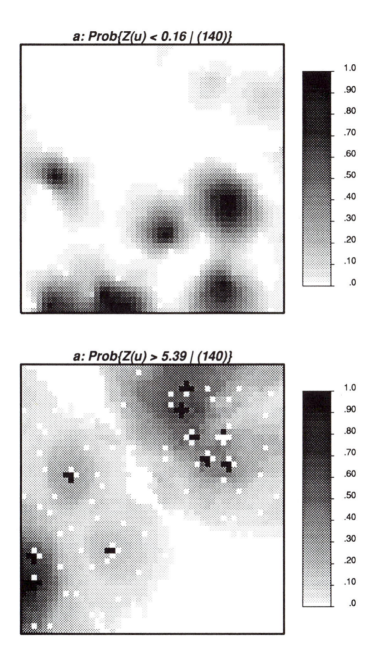

Figure A.51: Probability maps using 140 z-data. **a:** Probability of not exceeding $z_1 = .16$, **b:** Probability of exceeding $z_9 = 5.39$.

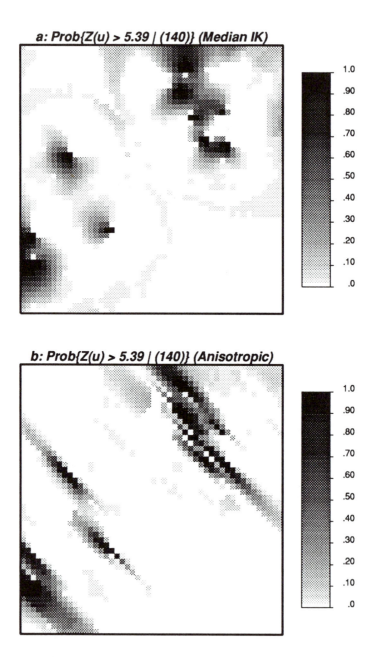

Figure A.52: Probability of exceeding $z_9 = 5.39$, **a:** with a median IK model, **b:** with a model indicating greater continuity of high z-values in direction NW-SE.

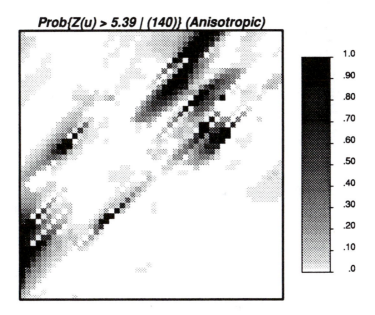

Figure A.53: Probability of exceeding $z_9 = 5.39$ under the same condition as Figure A.52-b but the direction of extreme-values continuity is NE-SW.

A.6   Problem Set Six: Simulations

There are two related objectives for stochastic simulation. The first is to create realizations that honor, up to statistical fluctuations, the histogram, variogram (or set of indicator variograms) and conditioning data. These realizations overcome the smoothing effect of kriging if used as estimated maps. The second objective is to assess spatial uncertainty through multiple realizations.

The gray scale used throughout most of this solution set is that considered for the reference images of Figures A.26 and A.27, i.e., between 0-white and 30-black.

Part A: Gaussian Simulation

As discussed in Chapter V the most straightforward algorithm for simulation of continuous variables is the sequential Gaussian algorithm. The 140 data in cluster.dat are used with declustering weights in **sgsim** to generate the two realizations shown at the top of Figure A.54. The parameter file for **sgsim** is shown in Figure A.55 and the first and last 10 lines of the corresponding output file are shown in Figure A.56.

A number of typical checks are also shown in Figure A.54. It is good practice to ensure that the univariate distribution and the normal scores variogram is reproduced within reasonable statistical fluctuations. The input isotropic variogram model is reproduced fairly well. The deviations from the model are explained by both statistical fluctuations and the fact that the input model (A.3) is not deduced from all the 140 conditioning data.

Sensitivity analysis to the input (normal score) variogram model can be done retaining the same 140 conditioning data. Figure A.57 shows three different input variograms corresponding to increasing spatial continuity and an example realization corresponding to each variogram.

A large number of realizations can be generated and used in conjunction with the **postsim** program to generate quantile maps and maps showing the probability of exceeding a cutoff. Note that these maps can also be generated as a direct result of multiGaussian kriging or indicator kriging; see Figure A.51.

The matrix approach to Gaussian simulation is particularly efficient for small scale simulations to be used for change of support problems; see section IV.1.13 and V.2.4. The program **lusim** can be used to simulate a grid of normal score point values within any block of interest; see parameter file and output in Figures A.58 and A.59. These normal score values are back-transformed (with program **backtr**) and averaged to generate block values. Sixteen point values discretizing a block were simulated 100 times and then averaged to create 100 simulated block values. The histogram of point and block values are shown in Figure A.60. Note that the linear averaging reduces the variance but leaves the overall mean unchanged.

Part B: Indicator Simulation

One advantage of the indicator formalism is that it allows a more complete specification of the bivariate distribution than afforded, e.g., by the multi-Gaussian model. Another advantage is that it allows the straightforward incorporation of secondary information and constraint intervals. An example of two realizations generated with **sisim** are shown in Figure A.61. The parameter file and a portion of the output file are shown in Figures A.62 and A.63. The apparent discontinuity of the very high simulated values is due to the fact that only three cutoffs were used, entailing artificially large within-class noise.

More cutoffs can be used to reduce the within-class noise. The variogram modeling, however, may then become tedious, order relation problems are likely to increase, and the computational requirements increase. As discussed in section IV.1.9, the median IK approximation is one way to increase the number of cutoffs without any of the disadvantages just mentioned. The price for the median IK approximation is that all indicator variograms are proportional, i.e., the standardized variograms are identical. Figure A.64 shows two realizations generated with **sisim** using 9 cutoffs (same as in problem set 5) and the median IK option (**mik=1**).

Once again it is good practice to ensure that the univariate distribution and the indicator variograms are reproduced within reasonable statistical fluctuations. Some of these checks are shown in Figure A.61. Quantile maps and maps of the probability of exceeding a cutoff can be derived if a large number of realizations have been generated.

Part C: Markov-Bayes Indicator Simulation

The Markov-Bayes simulation algorithm coded in program **mbsim** allows using soft prior information as provided by a secondary data set. A full indicator cokriging approach to handle the secondary variable would be quite demanding in that a matrix of cross covariance models must be inferred. This inference is avoided by making the Markov hypothesis; see section IV.1.12. Calibration parameters from the calibration scatterplot provide the linear rescaling parameters needed to establish the cross covariances.

Indicator cutoffs must be chosen for both the primary and secondary (soft) data. In this case the secondary data have essentially the same units as the primary data; therefore, the same five cutoffs were chosen for both: 0.5, 1.0, 2.5, 5.0, and 10.0. Based on the 140 declustered data these cutoffs correspond to cdf values of 0.30, 0.47, 0.72, 0.87, and 0.96. A parameter file for the calibration program **mbcalib** is shown in Figure A.65. The output from **mbcalib**, shown in Figure A.66, is directly input to the **mbsim** program.

The indicator variograms have been entered as standardized indicator variograms with the first and last cutoffs showing a higher (%15) relative nugget constant than the central cutoffs (%10). An example parameter file for the Markov-Bayes simulation program **mbsim** is shown in Figure A.67.

The first and last 10 lines of output are shown on Figure A.68.

Gray scale representations of the two realizations are shown in Figure A.69. Realizations generated by mbsim must be compared to the true distribution (see Figure A.26) and the distribution of the secondary variable (see Figure A.27).

Another approach to incorporate a secondary variable is the external drift formalism; see section IV.1.5. The Gaussian simulation program was modified (code not given) to allow the secondary variable to be used as an external drift in the kriging systems. Two realizations of this simulation procedure are shown in Figure A.70. When using the external drift constraint with Gaussian simulation the kriging variance or standard deviation is always larger than the theoretically correct simple kriging variance. This causes the final back-transformed values also to have too high a variance. In this case, however, the average kriging standard deviation increased by less than one percent. In other situations where the increase is more pronounced it may be necessary to solve the SK system separately and use the external drift estimate with the SK variance; for a similar implementation; see [86].

Part D: Indicator Principal Components Simulation

The typical application of the indicator principal components approach is to truly categorical variables. For this exercise the continuous data have been integer-coded to illustrate the use of ipcprep and ipcsim. An example parameter file for ipcprep and part of the output are shown in Figures A.71 and A.72. Computing the omnidirectional variograms on the resulting three indicator principal components yields the semivariograms shown in Figure A.73. Note that the first principal component shows the most spatial correlation and the third shows the least correlation (higher nugget effect).

An example parameter file for ipcsim is given in Figure A.74, the first and last 10 lines of output are shown in Figure A.75, and a gray scale map of the results is shown in Figure A.76. Note that no resolution is possible beyond the original four integer codes; thus, the map appears quite different from the continuous distribution as shown in Figure A.26; for this reason the integer coded true distribution is also shown in Figure A.76. The ipcsim realization fails to reproduce the gradual sequence of grays seen on the true image.

Part E: Categorical Variable Simulation

This section provides only a hint of the potential applications of Boolean and annealing-based simulation techniques.

The parameter file for ellipsim is shown in Figure A.77, the first and last 10 lines of output are shown in Figure A.78. The two corresponding output images are shown in Figure A.79. In this case the radius, anisotropy ratio, and direction of anisotropy were kept constant. As illustrated in Figure A.80 neither the radius, nor the angle, nor the anisotropy ratio need be kept constant.

The analytical bombing model described in the problem setting was approximated by a spherical variogram model. Program **sisimpdf** was then used to create two realizations; the parameter file is shown in Figure A.81 and the first and last 10 lines of output are shown in Figure A.82. The two realizations, shown in Figure A.83, reproduce the overall proportion of shales and the average aspect ratio. Nevertheless, they contain too much randomness, i.e., the shale shapes are not as well defined as in the **ellipsim** realizations of Figure A.79.

To demonstrate the use of the **anneal** program consider the post-processing of the **sisimpdf** realizations of Figure A.83 to honor more of the two-point statistics of the **ellipsim** realizations of Figure A.79. The parameter file for **anneal** is shown in Figure A.84. The control statistics were computed from the first **ellipsim** realization of Figure A.79. The two-point histogram for 10 lags in the two orthogonal directions aligned with the coordinate axes and for 5 lags in the two diagonal directions were retained. The first and last 10 lines of output from **anneal** are shown in Figure A.85. The two post-processed realizations are shown in Figure A.86. These realizations share more of the characteristics of the original **ellipsim** realizations.

Part F: Annealing Simulation

This section presents two possible realizations provided by the annealing simulation program **sasim**. The two realizations are shown at the top of Figure A.87. The parameter file for **sasim** is shown in Figure A.88 and the first and last 10 lines of the corresponding output file are shown in Figure A.89.

A number of typical checks are also shown in Figure A.87. The annealing algorithm leads to an excellent honoring of the univariate distribution (standard normal in this case) and the variogram model.

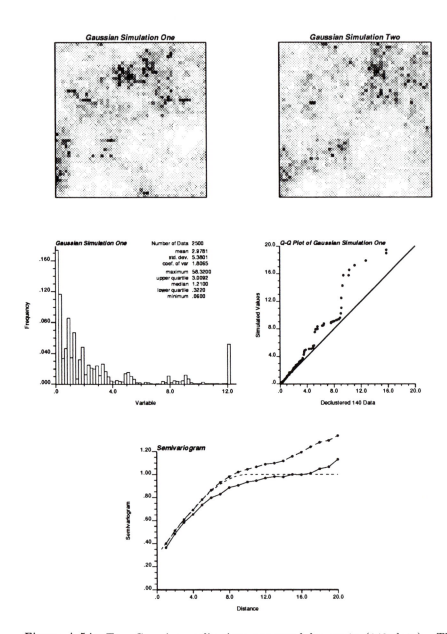

Figure A.54: Two Gaussian realizations generated by sgsim (140 data). The histogram, Q-Q plot, and NS and EW normal scores semivariogram of the first realization. The dotted line is the input isotropic model.

```
                    Parameters for SGSIM
                    ********************

START OF PARAMETERS:
cluster.dat                      \data file
1    2    0    3    5            \column: x,y,z,vr,wt
-1.0e21      1.0e21              \data trimming limits
0                                \0=transform the data, 1=don't
sgsim.trn                        \  output transformation table
0.0     30.0                     \  zmin,zmax(tail extrapolation)
1       0.0                      \  lower tail option, parameter
4       2.0                      \  upper tail option, parameter
sgsim.out                        \output File for simulation
1                                \debugging level: 0,1,2,3
sgsim.dbg                        \output File for Debugging
1711                             \random number seed
0                                \kriging type (0=SK, 1=OK)
2                                \number of simulations
50    0.5      1.0               \nx,xmn,xsiz
50    0.5      1.0               \ny,ymn,ysiz
1     0.0      1.0               \nz,zmn,zsiz
1                                \0=two part search, 1=data-nodes
0                                \max per octant(0 -> not used)
20.0                             \maximum search radius
 0.0   0.0   0.0   1.0   1.0     \sang1,sang2,sang3,sanis1,2
0     8                          \min, max data for simulation
16                               \number simulated nodes to use
1     0.3                        \nst, nugget effect
1    10.0     0.7                \it,   aa,   cc
0.0   0.0     0.0   1.0 1.0      \ang1,ang2,ang3,anis1,anis2:
```

Figure A.55: An example parameter file for **sgsim**.

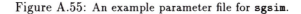

```
SISIM SIMULATIONS:   Clustered 140 primary and secondary data
1
Simulated Value
   1.9223
   1.6991
    .2366
    .0600
    .6339
    .0600
    .1080
    .0600
    .5555
    .1899
•  •  •
   1.9373
    .9900
    .5042
    .6232
   1.2812
    .9595
   1.9338
    .1831
    .4523
    .1149
```

Figure A.56: Output of **sgsim** corresponding to the parameter file shown in Figure A.55.

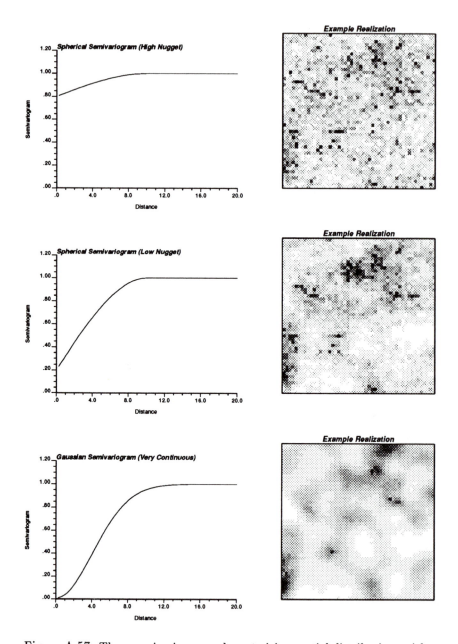

Figure A.57: Three semivariograms characterizing spatial distributions with varying degrees of spatial continuity are shown on the left. Example realizations corresponding to the variograms are shown on the right. The same 140 conditioning data were retained in all three cases.

```
                    Parameters for LUSIM
                    ********************

START OF PARAMETERS:
parta.dat                         \data file
1    2    0    4                  \column: x,y,z,vr - normal data
-1.0e21     1.0e21                \data trimming limits
lusim.out                         \output file for realization(s)
2                                 \debugging level: 0,1,2,3
lusim.dbg                         \output file for debugging
112063                            \random number seed
100                               \number of simulations
4    40.25    0.5                 \nx,xmn,xsiz
4    28.25    0.5                 \ny,ymn,ysiz
1     0.00    1.0                 \nz,zmn,zsiz
1     0.3                         \nst, nugget effect
1    10.0     0.7                 \it, aa, cc
0.0  0.0     0.0  1.0  1.0        \ang1,ang2,ang3,anis1,anis2:
```

Figure A.58: An example parameter file for **lusim**.

```
LUSIM Output
1
simulated values
 -.0800
  .7187
  .0444
  .3532
 -.1132
  .1197
-1.0783
 -.3101
  .2176
 -.8372
 • • •
  .7863
1.4875
1.9849
1.1984
1.2200
  .2193
  .7243
  .7333
1.3604
  .1456
```

Figure A.59: Output of **lusim** corresponding to the parameter file shown in Figure A.58.

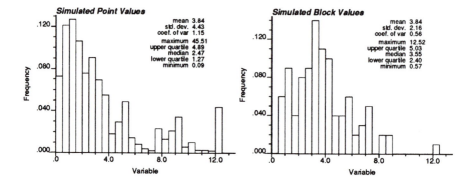

Figure A.60: The distribution of point values and the corresponding distribution of block values for a 2 by 2 unit block centered at x=41, y=29 generated by lusim.

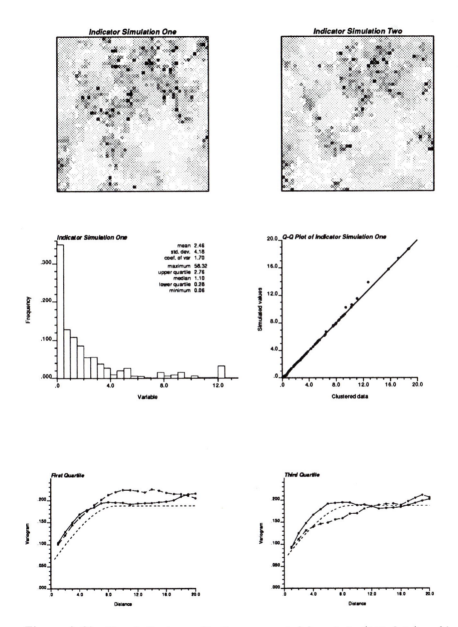

Figure A.61: Two indicator realizations generated by sisim (140 data), a histogram, Q-Q plot, and two directional indicator semivariograms from the first realization. The dotted line is the isotropic model.

```
                        Parameters for SISIM
                        ********************

START OF PARAMETERS:
cluster.dat                              \data file
1   2   0   3                            \column: x,y,z,vr
-1.0e21    1.0e21                        \data trimming limits
0.0   30.0                               \minimum and maximum data value
3       1.0                              \lower tail option and parameter
3       1.0                              \middle     option and parameter
3       1.0                              \upper tail option and parameter
cluster.dat                              \tabulated values for classes
3   5                                    \column for variable, weight
direct.ik                                \direct input of indicators
sisim.out                                \output file for simulation
2                                        \debugging level: 0,1,2,3
sisim.dbg                                \output File for Debugging
0                                        \0=standard order relation corr.
69069                                    \random number seed
2                                        \number of simulations
50    0.5    1.0                         \nx,xmn,xsiz
50    0.5    1.0                         \ny,ymn,ysiz
1     1.0   10.0                         \nz,zmn,zsiz
1                                        \0=two part search, 1=data-nodes
0                                        \  max per octant(0 -> not used)
20.0                                     \  maximum search radius
 0.0  0.0   0.0   1.0   1.0              \  sang1,sang2,sang3,sanis1,2
0    12                                  \  min, max data for simulation
12                                       \number simulated nodes to use
1     1.21                               \0=full IK, 1=med approx(cutoff)
0                                        \0=SK, 1=OK
3                                        \number cutoffs
0.34 0.25    1   0.30                    \cutoff, global cdf, nst, nugget
     1    9.0   0.70                     \        it, aa, cc
     0.0  0.0   0.0   1.0   1.0          \        ang1,ang2,ang3,anis1,2
1.21 0.50    1   0.25                    \cutoff, global cdf, nst, nugget
     1   10.0   0.75                     \        it, aa, cc
     0.0  0.0   0.0   1.0   1.0          \        ang1,ang2,ang3,anis1,2
2.75 0.75    1   0.35                    \cutoff, global cdf, nst, nugget
     1    8.0   0.65                     \        it, aa, cc
     0.0  0.0   0.0   1.0   1.0          \        ang1,ang2,ang3,anis1,2
```

Figure A.62: An example parameter file for **sisim**.

```
SISIM SIMULATIONS:  Clustered 140 primary and secondary data
1
Simulated Value
 13.1375
  5.1796
 10.2620
   .0834
   .1900
   .1000
   .1900
   .2099
   .2800
   .0874
● ● ●
   .3666
   .4426
  1.1090
  1.2100
  1.1875
  1.8182
   .5110
   .8897
   .3382
   .3222
```

Figure A.63: Output of sisim corresponding to the parameter file shown in Figure A.62.

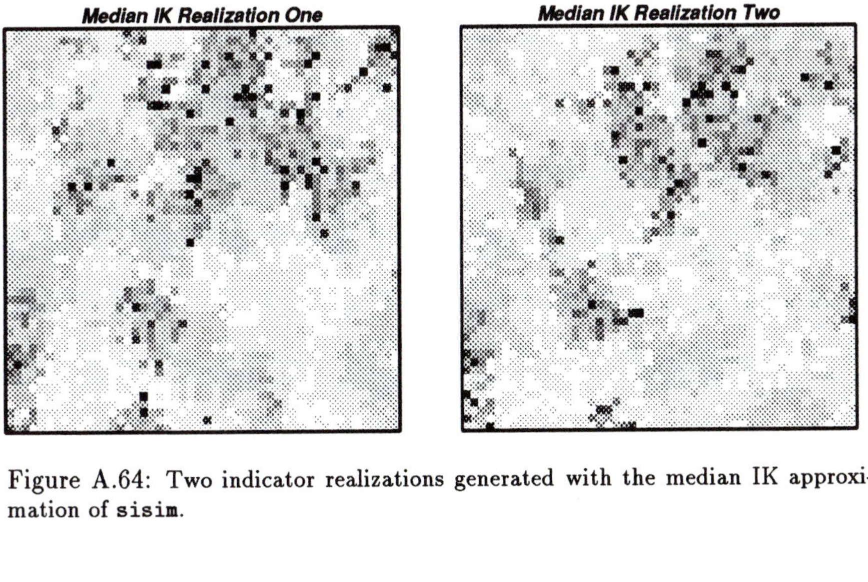

Figure A.64: Two indicator realizations generated with the median IK approximation of sisim.

```
              Parameters for MBCALIB
              ***********************

START OF PARAMETERS:
cluster.dat                        \Data File
3     4                            \column for u and v
-1.0e21                            \minimum acceptable value
mbsim.cal                          \Output File for MBSIM
mbcalib.rep                        \Output File for report
5                                  \number of cutoffs on u
0.50 1.00 2.50 5.00 10.0           \cutoffs on u
5                                  \number of cutoffs on v
0.50 1.00 2.50 5.00 10.0           \cutoffs on v
```

Figure A.65: Parameter file for mbcalib.

```
The local prior distribution table:
 .90476   .95238 1.00000 1.00000 1.00000
 .42105   .73684  .94737 1.00000 1.00000
 .03125   .31250  .78125  .90625 1.00000
 .06897   .06897  .31034  .79310  .96552
 .00000   .05000  .15000  .40000  .80000
 .00000   .00000  .00000  .05263  .52632
The calibration parameters B(i):
 .6066
 .5412
 .5633
 .5300
 .2935
```

Figure A.66: Output from mbcalib corresponding to the parameter file shown in Figure A.65.

```
                MBSIM: Indicator Sequential Simulation - Markov/Bayes Extension
                *****************************************************************

          START OF PARAMETERS:
          cluster.dat                       \HARD data file in GEOEAS format
          1    2    0    3                   \   column: x,y,z,vr
          -1.0e21      1.0e21                \data trimming limits
          mbsim.cal                         \SOFT data calibration File
          ydata.dat                         \SOFT data file in GEOEAS format
          1    2    0    4                   \   column: x,y,z,vr
          -1.0e21      1.0e21                \   trimming limits
          5                                 \   number of soft cutoffs
          0.5 1.0 2.5 5.0 10.0              \   soft cutoffs
          0.0  30.0                         \Minimum and maximum data value
          3    1.0                          \Lower tail option and parameter
          3    1.0                          \Middle      option and parameter
          3    1.0                          \Upper tail option and parameter
          cluster.dat                       \Tabulated values for classes
          3    5                            \   column for variable, weight
          mbsim.out                         \Output file for simulation
          2                                 \Debugging level: 1,2,3
          mbsim.dbg                         \Output file for debugging
          0                                 \0=standard order relation corr.
          69069                             \Random number seed
          2                                 \Number of simulations
          50    0.5      1.0                \nx,xmn,xsiz
          50    0.5      1.0                \ny,ymn,ysiz
          1     1.0      1.0                \nz,zmn,zsiz
          1                                 \0=two part search, 1=data-nodes
          0                                 \   max per octant (0-> not used)
          20.0                              \   maximum search radius
          0.0  0.0  0.0  1.0  1.0           \   sang1,sang2,sang3,sanis1,2
          0    12   12                      \   min, max hard, max soft data
          12                                \number simulated nodes to use
          0    2.5                          \0=full IK, 1=med approx(cutoff)
          0                                 \Kriging type (0=SK, 1=OK)
          5                                 \number of cutoffs
          0.5  0.12   1    0.15            \cutoff, global cdf, nst, nugget
               1    10.0   0.85            \        it, aa, cc
               0.0  0.0  0.0  1.0  1.0     \        ang1,ang2,ang3,anis1,2
          1.0  0.29   1    0.10            \cutoff, global cdf, nst, nugget
               1    10.0   0.90            \        it, aa, cc
               0.0  0.0  0.0  1.0  1.0     \        ang1,ang2,ang3,anis1,2
          2.5  0.50   1    0.10            \cutoff, global cdf, nst, nugget
               1    10.0   0.90            \        it, aa, cc
               0.0  0.0  0.0  1.0  1.0     \        ang1,ang2,ang3,anis1,2
          5.0  0.74   1    0.10            \cutoff, global cdf, nst, nugget
               1    10.0   0.90            \        it, aa, cc
               0.0  0.0  0.0  1.0  1.0     \        ang1,ang2,ang3,anis1,2
          10.0 0.88   1    0.15            \cutoff, global cdf, nst, nugget
               1    10.0   0.85            \        it, aa, cc
               0.0  0.0  0.0  1.0  1.0     \        ang1,ang2,ang3,anis1,2
```

Figure A.67: Parameter file for mbsim.

```
MBSIM SIMULATIONS:   Clustered 140 primary and secondary data
1
Simulated Value
    2.7504
    4.8685
    0.2827
    4.8370
    3.3570
    3.5627
    0.8748
    0.3111
    3.6130
    3.9541
 • • •
    3.0302
    4.0647
    2.7156
    4.9058
    3.7820
    4.9140
    2.7582
    0.1694
    0.1662
    0.2044
```

Figure A.68: Output of **mbsim** corresponding to the parameter file shown in Figure A.67.

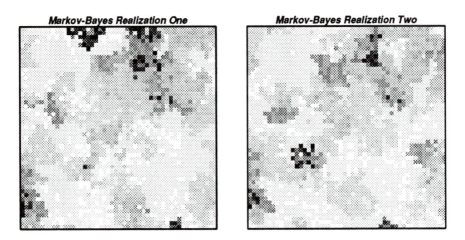

Figure A.69: Two output realizations of **mbsim**.

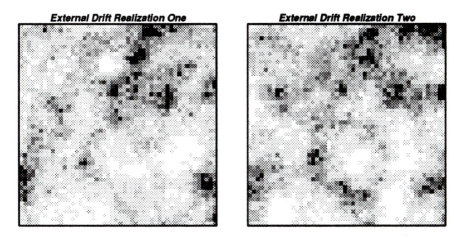

Figure A.70: Two output realizations of **sgsim** with an external drift variable. This result can be compared to that of **mbsim** shown in Figure A.69, the true distribution shown in Figure A.26, and the secondary variable distribution shown in Figure A.27.

```
              Parameters for IPCPREP
              **********************

START OF PARAMETERS:
ipcdat.dat                         \data file
1    2    0    3                   \column: x,y,z,vr
-1.0e21    1.0e21                  \data trimming limits
ipcprep.out                        \output file for trans. data
ipcprep.dbg                        \output file for debugging
ipcprep.svd                        \output file for orthog. matrix
4                                  \number of categories
1 2 3 4                            \categories
0.250 0.250 0.250 0.250            \''p'' values
```

Figure A.71: Parameter file for **ipcprep**.

```
PRINCIPAL COMPONENTS OF: IPCDAT: 140 integer coded data for GSLIB problems
          8
X
Y
Z
Original Value
IPC  1
IPC  2
IPC  3
IPC  4
     39.50        18.50        .00   1   .7461   .4398   .0000   .5000
      5.50         1.50        .00   1   .7461   .4398   .0000   .5000
     38.50         5.50        .00   1   .7461   .4398   .0000   .5000
     20.50         1.50        .00   1   .7461   .4398   .0000   .5000
     27.50        14.50        .00   1   .7461   .4398   .0000   .5000
     40.50        21.50        .00   1   .7461   .4398   .0000   .5000
     15.50         3.50        .00   1   .7461   .4398   .0000   .5000
      6.50        25.50        .00   1   .7461   .4398   .0000   .5000
     38.50        21.50        .00   1   .7461   .4398   .0000   .5000
     23.50        18.50        .00   1   .7461   .4398   .0000   .5000
• • •
      2.50        15.50        .00   3  -.6448   .5254   .2412   .5000
      3.50        14.50        .00   4  -.1567  -.3026  -.7961   .5000
     29.50        41.50        .00   4  -.1567  -.3026  -.7961   .5000
     30.50        40.50        .00   4  -.1567  -.3026  -.7961   .5000
     30.50        42.50        .00   4  -.1567  -.3026  -.7961   .5000
     31.50        41.50        .00   4  -.1567  -.3026  -.7961   .5000
     34.50        32.50        .00   4  -.1567  -.3026  -.7961   .5000
     35.50        31.50        .00   4  -.1567  -.3026  -.7961   .5000
     35.50        33.50        .00   4  -.1567  -.3026  -.7961   .5000
     36.50        32.50        .00   4  -.1567  -.3026  -.7961   .5000
```

Figure A.72: Output from **ipcprep** corresponding to the parameter file shown in Figure A.71.

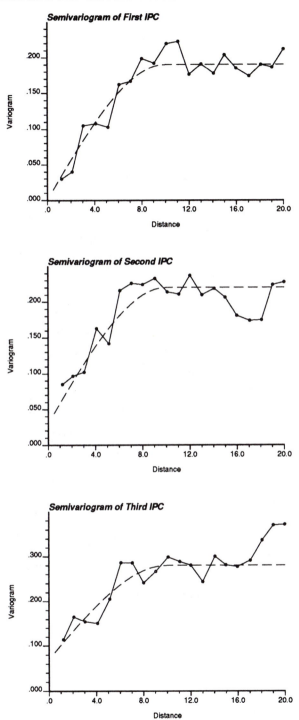

Figure A.73: Omnidirectional indicator principal component semivariograms and models.

```
                    Parameters for IPCSIM
                    *********************

START OF PARAMETERS:
ipcdat.dat                                  \data file
1    2    0    3                            \column: x,y,z,vr
-1.0e21    1.0e21                           \data trimming limits
ipcprep.svd                                 \file with the orthogonal matrix
ipcsim.out                                  \output file for simulation
1                                           \debugging level: 0,1,2,3
ipcsim.dbg                                  \output file for debugging
69069                                       \random number seed
1                                           \number of simulations
50   0.5     1.0                            \nx,xmn,xsiz
50   0.5     1.0                            \ny,ymn,ysiz
1    1.0    10.0                            \nz,zmn,zsiz
1                                           \0=two part search, 1=data-nodes
0                                           \max per octant (0-> not used)
25.0                                        \maximum search radius
 0.0 0.0     0.0  1.0  1.0                  \sang1,sang2,sang3,sanis1,2
0    12                                     \min, max data for simulation
12                                          \number simulated nodes to use
0                                           \kriging type (0=SK, 1=OK)
4    3                                      \# categories, # ipc's to krige
1    2    3    4                            \the categories
0.25 0.25 0.25 0.25                         \the ''p'' values
1 2 3                                       \ipc's to be kriged.
1  1   0.00                                 \ipc #,  nst, nugget
       1   10.0   0.19                      \       it, aa, cc
       0.0  0.0   0.0  1.0  1.0             \       ang1,ang2,ang3,anis1,2
2  1   0.03                                 \ipc #,  nst, nugget
       1   10.0   0.19                      \       it, aa, cc
       0.0  0.0   0.0  1.0  1.0             \       ang1,ang2,ang3,anis1,2
3  1   0.07                                 \ipc #,  nst, nugget
       1   10.0   0.21                      \       it, aa, cc
       0.0  0.0   0.0  1.0  1.0             \       ang1,ang2,ang3,anis1,2
```

Figure A.74: Parameter file for ipcsim.

```
IPCSIM REALIZATIONS:IPCDAT: 140 integer coded data for GSLIB problems
1
Simulated Value
 4
 4
 1
 4
 4
 4
 2
 1
 4
 4
• • •
 3
 3
 3
 3
 3
 3
 3
 3
 4
 4
```

Figure A.75: Output of `ipcsim` corresponding to the parameter file shown in Figure A.74.

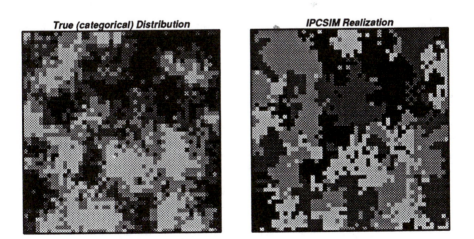

Figure A.76: The true image (coded with categorical indicators) and a realization obtained by indicator principal component simulation.

```
                    Boolean Simulation of Ellipses
                    ******************************
START OF PARAMETERS:
radius.dat                          \Radius File
1   0                               \columns: radius, weight
angles.dat                          \Angles File
1   0                               \columns: ang, weight
anis.dat                            \Anisotropy Ratios File
1   0                               \columns: anis, weight
ellipsim.out                        \Output File for simulation
69069                               \Random number seed
100     1.0                         \nx,xmn,xsiz
100     1.0                         \ny,ymn,ysiz
2                                   \number of simulations
0.25                                \Target proportion (in ellipses)
```

Figure A.77: Input parameter file for **ellipsim**.

```
Output from ELLIPSIM
1
code
0
0
0
0
0
0
0
0
0
0
• • •
1
1
1
1
1
1
1
1
1
1
```

Figure A.78: Output of **ellipsim** corresponding to the parameter file shown in Figure A.77.

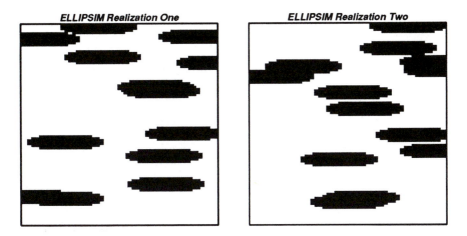

Figure A.79: Two ellipsim realizations.

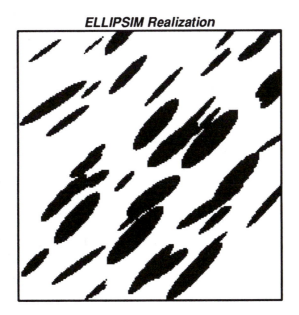

Figure A.80: An ellipsim realization with varying ellipse radii, anisotropy ratios and direction angles.

```
              Parameters for SISIMPDF
              **********************

START OF PARAMETERS:
nodata.dat                           \data file
1   2   0   3                        \column: x,y,z,vr
-1.0e21     1.0e21                    \data trimming limits
sisimpdf.out                         \output file for simulation
2                                    \debugging level: 0,1,2,3
sisimpdf.dbg                         \output File for Debugging
69069                                \random number seed
2                                    \number of simulations
100   0.5    1.0                     \nx,xmn,xsiz
100   0.5    1.0                     \ny,ymn,ysiz
1     1.0    10.0                    \nz,zmn,zsiz
1                                    \0=two part search, 1=data-nodes
0                                    \ max per octant(0 -> not used)
40.0                                 \ maximum search radius
90.0  0.0  0.0  0.2  1.0             \ sang1,sang2,sang3,sanis1,2
0    12                              \ min, max data for simulation
12                                   \number simulated nodes to use
0    2                               \0=full IK, 1=med approx(cat #)
0                                    \0=SK, 1=OK
2                                    \number of categories
0     0.75   1   0.00                \cat # , global pdf, nst, nugget
      1    40.0   1.00               \         it, aa, cc
      90.  0.0   0.0   0.2   1.0     \         ang1,ang2,ang3,anis1,2
1     0.25   1   0.00                \cat # , global pdf, nst, nugget
      1    40.0   1.00               \         it, aa, cc
      90.  0.0   0.0   0.2   1.0     \         ang1,ang2,ang3,anis1,2
```

Figure A.81: An example parameter file for **sisimpdf**.

```
SISIMPDF SIMULATIONS:
1
Simulated Value
 0
 0
 0
 0
 0
 0
 0
 0
 0
 0
```

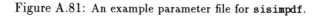

```
 0
 0
 0
 0
 0
 0
 0
 0
 0
 0
```

Figure A.82: Output of **sisimpdf** corresponding to the parameter file shown in Figure A.81.

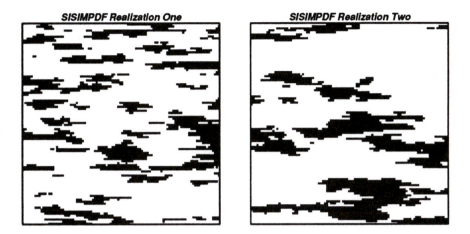

Figure A.83: Two sisim realizations with a variogram that approximately corresponds to the images shown in Figure A.79.

```
                    Parameters for ANNEAL
                    *********************

START OF PARAMETERS:
sisimpdf.out                        \input image(s)
ellipsim.out                        \training Image
anneal.out                          \output file for simulation
anneal.hst                          \output file for 2-pt histogram
3    10                             \debug level, reporting Interval
anneal.dbg                          \output file for debugging
69069                               \random number seed
10  0.000001                        \maximum iterations, tolerance
2                                   \number of simulations
100  100  1                         \nx, ny, nz
4                                   \ndir
1  0  0  10   1                     \ixl, iyl, izl, nlag, cross(1=y)
0  1  0  10   1                     \ixl, iyl, izl, nlag, cross(1=y)
1  1  0   5   0                     \ixl, iyl, izl, nlag, cross(1=y)
1 -1  0   5   0                     \ixl, iyl, izl, nlag, cross(1=y)
```

Figure A.84: Input parameter file for anneal.

```
ANNEAL SIMULATIONS: Output from ELLIPSIM
1
simulation
0
0
0
0
0
0
0
0
0
0
0
• • •
0
0
0
1
1
1
1
1
1
1
```

Figure A.85: Output of anneal corresponding to the parameter file shown in Figure A.84.

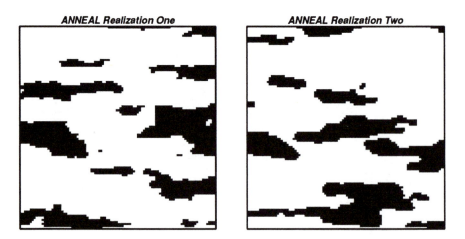

Figure A.86: Two anneal realizations starting from the sisim realizations shown in Figure A.83 and using control statistics taken from the left Figure A.79.

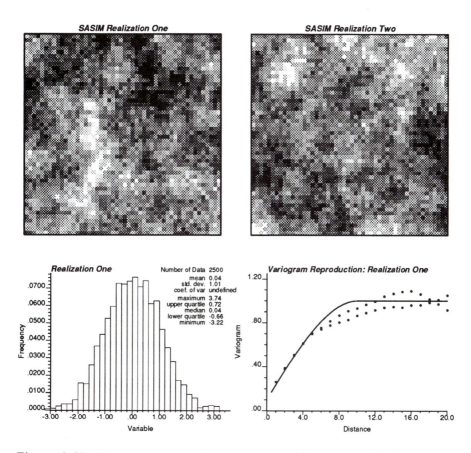

Figure A.87: Two annealing realizations generated by sasim. The histogram, experimental semivariogram (dots) in the NS and EW directions, and the model semivariogram (solid line) are shown for the first realization.

```
                    Parameters for SASIM
                    ********************

START OF PARAMETERS:
nodata                                      \conditioning data (if any)
1   2   0   3                               \columns: x,y,z,vr
-1.0e21    1.0e21                           \data trimming limits
1                                           \0=non parametric; 1=Gaussian
nodata                                      \non parametric distribution
3   5                                       \columns: vr,wt
0.0   90.0                                  \minimum and maximum data values
1     1.0                                   \lower tail option and parameter
4     2.0                                   \upper tail option and parameter
sasim.out                                   \output File for simulation
sasim.var                                   \output File for variogram
3   5000                                    \debug level, reporting interval
sasim.dbg                                   \output file for debugging
1                                           \annealing schedule? (0=auto)
1.0 0.1 75000 25000  3 0.0005   \manual schedule: t0,lambda,ka,k,e,Omin
1                                           \1 or 2 part objective function
69069                                       \random number seed
2                                           \number of simulations
50    0.5    1.0                            \nx,xmn,xsiz
50    0.5    1.0                            \ny,ymn,ysiz
1     0.0    1.0                            \nz,zmn,zsiz
24                                          \max lags for conditioning
1     0.1  1                                \nst, nugget, (1=renormalize)
1    10.0   0.9                             \it,aa,cc:        STRUCTURE 1
0.0   0.0   0.0   1.0   1.0                 \ang1,ang2,ang3,anis1,anis2:
```

Figure A.88: An example parameter file for **sasim**.

```
SASIM SIMULATIONS:
1
simulated value
   0.7472
   0.5026
  -0.0956
  -1.3585
  -0.9394
  -1.6167
  -0.3191
  -0.4137
  -0.2066
  -1.1922
• • •
  -1.6433
  -0.6361
  -1.8259
  -0.5939
  -1.3343
  -2.5655
  -2.1007
  -0.6607
  -1.1866
   0.2112
```

Figure A.89: Output of **sasim** corresponding to the parameter file shown in Figure A.88.

Appendix B

Software Installation

GSLIB does not provide executable programs and tutorials. All of the programs must be compiled prior to running them. This appendix provides some information and hints about loading and compiling the GSLIB programs.

The programs in this version of GSLIB have been developed and are used at Stanford in both UNIX and DOS environments. There are no restrictions on the type of operating system. Many of the comments implicitly assume either UNIX or DOS. The following information is necessarily general because of the many different possible operating systems.

As mentioned in Chapter I, the GSLIB programs are distributed without any support or warranty. The disclaimer shown on Figure B.1 is included in all 80 source code files (63 ".for" files and 17 ".inc" files) enclosed in the diistribution diskettes.

B.1   Installation

Choose a target computer with a Fortran compiler. The first step is to choose a computer to load GSLIB. In many cases this is evident; nevertheless, keep the intended applications in mind. Workstations or larger machines would be appropriate if large 3-D simulations and kriging runs are anticipated. A compiler for ANSI standard Fortran 77 (or any later release) must be available on the target computer.

Load the distribution software. The second step is to copy all of the files from the distribution diskettes to the target computer. Ensure that the directory structure is unchanged; there are some different files with the same name in different directories. The directory structure on the two 3.5 inch high-density 1.44mb DOS format distribution diskettes are shown on Table B.1. The files may be copied directly if the target computer is a IBM-type compatible PC with a high density diskette drive; otherwise, they will have to be transferred through an IBM-type

```
C%%%%%%%%%%%%%%%%%%%%%%%%%%%%%%%%%%%%%%%%%%%%%%%%%%%%%%%%%%%%%%%%%%%%%%%%
C                                                                       %
C Copyright (C) 1992 Stanford Center for Reservoir Forecasting.  All    %
C rights reserved.  Distributed with: C.V. Deutsch and A.G. Journel.    %
C ''GSLIB: Geostatistical Software Library and User's Guide,'' Oxford   %
C University Press, New York, 1992.                                     %
C                                                                       %
C The programs in GSLIB are distributed in the hope that they will be   %
C useful, but WITHOUT ANY WARRANTY.  No author or distributor accepts   %
C responsibility to anyone for the consequences of using them or for    %
C whether they serve any particular purpose or work at all, unless he   %
C says so in writing.  Everyone is granted permission to copy, modify   %
C and redistribute the programs in GSLIB, but only under the condition  %
C that this notice and the above copyright notice remain intact.        %
C                                                                       %
C%%%%%%%%%%%%%%%%%%%%%%%%%%%%%%%%%%%%%%%%%%%%%%%%%%%%%%%%%%%%%%%%%%%%%%%%
```

Figure B.1: The 16 line disclaimer included in all source code files.

PC to the target computer via some software transfer protocol.

The type of transfer protocol will be readily accessible from the local system expert or administrator. Some possibilities include a modem connection to the target computer, an ethernet card, or some other network connection.

File naming conventions. Different Fortran compilers require different file naming conventions, e.g., most UNIX implementations require Fortran source code files to be names with a `.f` extension, most DOS implementations require a `.for` extension, and some compilers require a `.ftn` extension. All of the Fortran code (named with a `.for` extension on the distribution diskettes) should be systematically renamed if appropriate. A UNIX shell script `mvsuff` is provided to change file suffixes, e.g., entering `mvsuff for f` will cause all files with a `.for` extension in the current working directory to be renamed with a `.f` extension.

Choose compiler options. The next step is to choose the compiler options and to document the changes that are required for the programs to be compiled. Compiler options vary considerably; the following should be kept in mind:

- The code can be optimized unless the use of a symbolic debugger is being considered. A high level of optimization on certain computers can introduce problems. For example, the `vargplt` program has failed to work properly when compiled with a high level of optimization on certain compilers.
- The source code is in mixed upper and lower case. In some cases a compiler option must be set or the source code may have to be systematically changed to upper case (the ANSI standard).
- Some compilers force each `do` loop to be executed at least once (i.e., the test for completion is at the end of the loop). This will

Disk	Directory	Subdirectory	size	files
Disk One	GAMLIB		160k	17 files
	KRIGLIB		323k	25 files
	SIMLIB		694k	36 files
	MISC		217k	20 files
	BIN		28k	19 files
Disk Two	POSTPLOT		221k	13 files
	SOLVE		30k	8 files
	DATA		98k	8 files
	SOLUTION	PROB1	40k	11 files
		PROB2	75k	8 files
		PROB4	170k	6 files
		PROB5	221k	4 files
		PROB6	451k	25 files

Table B.1: The directories, subdirectories, and number of files on each distribution diskette.

cause problems and must be avoided by additional coding (adding an **if** statement before entering a loop) or setting the appropriate option when compiling the programs.

Modifications required in one program will surely be required in other programs. It is good practice to modify systematically all programs at one time rather than when needed.

On UNIX machines a **makefile** can be used to facilitate source code management. If extensive modifications and upgrades are anticipated some type of source code control system could be considered.

Testing. A necessary but not sufficient check that the programs are working correctly is to reproduce the example run of each program.

B.2   Troubleshooting

This section describes some known installation pitfalls. When attempting to isolate a problem it would be a good idea to scan all of the following points:

APOLLO users note that unless extra steps are taken all of the **do** loops are executed once, which presently is not acceptable. The Fortran manual should be checked for the appropriate compiler option.

MAC users note that the quotes around the file name in the "include" statement must be removed.

SUN users will have to use the `cscale.old` source code rather than the `cscale.for` code for the color scale plotting program. The PostScript previewer (as of 1991) does not recognize the "colorimage" command.

VAX/VMS users note that the file opening convention will have to be modified in some cases. The VMS compiler does not accept the `status='UNKNOWN'` option when opening files.

Lahey Fortran users note that a variable that is in a common block can not be initialized by a "data" statement; a "block data" statement must be used.

File not found errors are often due to users forgetting that all file names are read as 40 characters. For example, accidentally removing some of the blank spaces before the start of a comment (e.g., `\data file`) will cause the program to fail.

Nonexistent file names are often used to inform the program that a particular operation is not needed, e.g., if the file specifying conditioning data, in the simulation programs, is missing then an unconditional simulation is constructed. It is good practice to explicitly specify a name that is never used, e.g., $\boxed{\text{nodata}}$.

Difficulty reproducing simulation results may be due to differences of numerical precision (machine representation of numbers) and a different sequence of pseudo-random numbers.

A DOS "end of line" is marked by two characters. Some file transfer protocols automatically remove the extra character or a special command is available on most computers (**dos2unix** on SUN computers, **to_unix** on SGI machines,...).

Appendix C

Programming Conventions

C.1 General

Many of these programs have been written and revised over the course of 12 years by numerous geostatisticians, most of whom were Stanford graduate students. Although they have been standardized as much as possible, some quirks from various programmers remain. Care has been taken to keep the code as streamlined and efficient as possible. The clarity and simplicity of the code has taken precedence over tricks for saving speed and storage.

Some other programming considerations:

- The names of frequently used variables have been standardized. For example, the number of variables is typically referred to as *nvar* and the variable values as an array named *vr*. Section C.2 contains a dictionary of commonly used variables.

- Most dimensioning parameters are specified in an include file to make it easy to compile versions with different storage capabilities. Note that the "include" statement is not standard Fortran, but is supported on virtually all compilers.

- Common blocks have been used to declare the variables used in a number of the more complex subroutines. This avoids potential errors with long argument lists and mismatched array sizes.

- Common blocks are typically defined in include files to facilitate quick modifications and avoid typographical errors.

- Maximum dimensioning parameters have been standardized as follows:

  MAXCTX: maximum x size of the covariance table
  MAXCTY: maximum y size of the covariance table
  MAXCTZ: maximum z size of the covariance table

MAXCUT: maximum number of cutoffs for an indicator-based method

MAXDAT: maximum number of data

MAXDIR: maximum number of directions in variogram programs

MAXDIS: maximum number of discretization points for a block

MAXLAG: maximum number of variogram lags

MAXNOD: maximum number of previously simulated nodes to use

MAXNST: maximum number of structures defining a variogram model

MAXSAM: maximum number of samples for kriging or simulation

MAXSBX: maximum number of nodes in x direction of super block network

MAXSBY: maximum number of nodes in y direction of super block network

MAXSBZ: maximum number of nodes in z direction of super block network

MAXVAR: maximum number of variables per datum

MAXSOF: maximum number of soft data

MAXX: maximum number of nodes in x direction

MAXY: maximum number of nodes in y direction

MAXZ: maximum number of nodes in z direction

MXJACK: maximum number of points to jackknife

MXVARG: maximum number of variograms to compute

Other less common dimensioning parameters, e.g., **MAXDT** maximum number of drift terms, are encountered in different include files.

- Some of the low level routines such as the sorting subroutine and matrix solvers are included in the programs that require them. This makes it easy to extract a single program from GSLIB, compile it, and execute the program for the desired results. An alternative approach for a regular user of this software would be to construct a library of compiled objects that is linked in with each main subroutine/program. Main programs are needed to test the routines and to provide a more advanced starting point for users who will be integrating these programs into a separate package.

C.2   Dictionary of Variable Names

a: left hand side matrix of covariance values

aa: range of a variogram nested structure

anis1, anis2: the anisotropy factors that apply after rotations using the angles $ang1, ang2, ang3$ (see Figure II.4 and the discussion in section II.3)

ang: azimuth angle measured in degrees clockwise from north.

ang1, ang2, ang3: angles defining the orientation of an ellipsoid in 3-D (see Figure II.4 and the discussion in section II.3)

atol: angular tolerance from azm

avg: average of the input data

azm: azimuth angle measured in degrees clockwise from north

bandw: the maximum acceptable distance perpendicular to the direction vector

c0: isotropic nugget effect

cbb: average covariance within a block

cc: variance contribution of a nested structure

cdf: array of cdf values for an indicator-based method

ccdf: array of ccdf values for an indicator-based method

close: array of pointers to the close data

cmax: maximum covariance value (used to convert semivariogram values to pseudo-covariance values)

covtab: table lookup array of covariance values

cut: an array of cutoff values for indicator kriging or simulation

datafl: the name of the input data file

dbgfl: the name of the debugging output file

dip: the dip angle measured in negative degrees down from horizontal

dis: an array used in the variogram programs to store the average distance for a particular lag

dtol: angular tolerance from dip

EPSLON: a small number used as a test to avoid division by zero

gam: an array used to store the computed semivariogram measures

gcdf: an array of global cdf values for indicator-based methods

gcut: an array of global cutoff values for indicator-based methods

h2: squared distance between two data

hm: an array used to store the mean of the "head" values in the variogram subroutines

hv: an array used to store the variance of the "head" values in the variogram subroutines

idbg: integer debugging level (0=none, 1=normal, 2=intermediate, 3=serious debugging)

idrif: integer flags specifying whether or not to estimate a drift term in KT

isim: a particular simulated realization of interest

isrot: pointer to the location of the rotation matrix for the anisotropic search

inoct: array containing the number of data in each octant

isk: flag specifying SK or not (0=no, 1=yes)

it: integer flag specifying type of variogram model

itrend: integer flag specifying whether to estimate the trend of the variable in KT

ivhead: the variable number for the head of a variogram pair (see section III.1)

ivtail: the variable number for the tail of a variogram pair (see section III.1)

ivtype: the integer code specifying the type of variogram to be computed (see section III.1)

ixinc, iyinc, izinc: the particular node offset currently being considered

ixd, iyd, izd: the node offset in each coordinate direction as used to define a direction in **gam2**, **gam3**, and **anneal**

ixnode, iynode, iznode: arrays containing the offsets for nodes to consider in a spiral search

KORDEI: order of ACORN random number generator, see [164]

ktype: kriging type (0=simple kriging, 1=ordinary kriging)

ldbg: Fortran logical unit number for debugging output

lin: Fortran logical unit number for input

lint: Fortran logical unit number for intermediate results

lout: Fortran logical unit number for output

ltail: indicator flag of lower tail option

ltpar: parameter for lower tail option

middle: indicator flag specifying option for interpolation within middle classes of a cdf

mik: indicator flag specifying median IK or not (0=no, 1=yes)

minoct: the minimum number of informed octants before estimating or simulating a point

nclose: the number of close data

nctx: the number x nodes in the covariance lookup table

ncty: the number y nodes in the covariance lookup table

nctz: the number z nodes in the covariance lookup table

nd: the number of data

ndmax: the maximum number of data used to estimate or simulate a point

ndmin: the minimum number of data required before estimating or simulating a point

ndir: the number of directions to be being considered in a variogram or simulated annealing simulation program

nlooku: the number of nodes to be searched in a spiral search

nlag: the number of lags to be computed in a variogram program or the number of lags to be matched in the simulated annealing simulation

nict: an array with the cumulative number of data in the super block and all other super blocks with a lower index

noct: the maximum number of data to retain from any one octant

np: an array used to store the number of pairs for a particular lag

nsbx: number of super blocks in x direction

nsby: number of super blocks in y direction

nsbz: number of super blocks in z direction

nsim: number of simulations to generate

nst: number of nested variogram structures (not including nugget)

nsxsb: number of super block nodes in x direction to search

nsysb: number of super block nodes in y direction to search

nszsb: number of super block nodes in z direction to search

nv: number of variables being considered

nx: number of blocks/nodes in x direction

nxdis: number of points to discretize a block in the x direction

ny: number of blocks/nodes in y direction

nydis: number of points to discretize a block in the y direction

nz: number of blocks/nodes in z direction

nzdis: number of points to discretize a block in the z direction

nvar: number of variables in cokriging or variogram subroutines

nvari: number of variables in an input data file

nvarg: number of variograms to compute

ortho: array containing the orthogonal decomposition of a covariance matrix

orthot: transpose of ortho

outfl: the name of the output file

r: right hand side vector of covariance values

rr: copy of right hand side vector of covariance values

rotmat: array with the rotation matrix for all variogram structures and the anisotropic search

radius: maximum radius for data searching

radsqd: maximum radius squared

s: vector of kriging weights

sang1,2,3: angles (see *ang*1, 2, 3) for anisotropic search

sanis1,2: anisotropy (see *anis*1, 2) for anisotropic search

seed: integer seed to the pseudo-random number generator

sill: array of sill values in the cokriging subroutine

skmean: mean used if performing SK

sstrat: indicator flag of search strategy (0=two part, 1=assign data to the nearest nodes)

ssq: sum of squares of the input data (used to compute the variance

title: a title for the current plot or output file

tm: an array used to store the mean of the "tail" values in the variogram subroutines

tmax: maximum allowable data value (values $\geq tmax$ are trimmed)

tmin: minimum data value (values$< tmin$ are accepted)

tv: an array used to store the variance of the "tail" values in the variogram subroutines

UNEST: parameter used to flag unestimated nodes

unbias: constant used in unbiasedness constraint

utail: indicator flag of upper tail option

utpar: parameter for upper tail option

vr: an array name used for data. There may be 2 or more dimensions for cokriging or some indicator implementations

vra: an array name used for data close to the point being considered

x: array of x locations of data

xa: array of x locations of data close to the point being considered

xlag: the lag distance used to partition the experimental pairs in variogram calculation

xltol: the half window tolerance for the lag distance $xlag$

xloc: x location of current node

xmn: location of first x node (origin of x axis)

xmx: location of last x node

xsbmn: x origin of super block search network

xsbsiz: x size of super blocks

xsiz: spacing of nodes or block size in x direction

y: array of y locations of data

ya: array of y locations of data close to the point being considered

yloc: y location of current node

ymn: location of first y node (origin of y axis)

ymx: location of last y node

ysbmn: y origin of super block search network

ysbsiz: y size of super blocks

ysiz: spacing of nodes or block size in y direction

z: array of z locations of data

za: array of z locations of data close to the point being considered

zloc: z location of current node

zmn: location of first z node (origin of z axis)

zmx: location of last z node

zsbmn: z origin of super block search network

zsbsiz: z size of super blocks

zsiz: spacing of nodes or block size in z direction

C.3   Reference Lists of Parameter Codes

The following codes are used in the variogram programs to indicate the type of variogram being computed:

ivtype	type of variogram
1	traditional semivariogram
2	traditional cross semivariogram
3	covariance
4	correlogram
5	general relative semivariogram
6	pairwise relative semivariogram
7	semivariogram of logarithms
8	semirodogram
9	semiMadogram
10	Indicator semivariogram

The following codes are used in the kriging, simulation, and utility programs to specify the type of variogram model being considered:

it	type of variogram model
1	spherical model (a=range, c=variance contribution)
2	exponential model (a=1/3 effective range, c=variance contribution)
3	Gaussian model ($a = 1/\sqrt{3}$ effective range, c=variance contribution)
4	power model (a=power, c=slope)

The following codes are used in the simulation and utility programs to specify the type of interpolation/extrapolation used for going beyond a discrete cdf:

ltail, middle, utail	option
1	linear interpolation between bounds
2	ω-power model interpolation between bounds
3	linear interpolation between tabulated quantiles
4	hyperbolic model extrapolation in upper tail

Appendix D

Alphabetical Listing of Programs

anneal: post-processing of simulated realizations by annealing

backtr: normal scores back transformation

bigaus: computes indicator semivariograms for a Gaussian model

cokb3d: cokriging

cscale: PostScript color scale map

declus: cell declustering

ellipsim: Boolean simulation of ellipses

gam2: computes variograms of 2-D regularly spaced data

gam3: computes variograms of 3-D regularly spaced data

gamv2: computes variograms of 2-D irregularly spaced data

gamv3: computes variograms of 3-D irregularly spaced data

gscale: PostScript gray scale map

histplt: PostScript plot of histogram with statistics

ik3d: indicator kriging

ipcprep: indicator principal components preparation

ipcsim: Indicator principal components simulation

ktb3d: 3-D Kriging (with various trend models)

lusim: LU simulation

mbcalib: Markov-Bayes calibration

mbsim: Markov-Bayes simulation

nscore: normal scores transformation

okb2d: 2-D ordinary kriging

postik: IK post processing

postsim: post-processing of multiple realizations

probplt: PostScript probability paper plot

qpplt: PostScript P-P or Q-Q plot

sasim: simulated annealing simulation

scatplt: PostScript plot of scattergram with statistics

sgsim: sequential Gaussian simulation

sisim: sequential indicator simulation (continuous variable)

sisimpdf: sequential indicator simulation (categorical variable)

tb3d: turning bands simulation

trans: general transformation program

vargplt: PostScript plot of variogram

vmodel: variogram file from model

xvkt3d: 3-D cross-validation/jackknifing

xvok2d: 2-D cross-validation/jackknifing

There is a total of 37 programs, not counting small utility programs such as linear system solvers and random number generators.

Appendix E

List of Acronyms and Notations

E.1 Acronyms

BLUE: best linear unbiased estimator

ccdf: conditional cumulative distribution function

cdf: cumulative distribution function

coIK: indicator cokriging

COK: cokriging

DK: disjunctive kriging

E-type: conditional expectation estimate

IK: indicator kriging

IPCK: indicator principal components kriging

IRF-k: intrinsic random functions of order k

Geo-EAS: geostatistical software released by the U.S. Environmental Protection Agency [50]

GSLIB: geostatistical software library (the software released with this guide)

KT: kriging with a trend model, also known as universal kriging

LS: least squares

M-type: conditional median estimate

Median IK: indicator kriging with the same indicator correlogram for all cutoff values

MG: multiGaussian (algorithm or model)

OK: ordinary kriging

pdf: probability density function

P-P plot: probability-probability plot

Q-Q plot: quantile-quantile plot

RF: random function

RV: random variable denoted by capital letter Z or Y

SK: simple kriging

sGs: sequential Gaussian simulation

sis: sequential Indicator simulation

E.2   Common Notation

$\forall$: whatever

a: range parameter

a_l: coefficient of component number l of the trend model

$B(z)$: Markov-Bayes calibration parameter

$C(\mathbf{0})$: covariance value at separation vector $\mathbf{h} = 0$ It is also the stationary variance of random variable $Z(\mathbf{u})$

$C(\mathbf{h})$: stationary covariance between any two random variables $Z(\mathbf{u})$, $Z(\mathbf{u} + \mathbf{h})$ separated by vector $\mathbf{h}$

$C(\mathbf{u}, \mathbf{u}')$: non-stationary covariance of two random variables $Z(\mathbf{u})$, $Z(\mathbf{u}')$

$C_k(\mathbf{h})$: nested covariance structure in the linear covariance model $C(\mathbf{h}) = \sum_{k=1}^{K} C_k(\mathbf{h})$

$C_I(\mathbf{h}; z_k, z_{k'})$: stationary indicator cross covariance for cutoffs z_k, $z_{k'}$; it is the cross covariance between the two indicator random variables $I(\mathbf{u}; z_k)$ and $I(\mathbf{u} + \mathbf{h}; z_{k'})$

$C_{ZY}(\mathbf{h})$: stationary cross-covariance between the two random variables $Z(\mathbf{u})$ and $Y(\mathbf{u} + \mathbf{h})$ separated by lag vector $\mathbf{h}$

$Circ(\mathbf{h})$: circular semivariogram function of separation vector $\mathbf{h}$

$E\{\cdot\}$: expected value

$E\{Z(\mathbf{u})|(n)\}$: conditional expectation of the random variable $Z(\mathbf{u})$ given the realizations of n other neighboring random variables (called data)

$Exp(\mathbf{h})$: exponential semivariogram model, a function of separation vector $\mathbf{h}$

$F(\mathbf{u}; z)$: non-stationary cumulative distribution function of random variable $Z(\mathbf{u})$

$F(\mathbf{u}; z|(n))$: non-stationary conditional cumulative distribution function of the continuous random variable $Z(\mathbf{u})$ conditioned by the realizations of n other neighboring random variables (called data)

$F(\mathbf{u}; k|(n))$: non-stationary conditional probability distribution function of the categorical variable $Z(\mathbf{u})$

$F(\mathbf{u}_1, \ldots, \mathbf{u}_K; z_1, \ldots, z_K)$: K-variate cumulative distribution function of the K random variables $Z(\mathbf{u}_1), \ldots, Z(\mathbf{u}_K)$

$F(z)$: cumulative distribution function of a random variable Z, or stationary cumulative distribution function of a random function $Z(\mathbf{u})$

$F^{-1}(p)$: inverse cumulative distribution function or quantile function for the probability value $p \in [0, 1]$

$\gamma(\mathbf{h})$: stationary semivariogram between any two random variables $Z(\mathbf{u})$, $Z(\mathbf{u} + \mathbf{h})$ separated by lag vector $\mathbf{h}$

$\gamma_I(\mathbf{h}; z)$: stationary indicator semivariogram for lag vector $\mathbf{h}$ and cutoff z: it is the semivariogram of the binary indicator random function $I(\mathbf{u}; z)$

$\gamma_I(\mathbf{h}; p)$: same as above, but the cutoff z is expressed in terms of p-quantile with $p = F(z)$

$\gamma_{ZY}(\mathbf{h})$: stationary cross semivariogram between the two random variables $Z(\mathbf{u})$ and $Y(\mathbf{u} + \mathbf{h})$ separated by lag vector $\mathbf{h}$

$G(y)$: standard normal cumulative distribution function

$G^{-1}(p)$: standard normal quantile function such that $G(G^{-1}(p)) = p \in [0, 1]$

$\mathbf{h}$: separation vector

$I(\mathbf{u}; z)$: binary indicator random function at location $\mathbf{u}$ and for cutoff z

$i(\mathbf{u}; z)$: binary indicator value at location $\mathbf{u}$ and for cutoff z

$j(\mathbf{u}; z)$: binary indicator transform arising from constraint interval

$f_l(\cdot)$: function of the coordinates used in a trend model

λ_α, $\lambda_\alpha(\mathbf{u})$: kriging weight associated to datum α for estimation at location $\mathbf{u}$

M: stationary median of the distribution function $F(z)$

m: stationary mean of the random variable $Z(\mathbf{u})$

$m(\mathbf{u})$: mean at location $\mathbf{u}$, expected value of random variable $Z(\mathbf{u})$; trend component model in the decomposition $Z(\mathbf{u}) = m(\mathbf{u}) + R(\mathbf{u})$, where $R(\mathbf{u})$ represents the residual component model

$m^*_{KT}(\mathbf{u})$: estimate of the trend component at location $\mathbf{u}$

$\mu, \mu(\mathbf{u})$: Lagrange parameter for kriging at location $\mathbf{u}$

$N(\mathbf{h})$: number of pairs of data values available at lag vector $\mathbf{h}$

$\prod_{i=1}^n y_i = y_1 \cdot y_2 \ldots y_n$: product

$q(p) = F^{-1}(p)$: quantile function, i.e., inverse cumulative distribution function for the probability value, $p \in [0, 1]$

$R(\mathbf{u})$: residual random function model in the decomposition $Z(\mathbf{u}) = m(\mathbf{u}) + R(\mathbf{u})$, where $m(\mathbf{u})$ represents the trend component model

ρ: correlation coefficient $\in [-1, +1]$

$\rho(\mathbf{h})$: stationary correlogram function $\in [-1, +1]$

$\sum_{i=1}^n y_i = y_1 + y_2 + \ldots + y_n$: summation

$\Sigma(\mathbf{h})$: matrix of stationary covariances and cross-covariances

σ^2: variance

$\sigma^2_{OK}(\mathbf{u})$: ordinary kriging variance of $Z(\mathbf{u})$

$\sigma^2_{SK}(\mathbf{u})$: simple kriging variance of $Z(\mathbf{u})$

$Sph(\mathbf{h})$: spherical semivariogram function of separation vector $\mathbf{h}$

$\mathbf{u}$: coordinates vector

$Var\{\cdot\}$: variance

$Y = \varphi(Z)$: transform function $\varphi(\cdot)$ relating two random variables Y and Z

$Z = \varphi^{-1}(Y)$: inverse transform function $\varphi(\cdot)$ relating random variables Z and Y

Z: generic random variable

$Z(\mathbf{u})$: generic random variable at location $\mathbf{u}$, or a generic random function of location $\mathbf{u}$

$Z^*_{COK}(\mathbf{u})$: cokriging estimator of $Z(\mathbf{u})$

$Z^*_{KT}(\mathbf{u})$: estimator of $Z(\mathbf{u})$ using some form of prior trend model

$Z^*_{OK}(\mathbf{u})$: ordinary kriging estimator of $Z(\mathbf{u})$

$Z^*_{SK}(\mathbf{u})$: simple kriging estimator of $Z(\mathbf{u})$

$\{Z(\mathbf{u}), \mathbf{u} \in A\}$: set of random variables $Z(\mathbf{u})$ defined at each location $\mathbf{u}$ of a zone A

$z(\mathbf{u})$: generic variable function of location $\mathbf{u}$

$z(\mathbf{u}_\alpha)$: z-datum value at location $\mathbf{u}$

z_k: k-th cutoff value

$z^{(l)}(\mathbf{u})$: l-th realization of the random function $Z(\mathbf{u})$ at location $\mathbf{u}$

$z_c^{(l)}(\mathbf{u})$: l-th realization conditional to some neighboring data

$z^*(\mathbf{u})$: an estimate of value $z(\mathbf{u})$

$[z(\mathbf{u})]^*_E$: E-type estimate of value $z(\mathbf{u})$, obtained as an arithmetic average of multiple simulated realizations $z^{(l)}(\mathbf{u})$ of the random function $Z(\mathbf{u})$

Bibliography

[1] E. Aarts and J. Korst. *Simulated Annealing and Boltzmann Machines.* John Wiley & Sons, New York, NY, 1989.

[2] M. Abramovitz and I. Stegun, editors. *Handbook of Mathematical Functions: with Formulas, Graphs, and Mathematical Tables.* Dover, New York, NY, 1972. 9th (revised) printing.

[3] Adobe Systems Incorporated. *PostScript Language Reference Manual.* Addison-Wesley, Menlo Park, CA, 1985.

[4] Adobe Systems Incorporated. *PostScript Language Tutorial and Cookbook.* Addison-Wesley, Menlo Park, CA, 1985.

[5] F. Alabert. The practice of fast conditional simulations through the LU decomposition of the covariance matrix. *Math Geology*, 19(5):369–386, 1987.

[6] F. Alabert. *Stochastic Imaging of Spatial Distributions Using Hard and Soft Information.* Master's thesis, Stanford University, Stanford, CA, 1987.

[7] F. Alabert and G. J. Massonnat. Heterogeneity in a complex turbiditic reservoir: stochastic modelling of facies and petrophysical variability. In *65th Annual Technical Conference and Exhibition*, pages 775–790, Society of Petroleum Engineers, September 1990.

[8] A. Almeida, F. Guardiano, and L. Cosentino. Generation of the Stanford 1 turbiditic reservoir. In *Report 5*, Stanford Center for Reservoir Forecasting, Stanford, CA, May 1992.

[9] T. Anderson. *An Introduction to Multivariate Statistical Analysis.* John Wiley & Sons, New York, NY, 1958.

[10] M. Armstrong. Improving the estimation and modeling of the variogram. In G. Verly et al., editors, *Geostatistics for natural resources characterization*, pages 1–20, Reidel, Dordrecht, Holland, 1984.

[11] P. Ballin, A. Journel, and K. Aziz. Prediction of uncertainty in reservoir performance forecasting. *JCPT*, 31(4), April 1992.

[12] R. Barnes and T. Johnson. Positive kriging. In G. Verly et al., editors, *Geostatistics for natural resources characterization*, pages 231–244, Reidel, Dordrecht, Holland, 1984.

[13] S. Begg and P. King. Modelling the effects of shales on reservoir performance: calculation of effective vertical permeability. 1985. SPE paper # 13529.

[14] J. Berger. *Statistical Decision Theory and Bayesian Analysis*. Springer Verlag, New York, NY, 1980.

[15] J. Besag. On the statistical analysis of dirty pictures. *J.R. Statistical Society B*, 48(3):259–302, 1986.

[16] L. Borgman. New advances in methodology for statistical tests useful in geostatistical studies. *Math Geology*, 20(4):383–403, 1988.

[17] L. Borgman and L. Frahme. A case study: multivariate properties of bentonite in NE Wyoming. In Guarascio et al., editors, *Advanced Geostatistics in the Mining Industry*, pages 381–390, Reidel, Dordrecht, Holland, 1976.

[18] L. Borgman, M. Taheri, and R. Hagan. Three-dimensional frequency-domain simulations of geological variables. In G. Verly et al., editors, *Geostatistics for natural resources characterization*, pages 517–541, Reidel, Dordrecht, Holland, 1984.

[19] S. Bozic. *Digital and Kalman Filtering*. E. Arnold, London, 1979.

[20] J. Bridge and M. Leeder. A simulation model of alluvial stratigraphy. *Sedimentology*, 26:617–644, 1979.

[21] P. Brooker. Kriging. *Engineering and Mining Journal*, 180(9):148–153, 1979.

[22] P. Brooker. Two-dimensional simulations by turning bands. *Math Geology*, 17(1):81–90, 1985.

[23] J. Carr and D. Myers. COSIM: a Fortran IV program for co-conditional simulation program. *Computers & Geosciences*, 11(6):675–705, 1985.

[24] G. Christakos. On the problem of permissible covariance and variogram models. *Water Resources Research*, 20(2):251–265, 1984.

[25] G. Christakos. *Random Field Modeling in Earth Sciences*. Academic Press, New York, NY, 1992.

[26] D. Cox and V. Isham. *Point Processes*. Chapman and Hall, New York, NY, 1980.

[27] N. Cressie and D. Hawkins. Robust estimation of the variogram. *Math Geology*, 12(2):115–126, 1980.

[28] M. Dagbert, M. David, D. Crozel, and A. Desbarats. Computing variograms in folded strata-controlled deposits. In G. Verly et al., editors, *Geostatistics for natural resources characterization*, pages 71–89, Reidel, Dordrecht, Holland, 1984.

[29] E. Damsleth, C. Tjolsen, K. Omre, and H. Haldorsen. A two-stage stochastic model applied to a north sea reservoir. In *65th Annual Technical Conference and Exhibition*, pages 791–802, Society of Petroleum Engineers, September 1990.

[30] M. David. *Geostatistical Ore Reserve Estimation*. Elsevier, Amsterdam, 1977.

[31] B. Davis. Indicator kriging as applied to an alluvial gold deposit. In G. Verly et al., editors, *Geostatistics for natural resources characterization*, pages 337–348, Reidel, Dordrecht, Holland, 1984.

[32] B. Davis. Uses and abuses of cross validation in geostatistics. *Math Geology*, 19(3):241–248, 1987.

[33] J. Davis. *Statistics and Data Analysis in Geology*. John Wiley & Sons, New York, NY, 2nd edition, 1986.

[34] M. Davis. Production of conditional simulations via the LU decomposition of the covariance matrix. *Math Geology*, 19(2):91–98, 1987.

[35] P. Delfiner. Linear estimation of non-stationary spatial phenomena. In Guarascio et al., editors, *Advanced Geostatistics in the Mining Industry*, pages 49–68, Reidel, Dordrecht, Holland, 1976.

[36] C. Deutsch. *Annealing Techniques Applied to Reservoir Modeling and the Integration of Geological and Engineering (Well Test) Data*. PhD thesis, Stanford University, Stanford, CA, 1992.

[37] C. Deutsch. DECLUS: a Fortran 77 program for determining optimum spatial declustering weights. *Computers & Geosciences*, 15(3):325–332, 1989.

[38] C. Deutsch. *A Probabilistic Approach to Estimate Effective Absolute Permeability*. Master's thesis, Stanford University, Stanford, CA, 1987.

[39] C. Deutsch and A. Journel. The application of simulated annealing to stochastic reservoir modeling. In *Report 4, Stanford Center for Reservoir Forecasting*, Stanford, CA, May 1991.

[40] C. Deutsch and R. Lewis. Advances in the practical implementation of indicator geostatistics. In *Proceedings of the 23rd International AP-COM Symposium*, pages 133–148, Society of Mining Engineers, Tucson, AZ, April 1992.

[41] L. Devroye. *Non-Uniform Random Variate Generation*. Springer Verlag, New York, NY, 1986.

[42] J. Doob. *Stochastic Processes.* John Wiley & Sons, New York, NY, 1953.

[43] P. Doyen. Porosity from seismic data: a geostatistical approach. *Geophysics*, 53(10):1263–1275, 1988.

[44] P. Doyen and T. Guidish. Seismic discrimination of lithology: a Bayesian approach. In *Geostatistics Symposium*, Calgary, AB, May 1990.

[45] P. Doyen, T. Guidish, and M. de Buyl. Monte carlo simulation of lithology from seismic data in a channel sand reservoir. 1989. SPE paper # 19588.

[46] O. Dubrule. A review of stochastic models for petroleum reservoirs. In Armstrong, editor, *Geostatistics*, pages 493–506, Kluwer, 1989.

[47] O. Dubrule and C. Kostov. An interpolation method taking into account inequality constraints. *Math Geology*, 18(1):33–51, 1986.

[48] B. Efron. *The Jackknife, the Bootstrap, and other Resampling Plans*. Soc. for Industrial and Applied Math, Philadelphia, 1982.

[49] A. Emanuel, G. Alameda, R. Behrens, and T. Hewett. Reservoir performance prediction methods based on fractal geostatistics. *SPE Reservoir Engineering*, 311–318, August 1989.

[50] E. Englund and A. Sparks. *Geo-EAS 1.2.1 User's Guide, EPA Report # 60018-91/008*. EPA-EMSL, Las Vegas, NV, 1988.

[51] C. Farmer. The generation of stochastic fields of reservoir parameters with specified geostatistical distributions. In S. Edwards and P. King, editors, *Mathematics in Oil Production*, pages 235–252, Clarendon Press, Oxford, 1988.

[52] C. Farmer. Numerical rocks. In J. Fayers and P. King, editors, *The Mathematical Generation of Reservoir Geology*, Oxford University Press, New York, NY, 1991.

[53] G. Fogg, F. Lucia, and R. Sengen. Stochastic simulation of inter-well scale heterogeneity for improved prediction of sweep efficiency in a carbonate reservoir. In L. Lake, H. Caroll, and P. Wesson, editors, *Reservoir Characterization II*, pages 355–381, Academic Press, 1991.

[54] R. Froidevaux. *Geostatistical Toolbox Primer, version 1.30.* FSS International, Troinex, Switzerland, 1990.

[55] I. Gelfand. Generalized random processes. pages 853–856, Dokl. Acad. Nauk., USSR, 1955.

[56] S. Geman and D. Geman. Stochastic relaxation, Gibbs distributions, and the Bayesian restoration of images. *IEEE Transactions on Pattern Analysis and Machine Intelligence*, PAMI-6(6):721–741, November 1984.

[57] D. J. Gendzwill and M. Stauffer. Analysis of triaxial ellipsoids: their shapes, plane sections, and plane projections. *Math Geology*, 13(2):135–152, 1981.

[58] C. Gerald. *Applied Numerical Analysis.* Addison-Wesley Publishing Co., Menlo Park, CA, 1978.

[59] A. Goldberger. Best linear unbiased prediction in the generalized linear regression model. *JASA*, 57:369–375, 1962.

[60] G. Golub and C. Van Loan. *Matrix Computations.* The Johns Hopkins University Press, Baltimore, 1990.

[61] J. Gómez-Hernández. *A Stochastic Approach to the Simulation of Block Conductivity Fields Conditioned upon Data Measured at a Smaller Scale.* PhD thesis, Stanford University, Stanford, CA, 1991.

[62] J. Gómez-Hernández and R. Srivastava. ISIM3D: an ANSI-C three dimensional multiple indicator conditional simulation program. *Computers & Geosciences*, 16(4):395–440, 1990.

[63] F. Guardiano and R. M. Srivastava. Borrowing complex geometries from training images: the extended normal equations algorithm. In *Report 5*, Stanford Center for Reservoir Forecasting, Stanford, CA, May 1992.

[64] A. Gutjahr. *Fast Fourier Transforms for Random Fields.* Technical Report No. 4-R58-2690R, Los Alamos, NM, 1989.

[65] H. Haldorsen, P. Brand, and C. Macdonald. Review of the stochastic nature of reservoirs. In S. Edwards and P. King, editors, *Mathematics in Oil Production*, pages 109–209, Clarendon Press, Oxford, 1988.

[66] H. Haldorsen and E. Damsleth. Stochastic modeling. *J. of Pet. Technology*, 404–412, April 1990.

[67] R. L. Hardy. Theory and application of the multiquadric-biharmonic method: 20 years of discovery: 1968-1988. *Computers Math. Applic*, 19(8-9):163–208, 1990.

[68] T. Hewett. Fractal distributions of reservoir heterogeneity and their influence on fluid transport. 1986. SPE paper # 15386.

[69] T. Hewett and R. Behrens. Conditional simulation of reservoir heterogeneity with fractals. *Formation Evaluation*, 300–310, 1990.

[70] M. Hohn. *Geostatistics and Petroleum Geology*. Van Nostrand, New York, NY, 1988.

[71] A. Householder. *Principles of Numerical Analysis*. McGraw-Hill Book Co. Inc., New York, NY, 1953.

[72] P. Huber. *Robust Statistics*. John Wiley & Sons, New York, NY, 1981.

[73] C. Huijbregts and G. Matheron. Universal kriging - An optimal approach to trend surface analysis. In *Decision Making in the Mineral Industry*, pages 159–169, Canadian Institute of Mining and Metallurgy, 1971. Special Volume 12.

[74] E. Isaaks. *The Application of Monte Carlo Methods to the Analysis of Spatially Correlated Data*. PhD thesis, Stanford University, Stanford, CA, 1990.

[75] E. Isaaks. *Risk Qualified Mappings for Hazardous Waste Sites: A Case Study in Distribution-Free Geostatistics*. Master's thesis, Stanford University, Stanford, CA, 1984.

[76] E. Isaaks and R. Srivastava. *An Introduction to Applied Geostatistics*. Oxford University Press, New York, NY, 1989.

[77] E. Isaaks and R. Srivastava. Spatial continuity measures for probabilistic and deterministic geostatistics. *Math Geology*, 20(4):313–341, 1988.

[78] M. Johnson. *Multivariate Statistical Simulation*. John Wiley & Sons, New York, NY, 1987.

[79] R. Johnson and D. Wichern. *Applied Multivariate Statistical Analysis*. Prentice Hall, Englewood Cliffs, NJ, 1982.

[80] T. Jones, D. Hamilton, and C. Johnson. *Contouring of Geological Surfaces with the Computer*. Van Nostrand Reinhold, New York, NY, 1986.

[81] A. Journel. Constrained interpolation and qualitative information. *Math Geology*, 18(3):269–286, 1986.

[82] A. Journel. *Fundamentals of Geostatistics in Five Lessons. Volume 8 Short Course in Geology*, American Geophysical Union, Washington, D.C., 1989.

[83] A. Journel. Geostatistics for conditional simulation of orebodies. *Economic Geology*, 69:673–680, 1974.

[84] A. Journel. *Geostatistics for the Environmental Sciences, EPA Project no. CR 811893*. Technical Report, US EPA, EMS Lab, Las Vegas, NV, 1987.

[85] A. Journel. Geostatistics: Models and tools for the earth sciences. *Math Geology*, 18(1):119–140, 1986.

[86] A. Journel. The lognormal approach to predicting local distributions of selective mining unit grades. *Math Geology*, 12(4):285–303, 1980.

[87] A. Journel. New distance measures: the route towards truly non-Gaussian geostatistics. *Math Geology*, 20(4):459–475, 1988.

[88] A. Journel. Non-parametric estimation of spatial distributions. *Math Geology*, 15(3):445–468, 1983.

[89] A. Journel. The place of non-parametric geostatistics. In G. Verly et al., editors, *Geostatistics for natural resources characterization*, pages 307–355, Reidel, Dordrecht, Holland, 1984.

[90] A. Journel. Recoverable reserves estimation - the geostatistical approach. *Mining Engineering*, 563–568, June 1985.

[91] A. Journel and F. Alabert. New method for reservoir mapping. *J. of Pet. Technology*, 212–218, February 1990.

[92] A. Journel and F. Alabert. Non-Gaussian data expansion in the earth sciences. *Terra Nova*, 1:123–134, 1989.

[93] A. Journel and J. Gómez-Hernández. Stochastic imaging of the Wilmington clastic sequence. 1989. SPE paper # 19857.

[94] A. Journel and C. J. Huijbregts. *Mining Geostatistics*. Academic Press, New York, NY, 1978.

[95] A. Journel and E. Isaaks. Conditional indicator simulation: application to a Saskatchewan uranium deposit. *Math Geology*, 16(7):685–718, 1984.

[96] A. Journel and D. Posa. Characteristic behavior and order relations for indicator variograms. *Math Geology*, 22(8):1011–1025, 1990.

[97] A. Journel and M. Rossi. When do we need a trend model in kriging? *Math Geology*, 21(7):715–739, 1989.

[98] A. Journel, W. Xu, and T. Tran. Integrating seismic data in reservoir modeling: the collocated cokriging alternative. In *Report 5, Stanford Center for Reservoir Forecasting*, Stanford, CA, May 1992.

[99] A. Journel and H. Zhu. Integrating soft seismic data: Markov-Bayes updating, an alternative to cokriging and traditional regression. In *Report 3, Stanford Center for Reservoir Forecasting*, Stanford, CA, May 1990.

[100] W. Kennedy Jr. and J. Gentle. *Statistical Computing*. Marcel Dekker, Inc, New York, NY, 1980.

[101] S. Kirkpatrick, C. Gelatt Jr., and M. Vecchi. Optimization by simulated annealing. *Science*, 220(4598):671–680, May 1983.

[102] A. Kolmogorov. *Grundbegriffe der Wahrschein-lichkkeitrechnung. Ergebnisse der Mathematik, English Translation: Foundations of the Theory of Probability*. Chelsea Publ., New York, NY, English translation: 1950 edition, 1933.

[103] G. Kunkel. *Graphic Design with PostScript*. Scott, Foreman and Company, Glenview, IL, 1990.

[104] I. Lemmer. Mononodal indicator variography - part 1: Theory, - part 2: Application to a computer simulated deposit. *Math Geology*, 18(7):589–623, 1988.

[105] D. Luenberger. *Optimization by Vector Space Methods*. John Wiley & Sons, New York, NY, 1969.

[106] G. Luster. *Raw Materials for Portland Cement: Applications of Conditional Simulation of Coregionalization*. PhD thesis, Stanford University, Stanford, CA, 1985.

[107] Y. Ma and J. Royer. Local geostatistical filtering application to remote sensing. In Namyslowska and Royer, editors, *Geomathematics and Geostatistics Analysis Applied to Space and Time data*, pages 17–36, Sciences de la Terre, Nancy, France, 1988.

[108] J. Mallet. Regression sous contraintes lineaires: application au codage des variables aleatoires. *Rev. Stat. Appl.*, 28(1):57–68, 1980.

[109] A. Mantoglou and J. Wilson. *Simulation of random fields with the turning band method*. Technical Report No. 264, Department of Civil Engineering, M.I.T., 1981.

[110] D. Marcotte and M. David. The biGaussian approach, a simple method for recovery estimation. *Math Geology*, 17(6):625–644, 1985.

[111] A. Marechal. Kriging seismic data in presence of faults. In G. Verly et al., editors, *Geostatistics for natural resources characterization*, pages 271–294, Reidel, Dordrecht, Holland, 1984.

[112] G. Marsaglia. The structure of linear congruential sequences. In S. Zaremba, editor, *Applications of Number Theory to Numerical Analysis*, pages 249–285, Academic Press, London, 1972.

[113] B. Matern. *Spatial Variation*. Volume 36 of *Lecture Notes in Statistics*, Springer Verlag, New York, NY, second edition, 1980. First edition published by Meddelanden fran Statens Skogsforskningsinstitut, Band 49, No. 5, 1960.

[114] G. Matheron. The internal consistency of models in geostatistics. In Armstrong, editor, *Geostatistics*, pages 21–38, Kluwer, 1989.

[115] G. Matheron. The intrinsic random functions and their applications. *Advances in Applied Probability*, 5:439–468, 1973.

[116] G. Matheron. La théorie des variables régionalisées et ses applications. 1971. Fasc. 5, Ecole Nat. Sup. des Mines, Paris.

[117] G. Matheron. A simple substitute for conditional expectation: the disjunctive kriging. In Guarascio et al., editors, *Advanced Geostatistics in the Mining Industry*, pages 221–236, Reidel, Dordrecht, Holland, 1976.

[118] G. Matheron. Traité de géostatistique appliquée. 1962. Vol. 1 (1962), Vol. 2 (1963), ed. Technip, Paris.

[119] G. Matheron, H. Beucher, H. de Fouquet, A. Galli, D. Guerillot, and C. Ravenne. Conditional simulation of the geometry of fluvio-deltaic reservoirs. 1987. SPE paper # 16753.

[120] N. Metropolis, A. Rosenbluth, M. Rosenbluth, A. Teller, and E. Teller. Equation of state calculations by fast computing machines. *J. Chem. Phys.*, 21(6):1087–1092, June 1953.

[121] M. Monmonier. *Computer-assisted Cartography: Principles and Prospects*. Prentice Hall, Englewood Cliffs, NJ, 1982.

[122] F. Mosteller and J. Tukey. *Data Analysis and Regression*. Addison-Wesley, Reading, MA, 1977.

[123] D. Myers. Cokriging - new developments. In G. Verly et al., editors, *Geostatistics for natural resources characterization*, pages 295–305, Reidel, Dordrecht, Holland, 1984.

[124] D. Myers. Matrix formulation of co-kriging. *Math Geology*, 14(3):249–257, 1982.

[125] R. Olea, editor. *Geostatistical Glossary and Multilingual Dictionary*. Oxford University Press, New York, NY, 1991.

[126] R. Olea, editor. *Optimum Mapping Techniques: Series on Spatial Analysis no. 2.* Kansas Geol. Survey, Lawrence, KA, 1975.

[127] H. Omre. The variogram and its estimation. In G. Verly et al., editors, *Geostatistics for natural resources characterization,* pages 107–125, Reidel, Dordrecht, Holland, 1984.

[128] H. Omre, K. Solna, and H. Tjelmeland. Calcite cementation description and production consequences. 1990. SPE paper # 20607.

[129] H. Parker. The volume-variance relationship: a useful tool for mine planning. In P. Mousset-Jones, editor, *Geostatistics,* pages 61–91, McGraw Hill, New York, NY, 1980.

[130] H. Parker, A. Journel, and W. Dixon. The use of conditional lognormal probability distributions for the estimation of open pit ore reserves in stratabound uranium deposits: a case study. In *Proceedings of the 16th International APCOM Symposium,* pages 133–148, Society of Mining Engineers, Tucson, AZ, October 1979.

[131] D. Posa. Conditioning of the stationary kriging matrices for some well-known covariance models. *Math Geology,* 21(7):755–765, 1988.

[132] W. Press, B. Flannery, S. Teukolsky, and W. Vetterling. *Numerical Recipes.* Cambridge University Press, New York, NY, 1986.

[133] J. M. Rendu. Normal and lognormal estimation. *Math Geology,* 11(4):407–422, 1979.

[134] B. Ripley. *Statistical Inference for Spatial Processes.* Cambridge University Press, New York, NY, 1988.

[135] B. Ripley. *Stochastic Simulation.* John Wiley & Sons, New York, NY, 1987.

[136] M. Rosenblatt. Remarks on a multivariate transformation. *Annals of Mathematical Statistics,* 23(3):470–472, 1952.

[137] F. Samper Calvete and J. Carrera Ramirez. *Geoestadistica: Aplicaciones a la Hidrologia Subterranea.* Centro Int. de Metodos Numericos en Ingenieria, Barcelona, Spain, 1990.

[138] L. Sandjivy. The factorial kriging analysis of regionalized data. its application to geochemical prospecting. In G. Verly et al., editors, *Geostatistics for natural resources characterization,* pages 559–571, Reidel, Dordrecht, Holland, 1984.

[139] H. Schwarz. *Numerical Analysis, A Comprehensive Introduction.* John Wiley & Sons, New York, NY, 1989.

[140] H. Schwarz, H. Rutishauser, and E. Stiefel. *Numerical Analysis of Symmetric Matrices.* Prentice-Hall, Englewood Cliffs, NJ, 1973.

[141] B. Silverman. *Density Estimation for Statistics and Data Analysis.* Chapman and Hall, New York, NY, 1986.

[142] A. Solow. *Kriging Under a Mixture Model.* PhD thesis, Stanford University, Stanford, CA, 1985.

[143] R. Srivastava. Minimum variance or maximum profitability? *CIM Bulletin,* 80(901):63–68, 1987.

[144] R. Srivastava. *A Non-ergodic Framework for Variogram and Covariance Functions.* Master's thesis, Stanford University, Stanford, CA, 1987.

[145] R. Srivastava and H. Parker. Robust measures of spatial continuity. In Armstrong, editor, *Geostatistics,* pages 295–308, Reidel, Dordrecht, Holland, 1988.

[146] M. Stein and M. Handcock. Some asymptotic properties of kriging when the covariance is unspecified. *Math Geology,* 21(2):171–190, 1989.

[147] D. Stoyan, W. Kendall, and J. Mecke. *Stochastic Geometry and its Applications.* John Wiley & Sons, New York, NY, 1987.

[148] G. Strang. *Linear Algebra and its Applications.* Academic Press, New York, NY, 1976. Revised edition 1980.

[149] J. Sullivan. Conditional recovery estimation through probability kriging: theory and practice. In G. Verly et al., editors, *Geostatistics for natural resources characterization,* pages 365–384, Reidel, Dordrecht, Holland, 1984.

[150] J. Sullivan. *Non-parametric Estimation of Spatial Distributions.* PhD thesis, Stanford University, Stanford, CA, 1985.

[151] V. Suro-Perez. Generation of a turbiditic reservoir: the Boolean alternative. In *Report 4,* Stanford Center for Reservoir Forecasting, Stanford, CA, May 1991.

[152] V. Suro-Perez. *Indicator Kriging and Simulation Based on Principal Component Analysis.* PhD thesis, Stanford University, Stanford, CA, 1992.

[153] V. Suro-Perez and A. Journel. Indicator principal component kriging. *Math Geology,* 23(5):759–788, 1991.

[154] V. Suro-Perez and A. Journel. Stochastic simulation of lithofacies: an improved sequential indicator approach. In Guerillot and Guillon, editors, *Proceedings of 2nd European Conference on the Mathematics of Oil Recovery,* pages 3–10, Technip Publ., Paris, September 1990.

[155] D. Tetzlaff and J. Harbaugh. *Simulating Clastic Sedimentation*. Reinhold Van Nostrand, New York, NY, 1989.

[156] J. Tukey. *Exploratory Data Analysis*. Addison-Wesley, Reading, MA, 1977.

[157] G. Verly. *Estimation of Spatial Point and Block Distributions: The MultiGaussian Model*. PhD thesis, Stanford University, Stanford, CA, 1984.

[158] G. Verly. The multiGaussian approach and its applications to the estimation of local reserves. *Math Geology*, 15(2):259–286, 1983.

[159] R. Voss. *The Science of Fractal Images*. Springer Verlag, New York, NY, 1988.

[160] H. Wackernagel. Geostatistical techniques for interpreting multivariate spatial information. In C. Chung et al., editors, *Quantitative Analysis of Mineral and Energy Resources*, pages 393–409, Reidel, Dordrecht, Holland, 1988.

[161] J. Westlake. *A Handbook of Numerical Matrix Inversion and Solution of Linear Equations*. John Wiley & Sons, New York, NY, 1968.

[162] P. Whittle. *Prediction and Regulation by Linear Least-Squares Methods*. University of Minnesota Press, MN, 1963. Revised edition 1983.

[163] N. Wiener. *Extrapolation, Interpolation, and Smoothing of Stationary Time Series*. MIT Press, Cambridge, 7th edition, 1966.

[164] R. Wikramaratna. ACORN - a new method for generating sequences of uniformly distributed pseudo-random numbers. *Journal of Computational Physics*, 83:16–31, 1989.

[165] Y. Wu, A. Journel, L. Abramson, and P. Nair. *Uncertainty Evaluation Methods for Waste Package Performance Assessment*. US Nuclear Reg. Commission, NUREG/CR-5639, 1991.

[166] H. Xiao. *A description of the behavior of indicator variograms for a bivariate normal distribution*. Master's thesis, Stanford University, Stanford, CA, 1985.

[167] A. Yaglom and M. Pinsker. Random processes with stationary increments of order n. pages 731–734, Dokl. Acad. Nauk., USSR, 1953.

[168] H. Zhu. *Modeling Mixture of Spatial Distributions with Integration of Soft Data*. PhD thesis, Stanford University, Stanford, CA, 1991.

[169] H. Zhu and A. Journel. Mixture of populations. *Math Geology*, 23(4):647–671, 1991.

Index